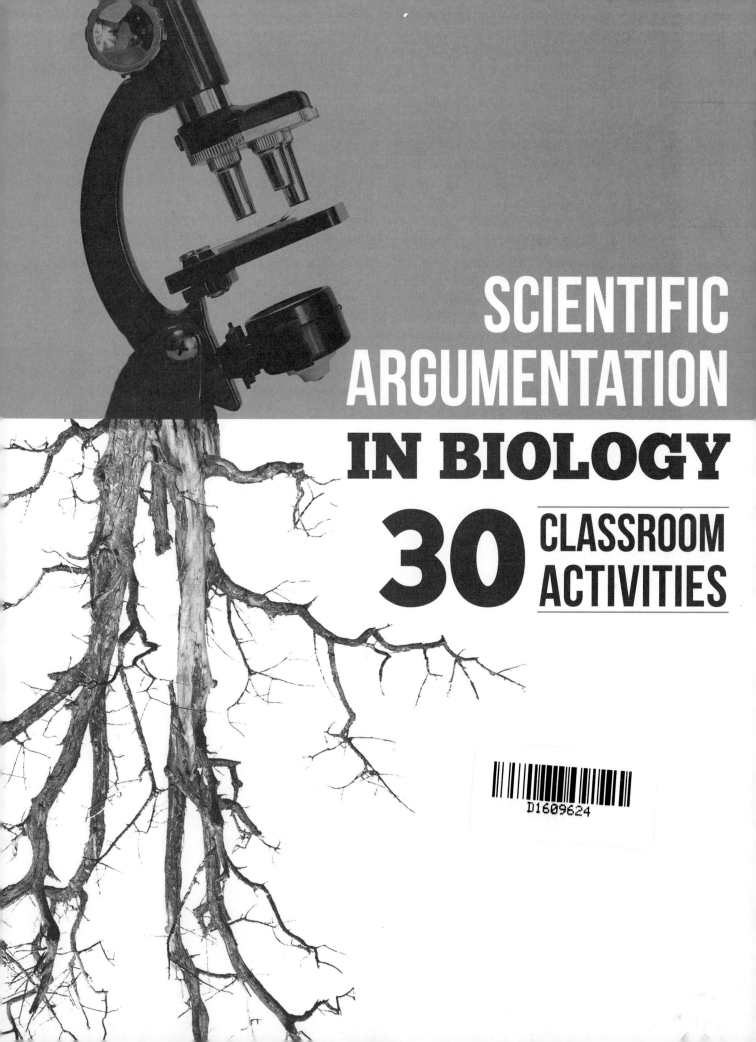

Claire Reinburg, Director
Jennifer Horak, Managing Editor
Andrew Cooke, Senior Editor
Wendy Rubin, Associate Editor
Agnes Bannigan, Associate Editor
Amy America, Book Acquisitions Coordinator

**ART AND DESIGN**
Will Thomas Jr., Director
Rashad Muhammad, Graphic Designer, Cover and Interior Design

**PRINTING AND PRODUCTION**
Catherine Lorrain, Director

**NATIONAL SCIENCE TEACHERS ASSOCIATION**
Gerald F. Wheeler, Executive Director
David Beacom, Publisher

1840 Wilson Blvd., Arlington, VA 22201
*www.nsta.org/store*
For customer service inquiries, please call 800-277-5300.

Copyright © 2013 by the National Science Teachers Association.
All rights reserved. Printed in the United States of America.
18 17 16 15     6 5 4 3

*NSTA is committed to publishing material that promotes the best in inquiry-based science education. However, conditions of actual use may vary, and the safety procedures and practices described in this book are intended to serve only as a guide. Additional precautionary measures may be required. NSTA and the authors do not warrant or represent that the procedures and practices in this book meet any safety code or standard of federal, state, or local regulations. NSTA and the authors disclaim any liability for personal injury or damage to property arising out of or relating to the use of this book, including any of the recommendations, instructions, or materials contained therein.*

**PERMISSIONS**
Book purchasers may photocopy, print, or e-mail up to five copies of an NSTA book chapter for personal use only; this does not include display or promotional use. Elementary, middle, and high school teachers may reproduce forms, sample documents, and single NSTA book chapters needed for classroom or noncommercial, professional-development use only. E-book buyers may download files to multiple personal devices but are prohibited from posting the files to third-party servers or websites, or from passing files to non-buyers. For additional permission to photocopy or use material electronically from this NSTA Press book, please contact the Copyright Clearance Center (CCC) (*www.copyright.com*; 978-750-8400). Please access *www.nsta.org/permissions* for further information about NSTA's rights and permissions policies.

Library of Congress Cataloging-in-Publication Data
Sampson, Victor, 1974-
  Scientific argumentation in biology : 30 classroom activities / by Victor Sampson and Sharon Schleigh.
    p. cm.
  Includes bibliographical references.
  ISBN 978-1-936137-27-5
  1. Qualitative reasoning. 2.  Biology.  I. Schleigh, Sharon, 1963- II. Title.
  Q339.25.S26 2012
  570.71'2--dc23
                      2012029423
eISBN 978-1-936959-56-3

# Contents

**PREFACE** ........................................................................................... ix

**INTRODUCTION** ............................................................................... xv

# Generate an Argument ....................................................... 1

## *Framework* Matrix .................................................................... 2

### **Activity 1:** Classifying Birds in the United States ............... 5
*(Species Concept)*

### **Activity 2:** Color Variation in Venezuelan Guppies ........... 19
*(Mechanisms of Evolution)*

### **Activity 3:** Desert Snakes ................................................... 29
*(Mechanics of Evolution)*

### **Activity 4:** Fruit Fly Traits .................................................. 45
*(Genetics)*

### **Activity 5:** DNA Family Relationship Analysis .................. 55
*(Genetics)*

### **Activity 6:** Evolutionary Relationships in Mammals ......... 67
*(Genetics and Evolution)*

### **Activity 7:** Decline in Saltwater Fish Populations ............. 81
*(Ecology and Human Impact on the Environment)*

### **Activity 8:** History of Life on Earth .................................. 103
*(Trends in Evolution)*

### **Activity 9:** Surviving Winter in the Dust Bowl ................ 113
*(Food Chains and Trophic Levels)*

### **Activity 10:** Characteristics of Viruses ............................ 123
*(Characteristics of Life)*

# Contents

## Evaluate Alternatives ... 133

*Framework* **Matrix** ... 134

**Activity 11:** Spontaneous Generation ... 137
*(Cell Theory)*

**Activity 12:** Plant Biomass ... 149
*(Photosynthesis)*

**Activity 13:** Movement of Molecules in or out of Cells ... 159
*(Osmosis and Diffusion)*

**Activity 14:** Liver and Hydrogen Peroxide ... 171
*(Chemical Reactions and Catalysts)*

**Activity 15:** Cell Size and Diffusion ... 181
*(Diffusion)*

**Activity 16:** Environmental Influence on Genotypes and Phenotypes ... 191
*(Genetics)*

**Activity 17:** Hominid Evolution ... 203
*(Macroevolution)*

**Activity 18:** Plants and Energy ... 219
*(Respiration and Photosynthesis)*

**Activity 19:** Healthy Diet and Weight ... 229
*(Human Health)*

**Activity 20:** Termite Trails ... 239
*(Animal Behavior)*

# Contents

## Refutational Writing — 249

*Framework* Matrix — 250

**Activity 21:** Misconception About Theories and Laws — 253
*(Nature of Science)*

**Activity 22:** Misconception About the Nature of Scientific Knowledge — 261
*(Nature of Science)*

**Activity 23:** Misconception About the Work of Scientists — 269
*(Nature of Science)*

**Activity 24:** Misconception About the Methods of Scientific Investigations — 277
*(Nature of Science)*

**Activity 25:** Misconception About Life on Earth — 285
*(Evolution)*

**Activity 26:** Misconception About Bacteria — 293
*(Microbiology)*

**Activity 27:** Misconception About Interactions That Take Place Between Organisms — 301
*(Ecology)*

**Activity 28:** Misconception About Plant Reproduction — 309
*(Botany)*

**Activity 29:** Misconception About Inheritance of Traits — 315
*(Genetics)*

**Activity 30:** Misconception About Insects — 321
*(Ecology)*

## ASSESSMENTS & STUDENT SAMPLES — 329

## APPENDIX — 361

## INDEX — 373

# Preface

## What Is Scientific Argumentation?

Scientific argumentation is an important practice in science. We define scientific argumentation as an attempt to validate or refute a claim on the basis of reasons in a manner that reflects the values of the scientific community (Norris, Philips, and Osborne 2007). A claim, in this context, is not simply an opinion or an idea; rather, it is a conjecture, explanation, or other conclusion that provides a sufficient answer to a research question. The term *reasons* is used to describe the support someone offers for a conclusion. The term *evidence* is often used to describe the reasons used by scientists, especially when the support is based on data gathered through an investigation. Yet reasons do not have to be based on measurements or observations to be viewed as scientific. Charles Darwin, for example, provided numerous reasons in *The Origin of Species* to support his claims that all life on Earth shares a common ancestor, biological evolution is simply descent with modification, and the primary mechanism that drives biological evolution is natural selection. Some of the reasons that Darwin used were theoretical in nature, such as appealing to population theory from Malthus and the ideas of uniformitarianism advocated by Lyell, while others were more empirical in nature, such as the appeals he made to the data that he gathered during his voyage to Central and South America. What made "Darwin's one long argument" (Mayr 1964, p. 459) so convincing and persuasive to others, however, was the way he was able to coordinate theory and evidence in order to validate his claims.

It is also important for teachers and students to understand how an argument (i.e., a written or spoken claim and support provided for it) in science is different than an argument that is used in everyday contexts or in other disciplines such as history, religion, or even politics. In order to make these differences explicit, we use the framework illustrated in Figure 1 (p. x).

In this framework, a claim is a conjecture, conclusion, explanation, or a descriptive statement that answers a research question. The evidence component of the argument refers to measurements, observations, or even findings from other studies that have been collected, analyzed, and then interpreted by the researchers. Biologists, for example, will often examine the data they collect in order to determine if there is (a) a trend over time, (b) a difference between groups or objects, or

*Figure 1. A Framework That Can Be Used to Illustrate the Components of a Scientific Argument and Some Criteria That Can and Should Be Used to Evaluate the Merits of a Scientific Argument*

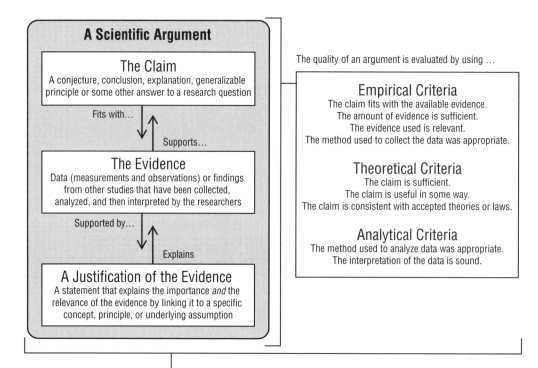

(c) a relationship between variables, and then they interpret their analysis in light of their research question, the nature of their study, and the available literature. Finally, the justification of the evidence component of the argument is a statement or two that explains the importance and the relevance of the evidence by linking it to a specific principle, concept, or underlying assumption.

It is also important for students to understand that some forms of evidence and some types of reasons are better than others in science. An important component of scientific argumentation involves the evaluation of the acceptability and sufficiency of the evidence or reasons that are used to support or challenge a claim. Therefore, in addition to the structural components of an argument, the framework in Figure 1 also highlights several empirical and theoretical criteria that students can and should use to evaluate the quality or merits of an argument in science. Empirical criteria include (a) how well the claim fits with all available evidence, (b) the sufficiency of the evidence included in the argument, (c) the quality of the evidence (i.e., validity and reliability), and (d) the predictive power of the claim. Theoretical criteria, on the other hand, refer to standards that are important in science but are not empirical in nature. These include criteria such as (a) the sufficiency of the claim (i.e., it includes everything it needs to), (b) the usefulness of the claim (e.g., it allows us to engage in new inquiries or understand a phenomenon), and (c) how consistent the claim and the reasoning is with other accepted theories, laws, or models. What counts as quality within these different categories, however, varies from discipline to discipline (e.g., physics, biology, geology) and within the fields that are found with a discipline (e.g., cell biology, evolutionary biology, genetics) due to differences in the types of phenomena investigated, what counts as an accepted mode of inquiry (e.g., experimentation vs. fieldwork), and the theory-laden nature of scientific inquiry. It is therefore important to keep in mind that the nature of scientific arguments and what counts as quality in science is discipline- and field-dependent.

## Why Integrate Argumentation Into the Teaching and Learning of Biology?

A major aim of science education in the United States is for all students to become proficient in science by the time they finish high school. Science proficiency consists of four interrelated aspects (Duschl, Schweingruber, and Shouse 2007). First, it requires an individual to know important scientific explanations about the natural world, to be able to use these explanations to solve problems, and to be able to understand new explanations when they are introduced. Second, it requires an individual to be able to generate and evaluate scientific explanations and scientific arguments. Third, individuals need to understand the nature of scientific knowledge and how scientific knowledge develops over time. Finally, and perhaps most importantly, individuals that are proficient in science should be able to understand the language of science and be able to participate in scientific practices (such as inquiry and argumentation). Empirical research, however, indicates that many students do not develop this knowledge or these abilities while in school (Duschl, Schweingruber, and Shouse 2007; NRC 2005, 2008).

One way to address this problem is to engage students in scientific argumentation as part of the teaching and learning of biology (Driver, Newton, and Osborne 2000; Duschl 2008; Duschl and Osborne

2002). In order to help students develop science proficiency by engaging them in scientific argumentation, however, the focus and nature of instruction inside biology classrooms will need to change from time to time. This change in focus, in part, will require teachers to place more emphasis on "how we know" in biology (i.e., how new knowledge is generated and validated) in addition to "what we know" about life on Earth (i.e., the theories, laws, and unifying concepts). Science teachers will also need to focus more on the abilities and habits of mind that students need to have in order to construct and support scientific knowledge claims through argument and to evaluate the claims or arguments developed by others.

In order to accomplish this goal, science teachers will need to design lessons that give students an opportunity to learn how to generate explanations from data, identify and judge the relevance or sufficiency of evidence, articulate and support an explanation in an argument, respond to questions or counterarguments, and revise a claim (or argument) based on the feedback they receive or in light of new evidence. Science teachers will also need to find a way to help students learn, adopt, and use the same criteria that biologists use to determine what counts as warranted scientific knowledge in a particular field of biology. This task, however, can be difficult for teachers to accomplish given the constraints of a science classroom without the development of new instructional strategies or techniques (Price Schleigh, Bosse, and Lee 2011). We have therefore used the available literature on argumentation in science education (e.g., Berland and Reiser 2009; Clark, Schleigh, and Menekse 2008; McNeill and Krajcik 2008a; Osborne, Erduran, and Simon 2004; Sampson and Clark 2008; Sandoval and Reiser 2004) to develop two different instructional models that teachers can use to promote and support student engagement in scientific argumentation in the biology classroom. We have also designed several stand-alone writing activities that teachers can use to help students learn how to write extended arguments that consist of multiple lines of reasoning that will help solidify their understanding of important biology content as part of the process.

All of these activities are designed so they can be used at different points during a biology course and in a variety of grade levels to help students learn how to generate a convincing scientific argument and to evaluate the validity or acceptability of an explanation or argument in science. In fact, we have used these activities included in this book to engage learners in scientific argumentation in middle school classrooms, high school classrooms, and in science teacher education programs. The activities in this book can also be used to help students understand the practices, crosscutting concepts, and core ideas found in *A Framework for K–12 Science Education* (NRC 2012) and develop the literacy in science skills outlined in the *Common Core State Standards for English Language Arts and Literacy* (NGA and CCSSO 2010).

## Development of the Activities

The integration of scientific argumentation into the teaching and learning of biology can be difficult for both the teachers and students. In fact, teachers often ask for specific instructional strategies and engaging activities based on these instructional activities that would allow students to learn how to engage in scientific argumentation as part of the inquiry process (see Sampson and Blanchard, forthcoming). We have also received many requests to help teachers develop the skills in facilitating this kind of activity inside the classroom. We have designed this book to satisfy these requests. This book's instructional strategies and the activities based on these strategies are grounded in not only current research on argumentation in science education (Berland and McNeill 2010; Clark et al. 2008; Driver, Newton, and Osborne 2000; Erduran and Jimenez-Aleixandre 2008; Jimenez-Aleixandre, Rodriguez, and Duschl 2000; McNeill and Krajcik 2008b; McNeill et al. 2006; Osborne, Erduran, and Simon 2004; Sampson and Blanchard, forthcoming; Sampson and Clark 2008, 2009; Sampson, Grooms, and Walker 2011) but also our experiences inside the classroom. Each activity has been field-tested in at least one middle school or high school (see Appendix A, p. 367, for a list of field test sites and teachers). The classrooms we used to test the activities were diverse and represented a wide range of student achievement levels (honors, general, advanced, and so on).

We used teacher comments and suggestions to refine the activities and to provide the guidance teachers need to implement the activities as Teacher Notes.

## References

Berland, L., and K. McNeill. 2010. A learning progression for scientific argumentation: Understanding student work and designing supportive instructional contexts. *Science Education* 94 (5): 765–793.

Berland, L., and B. Reiser. 2009. Making sense of argumentation and explanation. *Science Education* 93 (1): 26–55.

Clark, D., S. P. Schleigh, M. Menekse, and C. D'Angelo. 2008. Improving the quality of student argumentation through the initial structuring of online discussions. Paper presented at the proceedings of the American Educational Research Association (AERA) Annual Meeting.

Driver, R., P. Newton, and J. Osborne. 2000. Establishing the norms of scientific argumentation in classrooms. *Science Education* 84 (3): 287–313.

Duschl, R. 2008. Science education in three-part harmony: Balancing conceptual, epistemic, and social learning goals. *Review of Research in Education* 32: 268–291.

Duschl, R. A., and J. Osborne. 2002. Supporting and promoting argumentation discourse in science education. *Studies in Science Education* 38: 39–72.

Duschl, R., H. Schweingruber, and A. Shouse, eds. 2007. *Taking science to school: Learning and teaching science in grades K–8.* Washington, DC: National Academies Press.

Erduran, S., and M. Jimenez-Aleixandre, eds. 2008. *Argumentation in science education: Perspectives from classroom-based research.* Dordreht, Neth.: Springer Academic Publishers.

Jimenez-Aleixandre, M., M. Rodriguez, M., and R. A. Duschl. 2000. "Doing the lesson" or "doing science:" Argument in high school genetics. *Science Education* 84 (6): 757–792.

Mayr, E., ed. 1964. *On the origin of species by Charles Darwin: A facsimile of the first edition*. Cambridge, MA: Harvard University Press.

McNeill, K., and J. Krajcik. 2008a. Assessing middle school students' content knowledge and reasoning through written scientific explanations. In *Assessing science learning: perspectives from research and practice*, ed. J. Coffey, R. Douglas, and C. Stearns. Arlington, VA: National Science Teachers Association (NSTA) Press.

McNeill, K., and J. Krajcik. 2008b. Scientific explanations: Characterizing and evaluating the effects of teachers' instructional practices on student learning. *Journal of Research in Science Teaching* 45 (1): 53–78.

McNeill, K. L., D. J. Lizotte, J. Krajcik, and R. W. Marx. 2006. Supporting students' construction of scientific explanations by fading scaffolds in instructional materials. *The Journal of the Learning Sciences* 15 (2): 153–191.

National Governors Association Center (NGA) for Best Practices, and Council of Chief State School Officers (CCSSO). 2010. *Common core state standards for English language arts and literacy*. Washington, DC: National Governors Association for Best Practices, Council of Chief State School.

National Research Council (NRC). 2005. *America's lab report: Investigations in high school science*. Washington, DC: National Academies Press.

National Research Council (NRC). 2008. *Ready, set, science: Putting research to work in K–8 science classrooms*. Washington, DC: National Academies Press.

National Research Council (NRC). 2012. *A framework for K–12 science education: Practices, crosscutting concepts, and core ideas*. Washington, DC: National Academies Press.

Norris, S., L. Philips, and J. Osborne. 2007. Scientific inquiry: The place of interpretation and argumentation. In *Science as Inquiry in the Secondary Setting*, ed. J. Luft, R. Bell and J. Gess-Newsome. Arlington, VA: NSTA Press.

Osborne, J., S. Erduran, and S. Simon. 2004. Enhancing the quality of argumentation in science classrooms. *Journal of Research in Science Teaching* 41 (10): 994–1020.

Price Schleigh, S., M. Bosse, and T. Lee. 2011. Redefining curriculum integration and professional development: In-service teachers as agents of change. *Current Issues in Education* 14 (3).

Sampson, V., and M. Blanchard. Forthcoming. Science teachers and scientific argumentation: Trends in practice and views. *Journal of Research in Science Teaching*.

Sampson, V., and D. Clark. 2008. Assessment of the ways students generate arguments in science education: Current perspectives and recommendations for future directions. *Science Education* 92 (3): 447–472.

Sampson, V., and D. Clark. 2009. The effect of collaboration on the outcomes of argumentation. *Science Education* 93 (3): 448–484.

Sampson, V., J. Grooms, and J. Walker. 2011. Argument-driven inquiry as a way to help students learn how to participate in scientific argumentation and craft written arguments: An exploratory study. *Science Education* 95 (2): 217–257.

Sandoval, W. A., and B. J. Reiser. 2004. Explanation driven inquiry: Integrating conceptual and epistemic scaffolds for scientific inquiry. *Science Education* 88 (3): 345–372.

# Introduction

Many science educators view inquiry as a key component of any effort to help students develop science proficiency (Duschl, Schweingruber, and Shouse 2007; NRC 2008, 2012). Scientific inquiry refers to the diverse ways in which scientists study the natural world and propose explanations based on the evidence derived from their work. Inquiry refers to the understanding of how scientists study the natural world as well as the activities that students engage in when they attempt to develop knowledge and understanding of scientific ideas (NRC 1999). Students who learn science through inquiry are able to participate in many of the same activities and thinking processes as scientists do when they are seeking to expand our understanding of the natural world (NRC 2000). Yet educators seeking to engage students in inquiry inside the classroom do not always emphasize many of the activities and thinking processes used by scientists to generate and evaluate scientific knowledge.

Within the context of schools, scientific inquiry is often conceptualized as a straightforward process of "asking a question, devising a means to collect data to answer the question, interpreting this data, and then drawing a conclusion" (Sandoval and Reiser 2004, p. 345). Instruction, therefore, tends to focus on helping students master specific *skills* that are important to this process. Examples of such skills are formulating good research questions, designing controlled experiments, making careful observations, and organizing or graphing data. Although these types of skills are an important part of the inquiry process, they are often overemphasized at the expense of other important practices in inquiry such as proposing and testing alternatives, judging the quality or reliability of evidence, evaluating the potential viability of scientific claims, and constructing scientific arguments. As a result, typical science classrooms tend to place too much emphasis on individual exploration and the importance of experimentation in the inquiry process, which can cause students to develop an inaccurate understanding of how scientists study the natural world and how new knowledge is generated, justified, and evaluated by scientists (Duschl and Osborne 2002; Lederman and Abd-El-Khalick 1998; Osborne 2002; Sandoval 2005).

In light of this issue, *A Framework for K–12 Science Education: Practices, Crosscutting Concepts, and Core Ideas* highlights a

## INTRODUCTION

set of practices—such as asking questions, developing and using models, analyzing data, and communicating information—that students need to learn in order to be able to engage in inquiry (NRC 2012). The *Framework* also calls for explanation and argument to play a more central role in the teaching and learning of science. The *Framework* views explanation and argument as both the goal of an inquiry and the means to get there; that is, students construct explanations and supporting arguments in order to understand the phenomenon under investigation, and they also use explanation and argument as a guide to engage in the inquiry process (Bell and Linn 2000; Goldman et al. 2002; Sandoval and Reiser 2004). The National Research Council (NRC) made argumentation a foundation of the new framework because:

> All ideas in science are evaluated against alternative explanations and compared with evidence; acceptance of an explanation is ultimately an assessment of what data are reliable and relevant and a decision about which explanation is the most satisfactory. Thus knowing why the wrong answer is wrong can help secure a deeper and stronger understanding of why the right answer is right. Engaging in argumentation from evidence about an explanation supports students' understanding of the reasons and empirical evidence for that explanation, demonstrating that science is a body of knowledge rooted in evidence. (2012, p. 44)

In order to make engaging in argument from evidence an important practice *within* a science classroom, teachers need to help students develop the abilities and habits of mind needed to generate explanations and evaluate the conclusions or claims put forth by others. Teachers, therefore, need to give students opportunities to learn how to articulate a claim, support it with evidence, respond to critiques, and revise a claim based on feedback or new evidence. This type of focus supports learning by establishing a context for students that allows them to contrast varied forms of evidence, link evidence to methods, explore the criteria for selecting evidence, and reflect on the nature of scientific investigation (Abell, Anderson, and Chezem 2000). Driver et al. (1994) argue that these types of goals are not additional extraneous aspects of science but instead represent an essential element of science education. Jimenez-Aleixandre et al. emphasize the same idea:

> Argumentation is particularly relevant in science education since a goal of scientific inquiry is the generation and justification of knowledge claims, beliefs and actions taken to understand nature. Commitments to theory, methods, and aims are the outcome of critical evaluation and debates among communities of scientists. (2000, p. 758)

Current research in science education also supports calls to integrate argumentation in the teaching and learning. First, several studies have demonstrated that students who engage in argumentation as part of an inquiry often change or refine their image of science (Bell and Linn 2000; Price Schleigh, Bosse, and Lee 2011) or enhance their understanding of the nature of scientific knowledge (Yerrick 2000), because learners are able to experience the nature of science firsthand (Driver et al. 1994; Duschl 2000). Second, several studies have shown that students can learn to develop a better understanding of important content knowledge by engaging in argumentation (Bell and Linn 2000; Zohar and Nemet 2002). Third, current research indicates that argumentation encourages learners to develop different ways of thinking, because they have more opportunities to engage in the reasoning and discursive practices of scientists (Brown and Palincsar 1989; Kuhn 1993; Sandoval and Millwood 2005). Finally, research has demonstrated that opportunities to engage in argumentation as part of the inquiry process can improve students' investigative competencies (Sandoval and Reiser 2004; Tabak et al. 1996). Taken together, these studies provide strong support for efforts to integrate argumentation into science education.

There are a number of strategies or approaches that biology teachers can use to integrate argumentation into the teaching and learning of biology. One approach, which is frequently described in the science education literature, involves engaging students in the production and evaluation of scientific arguments. This approach frames the goal of inquiry as the construction of a good argument that provides and justifies a conclusion, explanation, or some other answer to a research question. Students develop one or more ways to investigate the phenomenon, make sense of the data they gather, and produce an argument that makes clear their understanding. The quality of these arguments then becomes the focal point of discussion in the classroom as students evaluate and critique methods, explanations, evidence, and reasoning (Erduran and Jimenez-Aleixandre 2008; Sandoval and Reiser 2004).

Another common framework for promoting and supporting scientific argumentation in classrooms has focused on designing activities or tasks that require students to examine and evaluate alternative theoretical interpretations of a particular phenomenon (Erduran and Jimenez-Aleixandre 2008; Monk and Osborne 1997; Osborne, Erduran, and Simon 2004). This type of approach provides opportunities for students to examine competing explanations, evaluate the evidence that does or does not support each perspective, and construct arguments justifying the case for one explanation or another.

Finally, teachers can also engage students in argumentation by requiring them to write a refutational essay. A refutational essay—which is designed to give students an opportunity to not only write the research question. To do this, each group of students need to be encouraged to first make sense of the provided measurements (e.g., size, temperature) or observations (e.g., appearance, location, behavior) by looking for trends over time, difference between groups, or relationships between variables. Once the groups have examined and analyzed the data, they are instructed to create a tentative argument that consists of (1) their answer to the research question, (2) their evidence (the data that has been analyzed and interpreted), and (3) a rationale (i.e., a statement that explains why the evidence they decided to use is important or relevant) on a medium that can be easily viewed by their classmates (see Figure 3). We recommend using a 2 ft. × 3 ft. whiteboard, such as the example shown in Figure 4, a large piece of butcher paper, or a digital display on a group computer.

The intention of this stage is to provide students with an opportunity to make sense of what they are seeing or doing. As students work together to create a tentative argument, they must talk with each other and determine how to analyze the data and how to best interpret the trends, difference, or relationships that they uncover. They must also decide if the evidence (i.e., data that have been analyzed and interpreted) they decide to include in their argument is relevant, sufficient, and convincing enough to support their claim. This, in turn, enables students to evaluate competing ideas and weed out any claim that is inaccurate,

*Figure 3. The Components of an Argument for Stage 2 of the Generate an Argument Instructional Model*

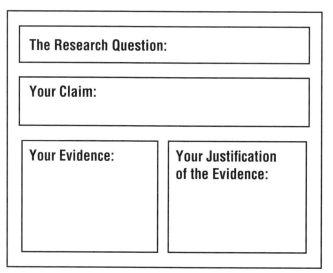

## Figure 4. An Example of an Argument Created by High School Students

> **Question:** Which of these Patients could have Cancer?
> **Claim:** Slide two is an example of a cancer patient.
>
> **Evidence**
>
> | Cell Type | Total Cell | Cell's in Mitosis | % of cells in Mitosis |
> |---|---|---|---|
> | Onion | 376 | 34 | 9% |
> | Parascarius | 36 | 4 | 11% |
> | Pat. 1 | 432 | 56 | 12% |
> | Pat. 2 | 331 | 69 | 20% |
>
> **Justification:**
> 1. Almost cells spend their time in interphase
> 2. If that is the case, we can infer that only a small % of healthy cells are in mitosis
> 3. After evaluating 2 different normal samples of healthy cells, we saw a 9% & 11% concentration of cells in Mitosis.
> 4. Patient 1 had a 12% Mitosis rate consistent w/ healthy cells, Patient 2 had double of the % of Mitosis rate.
> 5. Therefore, since we know cancer cells are the result of unchecked Mitosis, then patient 2 would the most probable candidate for Cancer.

contains contradictions, or does not fit with all the available data.

This stage is also designed to focus students' attention on the importance of argument in science. In other words, students need to understand that scientists must be able to support a conclusion, explanation, or an answer to a research question with appropriate evidence and then justify their use or choice of evidence with an adequate rationale. It also helps students develop new standards for what counts as high-quality evidence and a sufficient or adequate rationale (i.e., statements that explains why the evidence is important or relevant to the task at hand).

This stage of the model can be challenging for students because they are rarely asked to make sense of a phenomenon based on raw data. We therefore recommend that the classroom teacher circulate from group to group in order to act as a resource person for the students. It is the goal of the teacher at this stage of the model to ensure that students think about what they are doing and why. For example, teachers should ask students probing questions to help them remember the goal of the activity (e.g., *What are you trying to figure out?*), to encourage them to think about whether or not the data are relevant (e.g., *Why is that characteristic important?*), or to help them to remember to use rigorous criteria to evaluate the merits of an idea (e.g., *Does that fit with all the data or what we know about the solar system?*). It is also important to remember that students will struggle with this type of practical work at the beginning of the

*Figure 7. Example of a Round-Robin Argumentation Session*

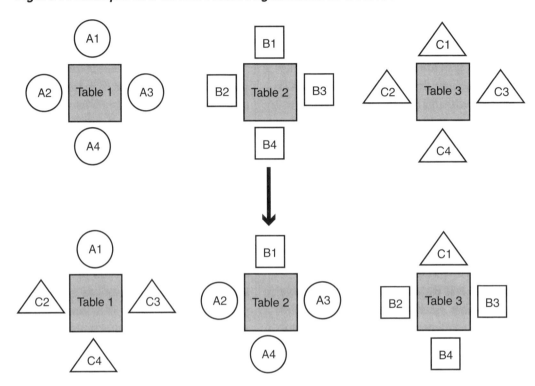

what they know and why some claims are more valid or acceptable in science. It is not the time to tell the students if they are right or wrong.

## Stage 4: A Reflective Discussion

The next stage in this instructional model is for the original groups to reconvene and discuss what they learned by interacting with individuals from the other groups. They should then modify their tentative argument as needed or conduct an additional analysis of the data. After the teacher gives the students a chance to debrief with their group, the teacher should lead a whole-class discussion. The teacher should encourage the students to explain what they learned about the phenomenon under investigation. This enables the classroom teacher to ensure the class reaches a scientifically acceptable conclusion and thinks about ways to improve the nature of their arguments in the future. The teacher can also discuss any issues that were a common challenge for the groups during the second and third stage of the activity.

## Stage 5: The Production of a Final Written Argument

In the final stage of the model, each student is required to make sense of his or her experience by producing a final argument in writing. This component is included in the instructional model because writing is an important part of doing science. Scientists must be able to read and understand the writing of others as well as evaluate its worth. They also must be able to share the results of their own research through writing. In addition, writing helps students learn how to articulate their thinking in a clear and concise manner; it encourages metacognition and improves student understanding of the content (Wallace, Hand, and Prain 2004). Finally, and perhaps most importantly, writing makes each student's thinking visible to the teacher (which facilitates assessment) and enables the teacher to provide students with the educative feedback they need to improve.

In order to help students learn how to write a persuasive and convincing scientific argument, we use the prompt provided in Figure 8. This prompt is designed to encourage students to think about what they know, how they know it, and why they accept it over alternatives. It is also designed to encourage students to think about the organization, sentence fluency, word choice, and writing conventions. Teachers can make a photocopy of the prompt for each student and have the student write his or her argument under the prompt. To reduce photocopies and paper usage, the teacher can also project the prompt on a screen by using a document camera, an overhead projector, or a computer for all students in the class to see and have students write their argument on their own piece of paper. In addition, teachers can have students write their arguments using a word processing application (or in another digital medium such as a wiki). A rubric for scoring these arguments is provided in Appendix B (p. 366). This rubric includes criteria that

## Figure 8. Writing Prompt for the Generate an Argument Instruction Model

In the space below, write an argument in order to persuade another biologist that your claim is valid and acceptable. As you write your argument, remember to do the following:

- State the claim you are trying to support
- Include genuine evidence (data + analysis + interpretation)
- Provide a justification of your evidence that explains why the evidence is relevant and why it provides adequate support for the claim
- Organize your argument in a way that enhances readability
- Use a broad range of words including vocabulary that we have learned
- Correct grammar, punctuation, and spelling errors

# INTRODUCTION

the various arguments in a small group format. We once again recommend that teachers use the round-robin structure so more students have an opportunity to determine if the data gathered by other groups is relevant, sufficient, and convincing enough to support one explanation over another.

## Stage 5: The Reflective Discussion

The next stage in this instructional model is for the original groups to reconvene and discuss what they learned by interacting with individuals from the other groups. Based on the discussion, they should then modify their tentative argument or collect and analyze additional data as needed. After the teacher gives the students a chance to debrief with their group, the teacher should lead a whole-class discussion. The teacher should, as in the Generate an Argument instructional model, encourage the students to explain what they learned about the phenomenon under investigation and to think about ways to improve the nature of their arguments in the future. The teacher should also pose questions to discuss ways to improve future investigations (e.g., *Why is it important to include a control? Why is it important that we conduct multiple trials?*).

## Stage 6: The Production of a Final Written Argument

In the last stage of the lesson, each student is required to produce a written argument in support of one of the explanations that also includes a challenge to an alternative explanation. The prompt provided in Figure 11 is included as part of the activity pages for each Evaluate Alternatives activity. This prompt is designed to encourage students to think about what they know, how they know it, and why one explanation is more valid or acceptable than the alternatives. It is also designed to encourage students to think about sentence fluency, word choice, and writing conventions. Perhaps more importantly, the writing prompt provides a summative assessment of student learning. Teachers can use the arguments and counterargument that students write to determine how well each student understands the content and how well he or she can provide evidence to support or challenge an explanation. A rubric for scoring the students' arguments is provided in Appendix C (p. 367).

## Refutational Writing Activities

This book, as discussed earlier, also includes several refutational writing activities (see Dlugokienski and Sampson 2008) that can be integrated into a unit. A refutational text introduces a common concept or idea; refutes it; offers an alternative concept, idea, or theory; and then attempts to show that this alternative way of thinking is more valid or acceptable (Guzzetti et al. 1997). An example of a refutation of the misconception that hypotheses become theories that in turn become laws can be seen in the following excerpt from a chapter written by William McComas, "The principal elements of the nature of science: Dispelling the myths of

*Figure 11. The Evaluate Alternatives Writing Prompt*

> In the space below, write a one- to three-paragraph argument to support the explanation that you think is the most valid or acceptable. Your argument must also include a challenge to one of the alternative explanations.
>
> As you write your argument, remember to do the following:
> - State the explanation you are trying to support
> - Include genuine evidence (data + analysis + interpretation)
> - Explain why the evidence is important and relevant
> - State the explanation you are trying to refute
> - Explain why the alternative explanation is invalid or unacceptable
> - Organize your argument in a way that enhances readability
> - Use a broad range of words including vocabulary that we have learned
> - Correct grammar, punctuation, and spelling errors

science." The key sentence that identifies this passage as refutational in nature is in in italics.

> [There is a] general belief that with increased evidence there is a developmental sequence through which scientific ideas pass on their way to final acceptance as mature laws. The implication is that hypotheses and theories are less secure than laws. A former U.S. president expressed his misunderstanding of science by saying that he was not troubled by the idea of evolution because it was, in his words, "just a theory." The president's misstatement is the essence of this myth; an idea is not worthy of consideration until "law-ness" has been bestowed upon it. *Theories and laws are very different kinds of knowledge, but the misconception portrays them as different forms of the same knowledge construct.* Of course there is a relationship between laws and theories, but it is not the case that one simply becomes the other—no matter how much empirical evidence is amassed. Laws are generalizations, principles, or patterns in nature and theories are the explanations of those generalizations. (Lederman and Abd-El-Khalick 1998, p. 56)

A text that is refutational in nature, such as the example provided, is one of three kinds of persuasive arguments that are often found in scientific writing (Hynd

# INTRODUCTION

2003). A one-sided persuasive argument only presents the concept, idea, or theory the author prefers a reader to adopt. Two-sided arguments can be nonrefutational or refutational. A two-sided, nonrefutational argument presents both sides of an issue but makes one side seem stronger by presenting more evidence, explaining it more logically, or in some other way making the argument more compelling yet without explicitly stating that the author prefers it. A refutational argument, in contrast, is more explicit than a nonrefutational argument about which is the preferred side.

Most textbooks and science trade books are written in an expository and authoritative style, and as a result usually do not include arguments. When they do, they often use one-sided arguments rather than refutational two-sided arguments. Thus, students are likely to be unfamiliar with this type of writing and will need explicit instruction, a great deal of practice, and good feedback in order to learn how to write in this manner. Science teachers, however, can help students learn to write a high-quality essay that is refutational in nature (and develop a better understanding of the content as part of the process) by using the refutational writing activities included in this book. These writing activities require students to produce an extended essay that refutes a common misconception related to an important biological concept (e.g., species do not evolve over time, or all bacteria cause disease) or to the nature of science (e.g., there is one scientific method, or theories turn into laws).

Each refutational writing prompt begins with a particular misconception to refute. It then outlines all the information a student will need such as the topic, the audience, the purpose, the form of the text, and reminders (Turner and Broemmel 2006). The reminders are designed to focus the writer's attention on important components of a quality refutational text that novices often forget or overlook in their writing. The prompt then concludes with information about the steps of the writing process that the student should follow (e.g., conducting research, creating an outline, producing a rough draft, editing, and publication). It also provides a space for the teacher to assign a due date for each step of the process. A rubric for scoring the argument is provided in Appendix D (p. 368).

We recommend that teachers treat these writing activities as opportunities for students to conduct literature reviews as part of the writing process. We also suggest that the essays are at least 100 words long, that students type their initial and final draft, and include properly formatted in-text citations. Students need to write to learn but also need to learn how to write in the context of science. The refutational writing activities provide students with an opportunity to do both inside the biology classroom.

## The Activities in This Book

This book includes 30 activities. These activities have been organized into three sections based on type. Ten of the activities are designed around the Generate an Argument model, and 10 are designed using the Evaluate Alternatives model. The remaining 10 activities are refutational writing activities. The investigations in many of these activities require safety considerations. Certain activities contain safety notes as needed, but before any activity, teachers should review NSTA's "Safety in the Science Classroom," which can be found at *http://www.nsta.org/pdfs/SafetyInTheScienceClassroom.pdf*.

Teachers can use these activities to integrate more scientific argumentation into the teaching and learning of biology. When teachers use several of these activities over the course of an academic year (e.g., two or three per semester), students will not only have an opportunity to learn important content (i.e., learn from scientific argumentation), but they will also learn more about scientific argumentation (i.e., what counts as evidence, how to support claims, how to evaluate scientific argument) in Biology. These activities can also be used to improve students' communication and critical-thinking skills.

## How to Use the Activities

The activities in this book are not designed to replace an existing curriculum but to supplement what teachers are already doing in the classroom. The teacher notes for each activity will suggest content that should be covered before, during, and after the activities in order to best foster student learning. The teacher notes also highlight the aspects of *A Framework for K–12 Science Education* that are aligned with each activity and the particular Common Core State Standards for English Language Arts and Literacy that the activity addresses. Lastly, the teacher notes also provide some suggestions for how to implement the activity in a particular context. It is suggested that teachers review the Curricular and Instructional Considerations section of each activity's teacher notes to best determine how the activity might supplement an existing curriculum. While we believe that the purpose of the activity is to help students understand important content and practices in science, teachers often need guidance about when to implement an activity and what to do before, during, and after a lesson. Reviewing this section will help teachers make these types of decisions.

The activities are flexible in that they can be used to at different points in the curriculum. A teacher can use these activities as a way to introduce students to new content or as a way to give students an opportunity to apply a theory, law, or unifying concept to a novel situation. Teachers can even use these activities as a way to allow students to demonstrate what they have learned after an instructional unit. To support student learning, we provide research related to misconceptions and suggestions to address the misconceptions.

In the Recommendations for Implementing the Activity section, we provide information about what teachers should look for while teaching and strategies

## INTRODUCTION

*Table 1. Generate an Argument Instructional Model Teacher Behaviors*

| Stage | What the teacher does that is ... | |
|---|---|---|
| | Consistent with the Generate an Argument instructional model | Inconsistent with the Generate an Argument instructional model |
| 1. The Identification of the Problem and the Research Question | • Sparks the students' curiosity<br>• Creates a need for the students to develop arguments<br>• Organizes the students into collaborative groups<br>• Supplies the students with the materials they will need<br>• Provides the students with hints | • Provides students with possible answers to the research question<br>• Allows students to organize into groups of existing consensus<br>• Tells students that there is one correct answer |
| 2. The Generation of a Tentative Argument | • Reminds students of the research questions and what counts as appropriate evidence in science<br>• Requires students to generate an argument that provides and supports a claim with genuine evidence<br>• Suggests that a model, diagram, or representation is created<br>• Asks students what opposing ideas or rebuttals they might anticipate<br>• Provides related theories and reference materials as tools | • Requires only one student to be prepared to discuss the argument<br>• Moves to groups to check on progress without asking students questions about why they are doing what they are doing<br>• Does not interact with students (uses the time to catch up on other responsibilities)<br>• Does not expect students to address validity or reliability of data collection<br>• Tells students which theories are best to support their ideas |
| 3. Argumentation Session | • Reminds students of appropriate and safe behaviors in the learning community<br>• Encourages students to ask peers the questions that the teacher asked in the previous stage<br>• Keeps the discussion focused on the evidence and data<br>• Encourages students to use appropriate criteria for determining what does and does not count | • Tells students when a good point was posed<br>• Allows students to negatively respond to others<br>• Asks questions about students' claims before other students can ask<br>• Allows students to be satisfied with ideas that are not supported by evidence<br>• Allows students to use inappropriate criteria for determining what does and does not count |
| 4. Reflective Discussion | • Encourages students to discuss what they learned about the content and how they know what they know<br>• Encourages students to discuss what they learned about the nature of science<br>• Encourages students to discuss ways in which they could be more productive in the future | • Provides a lecture on the content<br>• Provides a lecture about the nature of science<br>• Tells students what they should have learned or identifies what they should have figured out |
| 5. The Production of a Final Written Argument | • Provides an authentic purpose for the writing of the final argument<br>• Reminds students about the audience, topic, and purpose<br>• Provides a rubric in advance | • Places emphasis on spelling and grammar<br>• Moves on to the next activity or topic without providing feedback |

*Research in Science Teaching* 41 (10): 994–1020.

Price Schleigh, S., M. Bosse, and T. Lee. 2011. Redefining curriculum integration and professional development: In-service teachers as agents of change. *Current Issues in Education* 14 (3).

Sampson, V., and J. Grooms. 2009. Promoting and supporting scientific argumentation in the classroom: The evaluate alternatives instructional model. *Science Scope* 32 (10): 67–73.

Sampson, V., and J. Grooms. 2010. Generate an argument: An instructional model. *The Science Teacher* 77 (5): 33–37.

Sandoval, W. A. 2005. Understanding students' practical epistemologies and their influence on learning through inquiry. *Science Education* 89 (4): 634–656.

Sandoval, W. A., and K. Millwood. 2005. The quality of students' use of evidence in written scientific explanations. *Cognition and Instruction* 23 (1): 23–55.

Sandoval, W. A., and B. J. Reiser. 2004. Explanation driven inquiry: Integrating conceptual and epistemic scaffolds for scientific inquiry. *Science Education* 88 (3): 345–372.

Tabak, I., B. K Smith, W. A. Sandoval, and B. J. Reiser. 1996. Combining general and domain-specific strategic support for biological inquiry. In *Proceedings of the Third International Conference on Intelligent Tutoring Systems (ITS '96)*, eds. C. Frasson, G. Gauthier and A. Lesgold, Montreal, New York: Springer-Verlag.

Turner, T., and A. Broemmel. 2006. Fourteeen writing strategies. *Science Scope* (Dec): 27–31.

Wallace, C., B. Hand, and V. Prain, eds. 2004. *Writing and learning in the science classroom*. Boston: Kluwer Academic.

Yerrick, R. 2000. Lower track science students' argumentation and open inquiry instruction. *Journal of Research in Science Teaching* 37 (8): 807–838. Zohar, A., and F. Nemet. 2002. Fostering students' knowledge and argumentation skills through dilemmas in human genetics. *Journal of Research in Science Teaching* 39 (1): 35–62.

# GENERATE AN ARGUMENT

***Framework* Matrix** ........................................................... 2

**Activity 1:** Classifying Birds in the United States ................ 5
    *(Species Concept)*

**Activity 2:** Color Variation in Venezuelan Guppies ............ 19
    *(Mechanisms of Evolution)*

**Activity 3:** Desert Snakes ................................................. 29
    *(Mechanics of Evolution)*

**Activity 4:** Fruit Fly Traits ................................................ 45
    *(Genetics)*

**Activity 5:** DNA Family Relationship Analysis .................. 55
    *(Genetics)*

**Activity 6:** Evolutionary Relationships in Mammals .......... 67
    *(Genetics and Evolution)*

**Activity 7:** Decline in Saltwater Fish Populations .............. 81
    *(Ecology and Human Impact on the Environment)*

**Activity 8:** History of Life on Earth .................................. 103
    *(Trends in Evolution)*

**Activity 9:** Surviving Winter in the Dust Bowl .................. 113
    *(Food Chains and Trophic Levels)*

**Activity 10:** Characteristics of Viruses ............................ 123
    *(Characteristics of Life)*

# SECTION 1: GENERATE AN ARGUMENT

# FRAMEWORK MATRIX

| A Framework for K–12 Science Education | Classifying Birds in the United States | Color Variation in Venezuelan Guppies | Desert Snakes | Fruit Fly Traits | DNA Family Relationship Analysis | Evolutionary Relationships in Mammals | Decline in Saltwater Fish Populations | History of Life on Earth | Surviving Winter in the Dust Bowl | Characteristics of Viruses |
|---|---|---|---|---|---|---|---|---|---|---|
| **1. Scientific Practices** | | | | | | | | | | |
| Asking questions | | | | | | | | | | |
| Developing and using models | | | ■ | ■ | ■ | ■ | | | | |
| Planning and carrying out investigations | | | | | | | | | | |
| Using mathematics and computational thinking | | □ | ■ | ■ | | □ | □ | ■ | ■ | |
| Constructing explanations | ■ | ■ | ■ | ■ | ■ | ■ | ■ | ■ | ■ | ■ |
| Engaging in argument from evidence | ■ | ■ | ■ | ■ | ■ | ■ | ■ | ■ | ■ | ■ |
| Obtaining, evaluating, and communicating information | ■ | ■ | ■ | ■ | ■ | ■ | ■ | ■ | ■ | ■ |
| **2. Crosscutting Concepts** | | | | | | | | | | |
| Patterns | | ■ | ■ | ■ | ■ | ■ | | ■ | | ■ |
| Cause and effect: Mechanism and explanation | ■ | ■ | ■ | ■ | ■ | ■ | ■ | ■ | ■ | |
| Scale, proportion, and quantity | | ■ | | □ | | | | ■ | | |
| Systems and system models | | | | | | | | | ■ | |
| Energy and matter: Flows, cycles and conservation | | | | | | | □ | | ■ | |
| Structure and function | ■ | □ | ■ | ■ | | ■ | ■ | | | |
| Stability and change | | □ | | □ | | | | ■ | ■ | ■ |

■ = Strong alignment    □ = Weak alignment

# SECTION 1: GENERATE AN ARGUMENT

| A Framework for K–12 Science Education | Activities | | | | | | | | | |
|---|---|---|---|---|---|---|---|---|---|---|
| | Classifying Birds in the United States | Color Variation in Venezuelan Guppies | Desert Snakes | Fruit Fly Traits | DNA Family Relationship Analysis | Evolutionary Relationships in Mammals | Decline in Saltwater Fish Populations | History of Life on Earth | Surviving Winter in the Dust Bowl | Characteristics of Viruses |
| **3. Life Sciences Core Ideas** | | | | | | | | | | |
| From molecules to organisms: Structures and processes | | | | | ■ | ■ | | ■ | ■ | ■ |
| Ecosystems: Interactions, energy, and dynamics | | ☐ | ☐ | | | | ■ | | ■ | |
| Heredity: Inheritance and variation in traits | ■ | ■ | ■ | ■ | ■ | | | | | |
| Biological evolution: Unity and diversity | ■ | ■ | ■ | ☐ | ☐ | ■ | | ■ | | |
| **Common Core State Standards for English Language Arts and Literacy: Literacy in the Disciplines** | | | | | | | | | | |
| **1. Writing** | | | | | | | | | | |
| Text types and purposes | ■ | ■ | ■ | ■ | ■ | ■ | ■ | ■ | ■ | ■ |
| Production and distribution of writing | ■ | ■ | ■ | ■ | ■ | ■ | ■ | ■ | ■ | ■ |
| Research to build and present knowledge | ■ | ■ | ■ | ■ | ■ | ■ | ■ | ■ | ■ | ■ |
| Range of writing | ■ | ■ | ■ | ■ | ■ | ■ | ■ | ■ | ■ | ■ |
| **2. Speaking and Listening** | | | | | | | | | | |
| Comprehension and collaboration | ■ | ■ | ■ | ■ | ■ | ■ | ■ | ■ | ■ | ■ |
| Presentation of knowledge and ideas | ■ | ■ | ■ | ■ | ■ | ■ | ■ | ■ | ■ | ■ |

■ = Strong alignment   ☐ = Weak alignment

SECTION 1: GENERATE AN ARGUMENT

# CLASSIFYING BIRDS IN THE UNITED STATES (SPECIES CONCEPT) 1

Modern biological classification schemes generally contain a number of categories, each representing a group of organisms with a particular degree, or level, of relatedness to one another. Organisms that have the greatest number of shared characteristics are grouped together in the category of species. However, as important as the concept of a species is, the category itself is sometimes hard to define in practice. The following task is an example of this problem.

Figures 1.1–1.10 show 10 different birds that were recently observed in different parts of the United States.

*Figure 1.1. Bird A*

*Figure 1.2. Bird B*

*Figure 1.3. Bird C*

*Figure 1.4. Bird D*

*Figure 1.5. Bird E*

*Figure 1.6. Bird F*

*Figure 1.7. Bird G*

*Figure 1.8. Bird H*

*Figure 1.9. Bird I*

*Figure 1.10. Bird J*

All of these birds have very similar body shapes and coloration, but each one has a unique set of physical characteristics that can be used to distinguish it from the others (see Table 1.1, p. 7). As a result, some people think that these 10 birds represent 10 different species, while others think that these 10 birds represent one species consisting of many different varieties.

This has made many people wonder: **How many species do these 10 different birds represent?**

With your group, develop a claim that best answers this question. Once your group has developed your claim, prepare a whiteboard that you can use to share and justify your ideas. Your whiteboard should include all the information shown in the diagram on Figure 1.11(p. 6).

# SECTION 1: GENERATE AN ARGUMENT

## 1  CLASSIFYING BIRDS IN THE UNITED STATES

To share your work with others, we will be using a round-robin format. This means that one member of the group will stay at your workstation to share your groups' ideas while the other group members will go to the other groups one at a time in order to listen to and critique the arguments developed by your classmates.

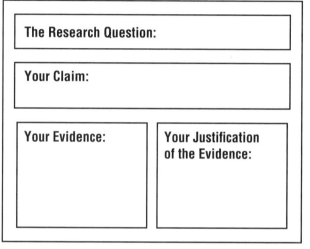

*Figure 1.11. Components of the Whiteboard*

To share your work with others, we will be using a round-robin format. This means that one member of the group will stay at your workstation to share your group's ideas while the other group members go to the other groups one at a time in order to listen to and critique the arguments developed by your classmates. Remember, as you critique the work of others, you need to decide if their conclusions are valid or acceptable based on the quality of their claim and how well they are able to support their ideas. In other words, you need to determine if their argument is convincing or not. One way to determine if their argument is convincing is to ask them some of the following questions:

- How did you analyze or interpret your data? Why did you decide to do it that way?
- How do you know that your analysis of the data is free from errors?
- Why does your evidence support your claim?
- Why did you decide to use that evidence? Why is your evidence important?
- How does your rationale fit with accepted scientific ideas?
- What are some of the other claims your group discussed before agreeing on your claim, and why did you reject them?

# SECTION 1: GENERATE AN ARGUMENT

## CLASSIFYING BIRDS IN THE UNITED STATES  1

### Table 1.1. Information About the 10 Birds

| Bird | Appearance | Characteristics |
|---|---|---|
| A | | **Habitat:** Deciduous woodlands and shade trees<br>**Range:** Washington, Oregon, California, Indiana, Nevada, Utah, Arizona, New Mexico, Texas, Montana, Wyoming, North Dakota, and South Dakota; Winters in tropics<br>**Gender:** Male<br>**Length:** 18–22 cm<br>**Diet:** Insectivorous but will eat fruit when available<br>**Song:** Clear and flutelike whistle; single or double notes in short, distinct phrases with much individual variation; also a rapid chatter<br>**Clutch Size:** Four to six grayish eggs<br>**Interactions:** Will not mate with Birds A, C, D, E, F, G, H, I, or J<br>**Behavior:** Creates a well-woven pendant bag nest that is made of plant fibers, bark, and string and is suspended from the tip of a branch |
| B | | **Habitat:** Deciduous woodlands and shade trees<br>**Range:** Washington, Oregon, California, Indiana, Nevada, Utah, Arizona, New Mexico, Texas, Montana, Wyoming, North Dakota, and South Dakota; Winters in tropics<br>**Gender:** Female<br>**Length:** 16–20 cm<br>**Diet:** Insectivorous but will eat fruit when available<br>**Song:** None<br>**Clutch Size:** Four to six grayish eggs<br>**Interactions:** Will not mate with Birds B, C, D, E, F, G, H, I, or J<br>**Behavior:** Lays eggs in a well-woven pendant bag of plant fibers, bark, and string and is suspended from the tip of a branch |
| C | | **Habitat:** Deciduous woodlands and shade trees<br>**Range:** North Dakota, South Dakota, Nebraska, Kansas, Oklahoma, Montana, Arizona, Texas, Louisiana, and Virginia; Winters in Florida and the southern Atlantic coast<br>**Gender:** Male<br>**Length:** 18–22 cm<br>**Diet:** Insectivorous but will eat fruit when available<br>**Song:** Clear and flutelike whistled single or double notes in short, distinct phrases with much individual variation<br>**Clutch Size:** Four to six grayish eggs<br>**Interactions:** Will not mate with Birds A, B, C, D, E, G, H, I, or J<br>**Behavior:** Creates a well-woven pendant bag nest that is made of plant fibers, bark, and string and is suspended from the tip of a branch |

*(continued)*

# SECTION 1: GENERATE AN ARGUMENT

## 1 CLASSIFYING BIRDS IN THE UNITED STATES

**Table 1.1.** *Information About the 10 Birds (continued)*

| Bird | Appearance | Characteristics |
|---|---|---|
| J | | **Habitat:** Tree plantations, city parks, and suburban areas with palm or eucalyptus trees and shrubbery<br>**Range:** California, Nevada, Arizona, New Mexico, and Texas<br>**Gender:** Female<br>**Length:** 18–20 cm<br>**Diet:** Insectivorous but will eat fruit when available<br>**Song:** None<br>**Clutch Size:** Three to five white eggs with dark brown and purple splotches<br>**Interactions:** Will not mate with A, B, C, E, F, G, H, I, or J<br>**Behavior:** Lays eggs in a basket nest of plant fibers with the entrance at the top, hanging from palm fronds or eucalyptus tree branches |

**Figure 1.12.** *A Map of the United States of America*

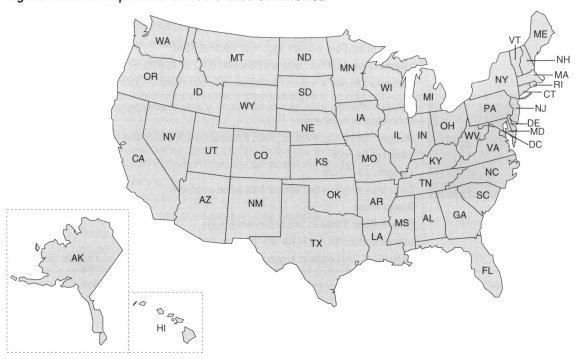

**SECTION 1: GENERATE AN ARGUMENT** 1

Name_____  Date_____

# CLASSIFYING BIRDS IN THE UNITED STATES:
## What Is Your Argument?

In the space below, write an argument in order to persuade another biologist that your claim is valid and acceptable. As you write your argument, remember to do the following:

- State the claim you are trying to support
- Include genuine evidence (data + analysis + interpretation)
- Provide a justification of your evidence that explains why the evidence is relevant and why it provides adequate support for the claim
- Organize your argument in a way that enhances readability
- Use a broad range of words including vocabulary that we have learned
- Correct grammar, punctuation, and spelling errors

SECTION 1: GENERATE AN ARGUMENT

# 1 CLASSIFYING BIRDS IN THE UNITED STATES
## TEACHER NOTES

## Purpose

The purpose of this activity is to help students understand (1) what counts as a species in the field of biology, (2) some of the various definitions for species that can be used by biologists, and (3) the challenges associated with biological classification. This activity also helps students learn how to engage in practices such as constructing explanations, arguing from evidence, and communicating information. This activity is also designed to give students an opportunity to learn how to write in science and develop their speaking and listening skills, which are important goals for literacy in science (see Standards Addressed in This Activity for a complete list of the practices, crosscutting concepts, core ideas, and literacy skills that are well-aligned with this activity).

## The Content and Related Concepts

A species can be defined as "a population or group of populations whose members have the potential to interbreed with one another in nature to produce viable, fertile offspring, but who cannot produce viable, fertile offspring with members of other species" (Campbell and Reece 2002, p. 465). This definition is known as the *biological species concept*. The basic principle underlying the biological species concept is simple: A species is a group of individuals that can exchange genetic information and is reproductively isolated from other groups of living things. A group of individuals can therefore be classified as a species when there are one or more factors that will prevent them from interbreeding with individuals from another group. These factors block genetic mixing and lead to reproductive isolation. These factors usually fall into one of two categories: Prezygotic barriers and postzygotic barriers. Prezygotic barriers hinder individuals from mating or prevent the fertilization of an egg if two individuals attempt to mate. Examples of prezygotic barriers include geographic isolation (i.e., individuals live in different regions), habitat isolation (i.e., individuals live in different habitats within the same region), temporal isolation (i.e., some organisms are only active during specific times of day or breed during specific seasons), mechanical isolation (i.e., anatomical differences that prevent copulation), and gametic isolation (i.e., egg and sperm fail to fuse to form a zygote). Postzygotic barriers, on the other hand, are factors that prevent a zygote from developing into a viable and fertile adult once sperm and egg fuse. The two most common postzygotic barriers are reduced hybrid viability (i.e., the zygote fails to develop) and reduced hybrid fertility (i.e., the offspring is sterile).

In nature, however, the biological species concept does not always work well. A bacterium, for example, reproduces by copying its

# SECTION 1: GENERATE AN ARGUMENT

## CLASSIFYING BIRDS IN THE UNITED STATES
### TEACHER NOTES 1

*Table 1.2. Classification of the 10 Birds*

| Rank | Name |
|---|---|
| Kingdom | Animalia |
| Phylum | Chordata |
| Class | Aves |
| Order | Passeriformes |
| Family | Icteridae |
| Genus | *Icterus* |

genetic material and then splitting (which is called binary fission). Therefore, defining a species as a group of interbreeding individuals only works with organisms that do not use an asexual form of reproduction. Most plants (and some animals) that use sexual reproduction can also self-fertilize, which makes it difficult to determine the boundaries of a species. Biologists are also unable to check for the ability to interbreed in extinct forms of organisms found in the fossil record. The biological species concept therefore has limitations. In order to address some of these limitations, many other species concepts have been proposed by scientists, such as the ecological species concept (which means a species is defined by its ecological niche or its role in a biological community), the morphological species concept (which means a species is defined using a unique set of shared structural features), and the genealogical species concept (which means a species is a set of organisms with a unique genetic history). The species concept that a scientist chooses to use will often reflect his or her research focus. Scientists, however, are expected to decide on a species concept, provide a rationale for doing so, and then use it consistently. Yet, scientists tend to use the biological species concept for most purposes and for communication with the general public.

All 10 birds in this activity are members of the same genus *Icterus*, or orioles (see Table 1.2 for more information about the way these birds are classified by biologists). When the biological species concept is used, the 10 birds represent six different species. Table 1.3 (p. 14) provides the species name for each bird. One of the most challenging aspects of classifying the birds is the fact that the female and male birds from the same species do not always have the same coloration. This is an example of sexual dimorphism or in this specific case, sexual dichromatism (different coloration). Sexual dichromatism in male and female birds results from sexual selection. The females tend to be most attracted to the brightest or flashiest males. Therefore, the brightest males tend to reproduce more than the dull males. The bright coloration, as a result, becomes more common in the population over time. The frequent occurrence of sexual dimor-

# SECTION 1: GENERATE AN ARGUMENT

## 1 CLASSIFYING BIRDS IN THE UNITED STATES
TEACHER NOTES

*Table 1.3. Names of the 10 Birds*

| Bird | Gender | Scientific Name | Common Name |
|---|---|---|---|
| A | Male | *Icterus bullockii* | Bullock's oriole |
| B | Female | *Icterus bullockii* | Bullock's oriole |
| C | Male | *Icterus galbula* | Baltimore oriole |
| D | Male | *Icterus cucullatus* | Hooded oriole |
| E | Male | *Icterus gularis* | Altamira oriole |
| F | Female | *Icterus galbula* | Baltimore oriole |
| G | Male | *Icterus parisorum* | Scott's oriole |
| H | Male | *Icterus pectoralis* | Spot-breasted oriole |
| I | Female | *Icterus parisorum* | Scott's oriole |
| J | Female | *Icterus cucullatus* | Hooded oriole |

phism and sexual dichromatism in nature is one reason why biologists cannot simply rely on appearance when attempting to define the boundaries of a species.

It is also important to note that the Bullock's oriole and the Baltimore oriole were once combined into a single species, called the northern oriole. This reclassification occurred after humans began planting trees on the Great Plains, which allowed the two different types of birds to extend their ranges and intermingle. At this point, the two types of birds began to interbreed, so the birds were combined into a single species. Now, it seems that in some places in the Central Plain, the birds are choosing mates of their own type (due to a behavioral prezygotic barrier). The birds are therefore considered two separate species again. This situation is an interesting example of how the biological species concept can be difficult to use in practice.

## Curriculum and Instructional Considerations

This activity can be used at several different points in a traditional biology curriculum. It can be used as part of a unit on classification, ecology, or evolution. It also may be used to either introduce students to the biological species concept or to give students a chance to apply their understanding of this concept in an unfamiliar context. If a teacher decides to use this activity as an introduction to the biological species concept, students do not need any additional information beyond what is supplied as part of the student pages in order to complete the activity. The teacher, however, will need to ask guiding questions, such as *Can*

# SECTION 1: GENERATE AN ARGUMENT

## CLASSIFYING BIRDS IN THE UNITED STATES
### TEACHER NOTES 1

*Table 1.4. Materials Needed to Implement the Activity in a Classroom of 28 Students*

| Material | Amount Needed With ... | |
|---|---|---|
| | Groups of 3 | Groups of 4 |
| Whiteboards (or chart paper)⁺ | 10 | 7 |
| Whiteboard markers (or permanent if using chart paper)⁺ | 20 | 14 |
| Copy of Student Pages (pp. 5–10)* | 28 | 28 |
| Copy of Student Page (p. 11)* | 10 | 7 |
| Copy of Appendix B (p. 366)* | 28 | 28 |

⁺ Teachers can also have students prepare their arguments in a digital medium (such as PowerPoint or Keynote).
* Teachers can also project these materials onto a screen in order to cut down on paper use.

*organisms look different and still be part of the same species?* and *What type of criteria should you use to determine if something is part of the same species?* as students attempt to make sense of the data and develop their tentative argument. The teacher will also need to explicitly discuss the concept and provide a working definition for the students as part of the reflective discussion stage of the lesson if the students are expected to develop a nuanced understanding of this important biological principle. On the other hand, if the activity is used as a way to allow students to apply their understanding of the biological species concept to an unfamiliar situation, then it will be important for the teacher to teach students about the concept before attempting to use this activity. The focus of the explicit discussion should then be on an aspect of nature of science or the nature of scientific inquiry. For example, a teacher could discuss how scientists use theories and laws to help make sense of their observations or the difference between data and evidence using what the students did during this activity as an illustrative example.

## Recommendations for Implementing the Activity

This activity takes approximately 100 minutes of instructional time to complete, but the amount of time devoted to each activity varies depending on how a teacher decides to spend time in class. See Appendix E for more information about how to implement this activity.

Table 1.4 provides information about the type and amount of materials needed to implement this activity in a classroom with 28 students with groups of four and groups of three.

## Assessment

The rubric provided in Appendix B can be used to assess the arguments crafted by each student at the end of the activity. To illustrate how

SECTION 1: GENERATE AN ARGUMENT

# COLOR VARIATION IN VENEZUELAN GUPPIES (MECHANISMS OF EVOLUTION) 2

When biologist John Endler began studying a species of wild guppy (*Poecilia reticulata*) in the 1970s, he was struck by the wide color variation among guppies from different streams and sometimes even among guppies living in different parts of the same stream. Guppies from one pool sported vivid blue and orange splotches along their sides, while those farther downstream carried only modest dots of color near their tails. The pictures in Figure 2.1 show how the coloration of guppies can range from drab to bright.

*Figure 2.1. Color Variation in Venezuelan Guppies*

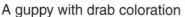

A guppy with drab coloration ⟶ A guppy with bright coloration

Endler photographed hundreds of guppies and carefully measured their size, color, and the placement of their spots. He began to see a pattern where guppies lived in a particular stream and whether the fish were bright or drab.

This led him to wonder: **What caused these trends in the coloration of the guppies?**

With your group, use the information that follows to develop a claim that best answers this question. Once your group has developed your claim, prepare a whiteboard that you can use to share and justify your ideas. Your whiteboard should include all the information shown in the diagram below.

To share your work with others, we will be using a round-robin format. This means that one member of the group will stay at your workstation to share your group's ideas while the other group members go to the other groups one at a time in order to listen to and critique the arguments developed by your classmates.

Remember, as you critique the work of others, you need to decide if their conclusions are valid or acceptable based on the quality of their claim and how well they are able to support their ideas.

## SECTION 1: GENERATE AN ARGUMENT

## 2  COLOR VARIATION IN VENEZUELAN GUPPIES

In other words, you need to determine if their argument is *convincing* or not. One way to determine if their argument is convincing is to ask them some of the following questions:

- How did you analyze or interpret your data? Why did you decide to do it that way?
- How do you know that your analysis of the data is free from errors?
- Why does your evidence support your claim?
- Why did you decide to use that evidence? Why is your evidence important?
- How does your justification of the evidence fit with accepted scientific ideas?
- What are some of the other claims your group discussed before agreeing on your claim, and why did you reject them?

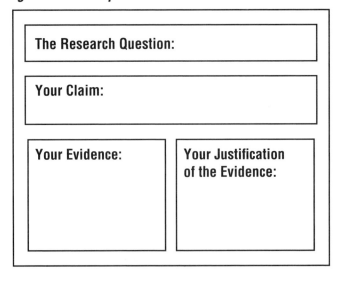

*Figure 2.2. Components of the Whiteboard*

# SECTION 1: GENERATE AN ARGUMENT

## COLOR VARIATION IN VENEZUELAN GUPPIES  2

*Table 2.1. Information About the Pools Where the Venezuelan Guppies Were Found*

| Characteristic | Pool 1 | Pool 2 | Pool 3 | Pool 4 |
| --- | --- | --- | --- | --- |
| Type | Deep (at a bend in the stream) | Deep (above a natural rock dam) | Shallow (at a bend in the stream) | Deep (above a 6 ft. waterfall) |
| Pool location (see Figure 2.3) | 50 m upstream from the river | 100 m upstream from the river | 150 m upstream from the river | 200 m upstream from the river |
| Turbidity of water (NTU) | Ranges between 27.50 and 36.25 | Ranges between 8.75 and 27.50 | Ranges between 3.00 and 8.75 | Ranges between 3.00 and 8.75 |
| Predatory fish in the pool (total) | 28 | 15 | 6 | 0 |
| Cichlids | 12 | 0 | 0 | 0 |
| Rivulus | 6 | 10 | 6 | 0 |
| Acara | 10 | 4 | 0 | 0 |
| Guppies found in the pool (total) | 102 | 165 | 187 | 231 |
| Bright males | 5 | 50 | 76 | 108 |
| Drab males | 41 | 19 | 10 | 5 |
| Bright females | 0 | 0 | 0 | 0 |
| Drab females | 56 | 96 | 101 | 118 |

*Note: Turbidity* is the cloudiness or haziness of a fluid. Nephelometric turbidity unit (NTU) range in value from 0 (completely clear) to 50 (no light passes through the fluid).

## SECTION 1: GENERATE AN ARGUMENT

## 2 COLOR VARIATION IN VENEZUELAN GUPPIES

*Figure 2.3. Map of the Pool Locations*

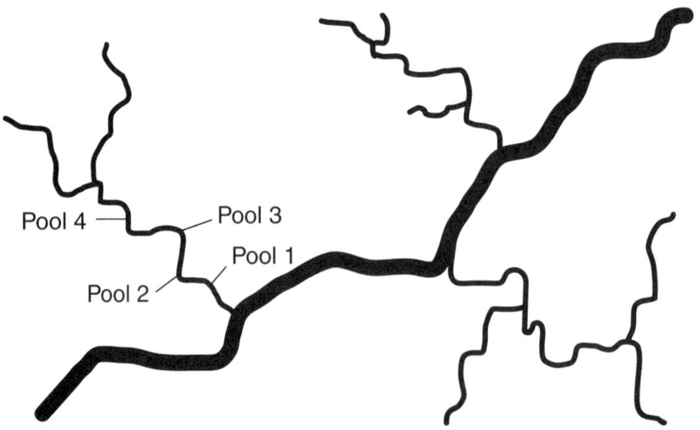

## Information About the Theory of Natural Selection

The fossil record provides convincing evidence that species evolve. In other words, the number of species found on Earth and the characteristics of these species have changed over time. However, these observations tell us little about the natural processes that drive evolution. A number of different explanations have been offered by scientists in an effort to explain why (or if) evolution occurs. One of these explanations is called natural selection. The basic tenets of natural selection are

- only a fraction of the individuals that make up a population survive long enough to reproduce;
- the individuals in a population are not all the same. Individuals have traits that make them unique;
- much but not all of this variation in traits is inheritable and can therefore be passed down from parent to offspring;
- the environment—including both abiotic (e.g., temperature and amount of water available) and biotic (e.g., amount of food and presence of predators) factors—determines which traits are favorable or unfavorable, because some traits increase an individual's chance of survival and others do not; and
- individuals with favorable traits tend to produce more offspring than those with unfavorable traits. Therefore, over time, favorable traits become more common within a population found in a particular environment (and unfavorable traits become less common). (Lawson 1995)

## Reference

Lawson, A. 1995. *Science teaching and the development of thinking.* Belmont, CA: Wadsworth.

SECTION 1: GENERATE AN ARGUMENT

Name_____ Date_____

# COLOR VARIATION IN VENEZUELAN GUPPIES:
## What Is Your Argument?

In the space below, write an argument in order to persuade another biologist that your claim is valid and acceptable. As you write your argument, remember to do the following:

- State the claim you are trying to support
- Include genuine evidence (data + analysis + interpretation)
- Provide a justification of your evidence that explains why the evidence is relevant and why it provides adequate support for the claim
- Organize your argument in a way that enhances readability
- Use a broad range of words including vocabulary that we have learned
- Correct grammar, punctuation, and spelling errors

## SECTION 1: GENERATE AN ARGUMENT

### 2 COLOR VARIATION IN VENEZUELAN GUPPIES
TEACHER NOTES

*Table 2.2. Materials Needed to Implement the Activity in a Classroom of 28 Students*

| Material | Amount Needed With ... | |
|---|---|---|
| | Groups of 3 | Groups of 4 |
| Whiteboards (or chart paper)+ | 10 | 7 |
| Whiteboard markers (or permanent if using chart paper)+ | 20 | 14 |
| Copy of Student Pages (pp. 19–22)* | 28 | 28 |
| Copy of Student Page (p. 23)* | 10 | 7 |
| Copy of Appendix B (p. 366)* | 28 | 28 |

+ Teachers can also have students prepare their arguments in a digital medium (such as PowerPoint or Keynote).
* Teachers can also project these materials onto a screen in order to cut down on paper use.

time devoted to each activity varies depending on how a teacher decides to spend time in class. For more information about how to implement the activity, see Appendix E on page 369.

Table 2.2 provides information about the type and amount of materials needed to implement this activity in a classroom with 28 students with groups of four and groups of three.

### Assessment

The rubric provided in Appendix B (p. 366) can be used to assess the arguments crafted by each student at the end of the activity. To illustrate how the rubric can be used to score an argument written by a student, consider the following example. This sample, which was written by a ninth-grade student, is an example of an argument that is weak in terms of content but adequate in terms of the writing mechanics.

> <u>Males are brighter in shallow pools because more sunlight can reach them.</u> **When the guppies are located above a natural rock dam there are 50 bright males and 19 drab males. Also it shows that 200 m upstream there are 234 guppies and 108 are bright males and 5 are drab males.** So depending on where they live the guppies can be bright or drab. This proves that the sunlight causes the males to be brighter.

The sample argument is poor for several reasons. The student's claim (underlined) is insufficient (0/1), because it does not answer the research question. It is also inaccurate (0/1). The student uses inadequate evidence (in bold) in the argument because she uses the supplied data to make comparisons of the characteristics of the guppies in different locations (2/2) but does not provide an interpretation of her analysis (0/1). The student also does not include a sufficient justification of the evidence in her argument; she does not explain why the evidence was important by linking it to a specific principle, concept, or underlying assumption (0/2). The author, however, uses scientific terms correctly (1/1) and phrases (e.g., "it

# SECTION 1: GENERATE AN ARGUMENT

## COLOR VARIATION IN VENEZUELAN GUPPIES
### TEACHER NOTES 2

shows") that reflect the nature of science (1/1). The organization of the argument is good, because the arrangement of the sentences does not distract from the development of the main idea (1/1). Finally, she uses appropriate grammar, spelling, punctuation, and capitalization (1/1). The overall score for the sample argument, therefore, is 6 out the 12 points possible.

## Standards Addressed in This Activity

This activity can be used to address the following dimensions outlined in *A Framework for K–12 Science Education* (NRC 2012):

## Scientific Practices

- Using mathematics and computational thinking
- Constructing explanations
- Engaging in argument from evidence
- Obtaining, evaluating, and communicating information

## Crosscutting Concepts

- Patterns
- Cause and effect: Mechanism and explanation
- Scale, proportion, and quantity

## Life Sciences Core Ideas

- Heredity: Inheritance and variation of traits
- Biological evolution: Unity and diversity

This activity can be used to address the following standards for literacy in science from the *Common Core State Standards for English Language Arts and Literacy* (NGA and CCSSO 2010):

## Writing

- Text types and purposes
- Production and distribution of writing
- Research to build and present knowledge
- Range of writing

## Speaking and Listening

- Comprehension and collaboration
- Presentation of knowledge and ideas

## References

Campbell, N., and J. Reece. 2002. *Biology*. 6th ed. San Francisco, CA: Benjamin Cummings.

Lawson, A. 1995. *Science teaching and the development of thinking*. Belmont, CA: Wadsworth.

National Governors Association Center (NGA) for Best Practices, and Council of Chief State School Officers (CCSSO). 2010. *Common core state standards for English language arts and literacy*. Washington, DC: National Governors Association for Best Practices, Council of Chief State School.

National Research Council (NRC). 2012. *A framework for K–12 science education: Practices, crosscutting concepts, and core ideas*. Washington, DC: National Academies Press.

SECTION 1: GENERATE AN ARGUMENT

# DESERT SNAKES (MECHANICS OF EVOLUTION) 3

There are numerous snakes that live in the deserts of the southwest United States. Many of these snakes have red, black, and yellow stripes that are easily seen against the colors of the environments in which they live. Take the following four species (Figures 3.1–3.4) as an example:

*Figure 3.1. Sonoran Coral Snake*

*Figure 3.2. Sonoran Mountain Kingsnake*

*Figure 3.3. Milk Snake*

*Figure 3.4. Sonoran Shovel-Nosed Snake*

This observation raises an interesting question: **Why do the Sonoran coral snake, the Sonoran Mountain kingsnake, the milk snake, and the Sonoran shovel-nosed snake look so similar?**

SCIENTIFIC ARGUMENTATION **IN BIOLOGY**: 30 CLASSROOM ACTIVITIES

## SECTION 1: GENERATE AN ARGUMENT

## 3 DESERT SNAKES

With your group, develop a claim that best answers this question. Once your group has developed your claim, prepare a whiteboard that you can use to share and justify your ideas. Your whiteboard should include all the information shown in Figure 3.5.

To share your work with others, we will be using a round-robin format. This means that one member of the group will stay at your workstation to share your group's ideas while the other group members go to the other groups one at a time in order to listen to and critique the arguments developed by your classmates.

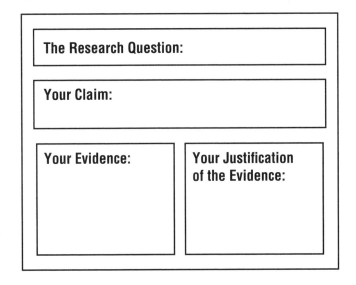

*Figure 3.5. Components of the Whiteboard*

Remember, as you critique the work of others, you need to decide if their conclusions are valid or acceptable based on the quality of their claim and how well they are able to support their ideas. In other words, you need to determine if their argument is *convincing* or not. One way to determine if their argument is convincing is to ask them some of the following questions:

- How did you analyze or interpret your data? Why did you decide to do it that way?
- How do you know that your analysis of the data is free from errors?
- Why does your evidence support your claim?
- Why did you decide to use that evidence? Why is your evidence important?
- How does your justification of the evidence fit with accepted scientific ideas?
- What are some of the other claims your group discussed before agreeing on your claim, and why did you reject them?

## SECTION 1: GENERATE AN ARGUMENT

### DESERT SNAKES  3

*Table 3.1. Information About the Desert Snakes*

| Name | Appearance | Distribution | Habitat | Diet | Reproduction | Venomous? |
|---|---|---|---|---|---|---|
| Sonoran coral snake<br><br>*Micruroides euryxanthus* | Photo © Randy Babb | | • Desert scrub<br>• Semidesert grasslands<br>• Woodlands | • Lizards<br>• Insects<br>• Scorpions<br>• Spiders<br>• Other invertebrates | Lays a clutch of 3 eggs | Yes |
| Sonoran Mountain kingsnake<br><br>*Lampropeltis pyromelana* | Photo © Tom Brennan | | • Heavily wooded, rocky slopes<br>• Steep canyon bottoms with abundant leaf-litter, fallen logs, and rocks<br>• Open rolling hills and grasslands | • Lizards<br>• Rodents<br>• Birds<br>• Bats | Lays a clutch of 9 eggs | No |
| Milk snake<br><br>*Lampropeltis triangulum* | | | • Open plains<br>• Low valleys,<br>• Semidesert Grassland | • Reptiles<br>• Mice<br>• Small mammals<br>• Birds<br>• Amphibians | Lays a clutch of 24 eggs | No |

*(continued)*

## SECTION 1: GENERATE AN ARGUMENT

## 3 DESERT SNAKES

*Table 3.1. Information About the Desert Snakes (continued)*

| Name | Appearance | Distribution | Habitat | Diet | Reproduction | Venomous? |
|---|---|---|---|---|---|---|
| Sonoran shovel-nosed snake *Chionactis palarostris* | Photo © Tom Brennan | | • Desert scrub | • Insects<br>• Scorpions<br>• Spiders<br>• Centipedes<br>• Other invertebrates | Lays a clutch of 5 eggs | No (has mildly toxic saliva) |
| Terrestrial garter snake *Thamnophis elegans* | | | • Desert scrub<br>• Grassland<br>• Woodlands<br>• Conifer Forests<br>• Usually near water | • Fish<br>• Amphibians<br>• Lizards<br>• Small mammals<br>• Birds<br>• Insects<br>• Snails<br>• Worms | Lays a clutch of 24 eggs | No |
| Western patch-nosed snake *Salvadora hexalepis* | Photo © Russ Jones | | • Desert scrub<br>• Woodlands | • Lizards<br>• Mice and other small mammals<br>• Reptile eggs<br>• Birds | Lays a clutch of 12 eggs | No |

## SECTION 1: GENERATE AN ARGUMENT
### DESERT SNAKES 3

*Table 3.2. Population Density*

| Species | Individuals/mi$^2$ |
|---|---|
| Milk snake (*Lampropeltis triangulum*) | 0.8 |
| Sonoran Mountain kingsnake (*Lampropeltis pyromelana*) | 0.3 |
| Sonoran shovel-nosed snake (*Chionactis palarostris*) | 2.7 |
| Sonoran coral snake (*Micruroides euryxanthus*) | 2.2 |
| Terrestrial garter snake (*Thamnophis elegans*) | 10.1 |
| Western patch-nosed snake (*Salvadora hexalepis*) | 7.5 |

## SECTION 1: GENERATE AN ARGUMENT

### 3 DESERT SNAKES

## Primary Snake Predators

### Raptors

Raptors such as the ferruginous hawk (Figure 3.6), the Cooper's hawk (Figure 3.7), and the red-tailed hawk (Figure 3.8) are widespread in the southwest United States.

*Figure 3.6.*
*Ferruginous Hawk*

*Figure 3.7.*
*Cooper's Hawk*

*Figure 3.8.*
*Red-Tailed Hawk*

Raptors are very agile and lively and spend a large part of their day searching for food. They are visual predators, relying on seeing movement and coloration to help them identify their prey. Raptors mainly eat birds such as mourning doves, rock pigeons, robins, several kinds of jays, Northern flicker, quail, pheasants, grouse, and chickens. Raptors eat reptiles such as lizards or snakes and small mammals such as chipmunks, hares, mice, squirrels, and bats. Reptiles and mammals are more common in diets of raptors in the West.

| Raptor Dietary Components | Units Consumed/Day |
|---|---|
| Birds | 2.1 |
| Lizards | 3.4 |
| Chipmunks and squirrels | 1.2 |
| Hares | 0.4 |
| Mice | 3.1 |
| Bats | 0.01 |
| Garter snake | 1.6 |
| Milk snake | 0.02 |
| Kingsnake | 0.01 |
| Shovel-nosed snake | 0.03 |
| Coral snake | 0.01 |
| Patch-nosed snake | 2.1 |

# SECTION 1: GENERATE AN ARGUMENT

## DESERT SNAKES 3

### Long-Tailed Weasel

The long-tailed weasel (Figure 3.9) is the only species of weasel that is found in Arizona. It is readily identifiable by its dark brown coat and orangish under parts. Some white is often present on the head, and some animals may turn all white in winter. The animals range in length from 8 to 10½ inches with the black-tipped tail adding another 4 to 6 inches.

Weasels are voracious predators. They eat cottontail rabbits, hares, and rodents. They also eat birds, snakes, and lizards.

*Figure 3.9 Long-Tailed Weasel*

| Weasel Dietary Components | Units Consumed/Day |
|---|---|
| Rabbits | 1.0 |
| Hares | 1.3 |
| Chipmunks and squirrels | 2.4 |
| Mice | 4.3 |
| Lizards | 3.6 |
| Birds | 1.2 |
| Garter snake | 3.2 |
| Milk snake | 0.02 |
| Kingsnake | 0.01 |
| Shovel-nosed snake | 0.03 |
| Coral snake | 0.04 |
| Patch-nosed snake | 5.1 |

## SECTION 1: GENERATE AN ARGUMENT

### 3 DESERT SNAKES

### Badgers

The badger is readily recognized by its grizzled gray, white, and black fur, cheek stripes, short legs, long claws, and the white stripe down its head and back (Figure 3.10). Adults may weigh from approximately 10 pounds up to 20 pounds and are approximately 26 inches long. Widely distributed, the badger is found almost everywhere in Arizona.

Badgers feed primarily on burrowing rodents such as prairie dogs and ground squirrels but also eat snakes, lizards, and even insects on occasion. Mating for these usually solitary animals takes place in the summer, the young being born the following spring due to delayed implantation. Primarily a nocturnal animal, badgers are sometimes encountered during the early morning hours.

*Figure 3.10. Badger*

| Badger Dietary Components | Units Consumed/Day |
|---|---|
| Prairie dogs | 2.5 |
| Lizards | 1.9 |
| Squirrels | 3.7 |
| Lizards | 2.3 |
| Mice | 4.1 |
| Insects | 0.3 |
| Garter snake | 0.4 |
| Milk snake | 0.01 |
| Kingsnake | 0.01 |
| Shovel-nosed snake | 0.00 |
| Coral snake | 0.02 |
| Patch-nosed snake | 0.5 |

## SECTION 1: GENERATE AN ARGUMENT

### DESERT SNAKES  3

## The Theory of Natural Selection

The fossil record provides convincing evidence that species evolve. In other words, the number of species found on Earth and the characteristics of these species has changed over time. However, these observations tell us little about the natural processes that drive evolution. A number of different explanations have been offered by scientists in an effort to explain why (or if) evolution occurs. One of these explanations is called natural selection. The basic tenets of natural selection are

- only a fraction of the individuals that make up a population survive long enough to reproduce;
- the individuals in a population are not all the same. Individuals have traits that make them unique;
- much, but not all, of this variation in traits is inheritable and can therefore be passed down from parent to offspring;
- the environment—including both *abiotic* (e.g., temperature, amount of water available) and *biotic* (e.g., amount of food, presence of predators) factors—determines which traits are favorable or unfavorable because some traits increase an individual's chance of survival and others do not; and
- individuals with favorable traits tend to produce more offspring than those with unfavorable traits. Therefore, over time, favorable traits become more common within a population found in a particular environment (and unfavorable traits become less common). (Lawson 1995)

## Reference

Lawson, A. 1995. *Science teaching and the development of thinking*. Belmont, CA: Wadsworth.

## SECTION 1: GENERATE AN ARGUMENT

Name_____  Date_____

# DESERT SNAKES:
## What Is Your Argument?

In the space below, write an argument in order to persuade another biologist that your claim is valid and acceptable. As you write your argument, remember to do the following:

- State the claim you are trying to support
- Include a sufficient amount of genuine evidence
- Provide a justification of your evidence that explains why the evidence is important and relevant by linking it a specific concept, principle, or an underlying assumption
- Organize your paper in a way that enhances readability
- Use a broad range of words including vocabulary that we have learned
- Make sure your writing has an easy flow and rhythm
- Correct grammar, punctuation, and spelling errors

SECTION 1: GENERATE AN ARGUMENT

# DESERT SNAKES
## TEACHER NOTES 3

## Purpose

The purpose of this activity is to help students understand how natural selection can shape the traits of a population found in different habitats over time and how *aposematic coloration* (i.e., the bright colors of bad tasting or venomous organisms) and *Batesian mimicry* (i.e., harmless organisms that look like dangerous ones) are adaptations that serve as a defense against predators. This activity also helps students learn how to engage in practices such as using models, constructing explanations, arguing from evidence, and communicating information. This activity is also designed to give students an opportunity to learn how to write in science and develop their speaking and listening skills, which are important goals for literacy in science (see Standards Addressed in this Activity for a complete list of the practices, crosscutting concepts, core ideas, and literacy skills that are aligned with this activity).

## The Content and Related Concepts

Natural selection is a mechanism that drives evolution. Biological evolution is defined as descent with modification. This definition includes both small-scale evolution, which refers to a change in gene frequency from one generation to the next, and large-scale evolution, which refers to the descent of different species from a common ancestor over many generations. Natural selection is a relatively simple process and consists of five basic components (Lawson 1995): (1) In a population of organisms, only a fraction of the individuals that make up a population survive long enough to reproduce; (2) the individuals in that population are not all the same. Individuals have traits that make them unique; (3) most of these traits are inheritable and can therefore be passed down from parent to offspring; (4) factors in the environment, such as temperature, amount of food available, and presence of predators, determine which traits are favorable or unfavorable because some traits increase an individual's chance of survival and others do not; (5) therefore, over time, favorable traits become more common within a population found in a particular environment (and unfavorable traits become less common). These traits are often referred to as adaptation.

One type of adaptation is called aposematic coloration (Campbell and Reece 2002). Aposematic coloration is also known as *warning coloration*. Aposematic coloration is considered an adaptation, because predators are often more cautious when dealing with prey that has bright color patterns, perhaps because so many aposematic animals *are* dangerous prey. A species of prey may gain significant protection through mimicry, an adaptation in which one species mimics the appearance of another. In Batesian mimicry, a palatable or harmless

# SECTION 1: GENERATE AN ARGUMENT

## 3 DESERT SNAKES
### TEACHER NOTES

species mimics an unpalatable, venomous, or otherwise harmful species. In Müllerian mimicry, two or more unpalatable species resemble one another. This form of mimicry is advantageous, because predators learn more quickly to avoid any prey with a particular appearance.

The scenario in Activity 3 is an example of how the process of natural selection can result in aposematic coloration and Batesian mimicry. First and foremost, coloration is a highly inheritable trait in the four species of desert snakes, which means that coloration is passed down from generation to generation. The presence of predators serves as the major selective pressure in the desert environment. The Sonoran coral snake is a highly venomous snake, and its bright coloration serves as a warning signal to would-be predators. Therefore, individuals with the yellow, black, and red banding pattern have a survival advantage over individuals without this type of coloration because predators such as weasels, hawks, and badgers know to stay away from snakes that are brightly colored. The Sonoran Mountain kingsnake, the milk snake, and the Sonoran shovel-nosed snake all live within the range of the Sonoran coral snake. These species are therefore able to take advantage of Batesian mimicry. These species have a similar yellow, black, and red banding pattern although they are not venomous. These snakes, however, have a survival advantage in the deserts of Arizona because they have the same aposematic coloration of the venomous Sonoran coral snake.

## Curriculum and Instructional Considerations

This activity is best used as part of a unit on evolution. It can also be used to either introduce students to the process of natural selection and the concepts of aposematic coloration and Batesian mimicry or to give students a chance to apply their understanding of them in an unfamiliar context. If a teacher decides to use this activity as an introduction to natural selection, aposematic coloration, and Batesian mimicry, students do not need any additional information beyond what is supplied in the student pages in order to complete the activity. In this case, the teacher should not give the students the part of the handout labeled *The Theory of Natural Selection*, because it will be unfamiliar to the students. The teacher, however, will need to explicitly discuss the concept and provide a working definition for the students as part of the reflective discussion stage of the lesson if the students are expected to develop a nuanced understanding of this important biological principle. On the other hand, if the teacher uses the activity as a way to give students an opportunity to apply their understanding of natural selection, aposematic coloration, and Batesian mimicry in an unfamiliar situation, then it will be important for the teacher to teach students about the concept before attempting to use this activity. The focus of the explicit discussion should then be on an aspect of the nature of science or the nature of scientific inquiry. For example, a teacher could discuss how all scientific knowledge is, in principle, subject to change as new evidence

## SECTION 1: GENERATE AN ARGUMENT

### DESERT SNAKES
### TEACHER NOTES 3

becomes available or how scientific explanation must be consistent with observational evidence about nature and must make accurate predictions, when appropriate, about systems being studied using what the students did during this activity as an illustrative example.

## Recommendations for Implementing the Activity

This activity takes approximately 100 minutes of instructional time to complete, but the amount of time devoted to each activity varies depending on how a teacher decides to spend time in class. For more information on how to implement the activity, see Appendix E on page 369.

Table 3.3 provides information about the type and amount of materials needed to implement this activity in a classroom with 28 students with groups of four and groups of three.

## Assessment

The rubric provided in Appendix B (p. 366) can be used to assess the arguments crafted by each student at the end of the activity. To illustrate how the rubric can be used to score an argument written by a student, consider the following example. This sample argument, which was written by an eighth-grade student, is an example of an argument that is weak in terms of content but adequate in terms of writing mechanics.

> All four of the Arizona-native snakes look similar for two main reasons. The first reason is because they are all commonly found in the same type of rocky habitat. Therefore, they need to be built for the same kind of lifestyle. The second reason they

*Table 3.3. Materials Needed to Implement the Activity in a Classroom of 28 Students*

| Material | Amount Needed With ... | |
|---|---|---|
| | Groups of 3 | Groups of 4 |
| Whiteboards (or chart paper)+ | 10 | 7 |
| Whiteboard markers (or permanent if using chart paper)+ | 20 | 14 |
| Copy of Student Pages (pp. 29–37)* | 28 | 28 |
| Copy of Student Page (p. 38)* | 10 | 7 |
| Copy of Appendix B (p. 366)* | 28 | 28 |

+ Teachers can also have students prepare their arguments in a digital medium (such as PowerPoint or Keynote).
* Teachers can also project these materials onto a screen in order to cut down on paper use.

## SECTION 1: GENERATE AN ARGUMENT

### 3 DESERT SNAKES
#### TEACHER NOTES

look so similar is because one of the snakes is venomous. The other three developed the same colors so that predators are less likely to eat them. We also have quite a bit of evidence to prove that we are correct about our claim. **For starters, all four of the Arizona-native snakes do live in the same type of habitat. Which means that they would need to have similar adaptations in order to live in that environment. They are all marked with black bands, or rings, going all the way down their bodies. This leads to another piece of evidence that we discovered, which is a well known quote, "red and black is a friend of Jack but red and yellow will kill a fellow." We believe this to be evidence because of the snakes colors.** Our rationale goes as followed: All the snakes live in an open plain-like rocky areas, so they have to learn to adapt to the same conditions. All these snakes are also nocturnal and all similar in size. Finally, al the snakes are also non-venomous, with the exception of the coral snake.

The example argument is weak for several reasons. The student's claim (underlined) is sufficient (1/1) because it provides a complete answer the research question, but it is inaccurate (0/1) because it implies that some of the species snakes changed colors in order to look more like venomous snakes (which is a Lamarckian explanation rather than a Darwinian explanation for adaptations). The student also does not use genuine evidence (in bold) because she does not use the supplied data to support the idea that similar bright coloration provides a selective advantage. The evidence comprises several unsubstantiated inferences (0/3). The student also does not include a sufficient justification of the evidence in her argument because she does not explain why the evidence was important by linking it to a specific principle, concept, or underlying assumption (0/2). Instead, she simply restates her original claim (which, again, is based on a need or choice-based reason for an adaptation). The author uses scientific terms correctly (1/1) but uses phrases (e.g., "to prove we are correct", "evidence we discovered") that misrepresent the nature of science (0/1). The organization of the argument is sufficient (1/1) because the arrangement of the sentences does not distract from the development of the main idea (0/1), but there are significant punctuation (0/1) and grammatical errors (0/1) in the argument. The overall score for the sample argument, therefore, is 3 out the 12 points possible.

## Standards Addressed in This Activity

This activity can be used to address the following dimensions outlined in *A Framework for K–12 Science Education* (NRC 2012):

### Scientific Practices

- Developing and using models
- Using mathematics and computational thinking

## SECTION 1: GENERATE AN ARGUMENT

### DESERT SNAKES
#### TEACHER NOTES 3

- Constructing explanations
- Engaging in argument from evidence
- Obtaining, evaluating, and communicating information

## Crosscutting Concepts

- Patterns
- Cause and effect: Mechanism and explanation
- Structure and function

## Life Sciences Core Ideas

- Heredity: Inheritance and variation of traits
- Biological evolution: Unity and diversity

This activity can be used to address the following standards for literacy in science from the *Common Core State Standards for English Language Arts and Literacy* (NGA and CCSSO 2010):

## Writing

- Text types and purposes
- Production and distribution of writing
- Research to build and present knowledge
- Range of writing

## Speaking and Listening

- Comprehension and collaboration
- Presentation of knowledge and ideas

## References

Campbell, N., and J. Reece. 2002. *Biology*. 6th ed. San Francisco, CA: Benjamin Cummings.

Lawson, A. 1995. *Science teaching and the development of thinking*. Belmont, CA: Wadsworth.

National Governors Association Center (NGA) for Best Practices, and Council of Chief State School Officers (CCSSO). 2010. *Common core state standards for English language arts and literacy*. Washington, DC: National Governors Association for Best Practices, Council of Chief State School.

National Research Council (NRC). 2012. *A framework for K–12 science education: Practices, crosscutting concepts, and core ideas*. Washington, DC: National Academies Press.

SECTION 1: GENERATE AN ARGUMENT

# FRUIT FLY TRAITS (GENETICS) 4

There are many different ways a trait can be passed down from a set of parents to their offspring. A trait, for example, can be determined by one gene that consists of two alleles, with one allele being dominant over the other. The two alleles of a gene can also be co-dominant or one allele can have incomplete dominance over the other allele. Sometimes, a gene can have more than two alleles as in the case of multiple allele traits. To complicate things even further, traits can also be determined on the basis of sex. Finding out how a trait is inherited (i.e., the mode of inheritance) is often difficult. To illustrate this problem, consider the following situation.

Fruit flies (*Drosophila melanogaster*) are small insects that are often found in homes, restaurants, supermarkets, and wherever else food could be left out or rotten. Fruit flies lay their eggs near the surface of any moist organic material. Upon emerging, the tiny larvae continue to feed. The reproductive potential of fruit flies is enormous; given the opportunity, they will lay approximately 500 eggs, and the entire lifecycle from egg to adult can be completed in about a week.

Typical, or wild-type fruit flies have brick-red eyes, are yellow-brown in color, have transverse black rings across their abdomen, and wings that extend beyond their abdomen (Figure 4.1). However, some fruit flies have an abnormal trait. Some fruit flies, for example, have a light yellow body (Figure 4.2) or have nonfunctional wings that are curled (Figure 4.3).

*Figure 4.1.*
*A Typical or Wild-Type Drosophila*

*Figure 4.2.*
*A Mutant Drosophila With a Yellow Body*

*Figure 4.3.*
*A Mutant Drosophila With Curly Wings*

Artwork © The Exploratorium

This has made many scientists wonder: **Which mode of inheritance, if any, do the yellow body trait and the curly wing trait follow?**

SCIENTIFIC ARGUMENTATION **IN BIOLOGY: 30 CLASSROOM ACTIVITIES**

## SECTION 1: GENERATE AN ARGUMENT

## 4  FRUIT FLY TRAITS

With your group, use the data provided to develop a claim that best answers this question. Once your group has developed your claim, prepare a whiteboard that you can use to share and justify your ideas. Your whiteboard should include all the information shown in Figure 4.4.

To share your work with others, we will be using a round-robin format. This means that one member of the group will stay at your workstation to share your group's ideas while the other group members go to the other groups one at a time in order to listen to and critique the arguments developed by your classmates.

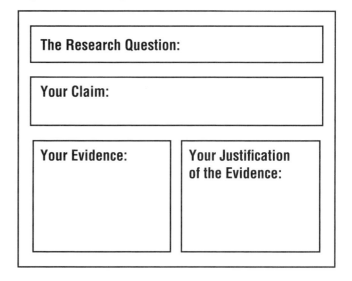

*Figure 4.4. Components of the Whiteboard*

Remember, as you critique the work of others, you need to decide if their conclusions are valid or acceptable based on the quality of their claim and how well they are able to support their ideas. In other words, you need to determine if their argument is *convincing* or not. One way to determine if their argument is convincing is to ask them some of the following questions:

- How did you analyze or interpret your data? Why did you decide to do it that way?
- How do you know that your analysis of the data is free from errors?
- Why does your evidence support your claim?
- Why did you decide to use that evidence? Why is your evidence important?
- How does your justification of the evidence fit with accepted scientific ideas?
- What are some of the other claims your group discussed before agreeing on your claim, and why did you reject them?

## Information About the Results of the Various Fruit Fly Crosses

In order to determine a mode of inheritance for a specific trait (such as dominant-recessive, co-dominance, incomplete dominance, multiple allele, or sex-linked), scientists must breed organisms together and then track the number of individuals with and without the trait over several generations. The information found in Tables 4.1–4.6 are the results of several different crosses that were done by a group of researchers in order to determine how the yellow body trait and the curly wing

## SECTION 1: GENERATE AN ARGUMENT

### FRUIT FLY TRAITS  4

trait are passed down from parent to offspring. Some of this information may be helpful to you and your group, and some of it may not.

The flies that were used in each parental (P) generation were true breeders. This means that the flies are homozygous for their traits; for those traits they possess only type of allele. So if you breed a fly with red (wild-type) eyes with another fly with red eyes, all of their offspring will have red eyes. This is true for each of the crosses.

*Table 4.1. Cross A: Wild-Type Male × Yellow Female*

| Generation | Male | | Female | |
|---|---|---|---|---|
| | Wild | Yellow | Wild | Yellow |
| 1–P (Parental) | 1 | 0 | 0 | 1 |
| 2–F1 (First Offspring) | 0 | 61 | 44 | 0 |
| 3–F2 (Second Offspring)* | 68 | 63 | 57 | 62 |

* The offspring of one yellow male and one wild-type female from the F1 generation

*Table 4.2. Cross B: Yellow Male × Wild-Type Female*

| Generation | Male | | Female | |
|---|---|---|---|---|
| | Wild | Yellow | Wild | Yellow |
| 1–P (Parental) | 0 | 1 | 1 | 0 |
| 2–F1 (First Offspring) | 52 | 0 | 57 | 0 |
| 3–F2 (Second Offspring)* | 88 | 43 | 114 | 0 |

* The offspring of one wild-type male and one wild-type female from the F1 generation

*Table 4.3. Cross C: Yellow Male × Yellow Female*

| Generation | Male | | Female | |
|---|---|---|---|---|
| | Wild | Yellow | Wild | Yellow |
| 1–P (Parental) | 0 | 1 | 0 | 1 |
| 2–F1 (First Offspring) | 0 | 84 | 0 | 91 |
| 3–F2 (Second Offspring)* | 0 | 89 | 0 | 75 |

* The offspring of one yellow male and one yellow female from the F1 generation

# SECTION 1: GENERATE AN ARGUMENT

## 4 FRUIT FLY TRAITS

### Table 4.4. Cross D: Wild-Type Male × Curly Wings Female

| Generation | Male | | Female | |
|---|---|---|---|---|
| | Wild | Curly | Wild | Curly |
| 1–P (Parental) | 1 | 0 | 0 | 1 |
| 2–F1 (First Offspring) | 0 | 43 | 0 | 52 |
| 3–F2 (Second Offspring)* | 12 | 40 | 8 | 37 |

* The offspring of one wild-type male and one wild-type female from the F1 generation

### Table 4.5. Cross E: Curly Wings Male × Wild-Type Female

| Generation | Male | | Female | |
|---|---|---|---|---|
| | Wild | Curly | Wild | Curly |
| 1–P (Parental) | 0 | 1 | 1 | 0 |
| 2–F1 (First Offspring) | 0 | 64 | 0 | 71 |
| 3–F2 (Second Offspring)* | 34 | 67 | 21 | 102 |

* The offspring of one wild-type male and one wild-type female from the F1 generation

### Table 4.6. Cross F: Curly Wings Male × Curly Wings Female

| Generation | Male | | Female | |
|---|---|---|---|---|
| | Wild | Curly | Wild | Curly |
| 1–P (Parental) | 0 | 1 | 0 | 1 |
| 2–F1 (First Offspring) | 0 | 27 | 0 | 22 |
| 3–F2 (Second Offspring)* | 0 | 13 | 0 | 29 |

* The offspring of one curly wing male and one curly wing female from the F1 generation

## SECTION 1: GENERATE AN ARGUMENT

Name_____ Date_____

# FRUIT FLY TRAITS:
## What Is Your Argument?

In the space below, write an argument in order to persuade another biologist that your claim is valid and acceptable. As you write your argument, remember to do the following:

- State the claim you are trying to support
- Include genuine evidence (data + analysis + interpretation)
- Provide a justification for your evidence that explains why the evidence is relevant and why it provides adequate support for the claim
- Organize your argument in a way that enhances readability
- Use a broad range of words including vocabulary that we have learned
- Correct grammar, punctuation, and spelling errors

SECTION 1: GENERATE AN ARGUMENT

# 4 FRUIT FLY TRAITS
## TEACHER NOTES

## Purpose

The purpose of this activity is to help students understand Mendelian genetics (including the law of segregation and the law of independent assortment) and several different patterns of inheritance (such as dominant-recessive, co-dominance, incomplete dominance, multiple allele, sex-linked, and polygenic). This activity also helps students learn how to engage in practices such as using models, constructing explanations, arguing from evidence, and communicating information. This activity is also designed to give students an opportunity to learn how to write in science and develop their speaking and listening skills, which are important goals for literacy in science (see Standards Addressed in This Activity for a complete list of the practices, crosscutting concepts, core ideas, and literacy skills that are aligned with this activity).

## The Content and Related Concepts

Mendelian genetics is the basis for modern research on inheritance. This important model can be broken down into five interrelated ideas (Campbell and Reece 2002). First and foremost, the fundamental unit of inheritance is the gene, and alternative versions of a gene account for variations in inheritable characteristics. The gene for a particular inherited characteristic, such as eye color in fruit flies, resides at a specific locus (position) on a specific chromosome (Figure 4.5). Alleles are variants of a particular gene. In Figure 4.5, for example, the eye color gene exists in two versions: the allele for red eyes and the allele for white eyes. Second, an organism inherits two alleles for each character, one from each parent. This occurs because individuals inherit one chromosome for each homologous pair from each parent (see Figure 4.5). Third, if the two alleles differ, then one is fully expressed in the organism's appearance (this version of the gene is called the dominant allele), while the other one has no noticeable effect on the organism's appearance (this version of the gene is called the recessive allele). Fourth, the two alleles for each character segregate (or separate) during gamete production. Therefore, an egg or a sperm cell only gets one of the two alleles that are present in the somatic cells of the organism. This idea is known as the law of segregation and is the result from the distribution of homologous chromosomes in different gametes during the process of meiosis.

There are also a number of additional patterns or modes of inheritance that follow the basic principles of Mendelian genetics but involve different interactions between the alleles. The first mode of inheritance is called incomplete dominance. In contrast to the dominant-recessive pattern of inheritance described earlier, traits that follow an incomplete dominance pattern produce hybrids

*Figure 4.5. Allele or Alternative Versions of a Gene*

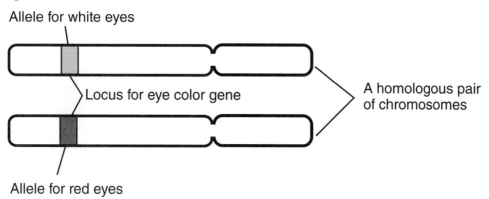

with an appearance somewhere in between the phenotypes of the two parental varieties. For example, when snapdragons that have red flowers ($C^RC^R$) are crossed with snapdragons that have white flowers ($C^WC^W$), all the snapdragons in the next generation will have pink flowers ($C^RC^W$), because one version of the allele does not have complete dominance over the other.

Another mode of inheritance is called co-dominance. In this situation, both alleles affect the phenotype of the individual in separate and distinguishable ways. For example, the human blood groups M, N, and MN are an example of the co-dominant mode of inheritance. Individuals with the blood group M are homozygous for the M allele and have M-type antigens on the surface of their red blood cells. Individuals with the blood group N are homozygous for the N allele and have N-type antigens on the surface of their red blood cells. Individuals with the blood group MN inherit a copy of each allele. The MN phenotype, however, is not an intermediate between the M and N phenotype; individuals with the MN phenotype actually have both the M- and the N-type antigens on their red blood cells.

A third mode of inheritance is called multiple allele. A multiple allele mode of inheritance simply means that there are more than two versions of a gene for a given trait with a population. The four different ABO blood types in humans are an example of a trait that follows the multiple allele mode of inheritance, because there are three alleles in the population: A, B, and O (A and B are co-dominant, and O is recessive to both A and B).

A fourth mode of inheritance is called polygenetic. In polygenetic inheritance, an additive effect of two or more genes determines the phenotype of an individual.

The fifth and final mode of inheritance that will be discussed is called sex-linked. Genes located on sex chromosomes are called sex-linked genes. Females and males differ in the number of genes they inherit when the gene is found on the sex chromosome; one gender will inherit two copies of the gene, while the other

## SECTION 1: GENERATE AN ARGUMENT

## 4 FRUIT FLY TRAITS
TEACHER NOTES

gender only inherits one (in humans, females will inherit two copies of a sex-linked gene, because they inherit two X chromosomes, while men only inherit one, because they inherit one X chromosome and one Y chromosome).

In this activity, the students are asked to determine the mode of inheritance for the yellow body trait and the curly wing trait. The yellow body color trait is sex-linked. Yellow flies have an alteration in the "yellow" allele, which is located on the X chromosome. Male flies with the genotype $X^y$ and female flies with the genotype $X^yX^y$ cannot produce black pigment as normal flies do, and as a result, appear yellow. Male flies with the genotype X and female flies with the genotype XX or $XX^y$ appear normal because they inherited a least one copy of a yellow allele that was not altered. The curly wing trait, on the other hand, follows a basic dominant-recessive inheritance pattern. Curly wing flies have an alteration in their "curly" gene, which is located on the second chromosome. The curly allele, however, is a dominant mutation. Flies with the genotype CC or Cc have curly wings, and flies with genotype cc have normal wings.

## Curriculum and Instructional Considerations

This activity is best used as part of a unit on genetics. However, it should only be used to give students a chance to apply their understanding of Mendelian inheritance in an unfamiliar context. It is therefore important for the teacher to introduce students to the concept of genes and the law of segregation as well as several different modes of inheritance such as co-dominance, incomplete dominance, multiple allele, polygenic, and sex-linked. Students should also understand how to use Punnett squares to predict the phenotypic ratios of offspring from a cross. Students will need a basic understanding of these ideas in order to be able to analyze and interpret the data that will be supplied to them during the activity. Teachers can also use this activity to introduce students to statistical hypothesis testing and the chi-square test (which is employed to test the difference between an actual sample and an expected sample).

The focus of the explicit discussion at the end of the activity should focus on an aspect of the nature of science or the nature of scientific inquiry. For example, a teacher could discuss how scientific explanation must be consistent with observational evidence about nature and must make accurate predictions, when appropriate, about systems being studied or how mathematics plays an important role in scientific inquiry using what the students did as an illustrative example.

## Recommendations for Implementing the Activity

This activity takes approximately 100 minutes of instructional time to complete but the amount of time devoted to each activity varies depending on how a teacher decides to spend time in class. For more information about how to implement the activity, see Appendix E on page 369.

Table 4.7 provides information about the type and amount of materials needed to imple-

## SECTION 1: GENERATE AN ARGUMENT

### FRUIT FLY TRAITS
### TEACHER NOTES 4

*Table 4.7. Materials Needed to Implement the Activity in a Classroom of 28 Students*

| Material | Amount Needed With … | |
| --- | --- | --- |
|  | Groups of 3 | Groups of 4 |
| Whiteboards (or chart paper)+ | 10 | 7 |
| Whiteboard markers (or permanent if using chart paper)+ | 20 | 14 |
| Copy of Student Pages (pp. 45–48)* | 28 | 28 |
| Copy of Student Page (p. 49)* | 10 | 7 |
| Copy of Appendix B (p. 366)* | 28 | 28 |

+ Teachers can also have students prepare their arguments in a digital medium (such as PowerPoint or Keynote).
* Teachers can also project these materials onto a screen in order to cut down on paper use.

ment this activity in a classroom with 28 students with groups of four and groups of three.

## Assessment

The rubric provided in Appendix B can be used to assess the arguments crafted by each student at the end of the activity. To illustrate how the rubric can be used to score an argument, consider the following example. This sample argument, which was written by an eighth-grade student, focuses only on the yellow body trait and ignores the curly wing trait. This argument is weak in terms of content but adequate in terms of mechanics (although it is rather brief).

> The mode of inheritance that the yellow body trait follows is sex-linked. We came to this conclusion by studying the information about the file crosses and then finding the following evidence. **When a male wild-type flies mates with a yellow body female it results in yellow and wild-type offspring. When a yellow male fly mates with a wild type female, on the other hand, it results in all wild-type offspring.** This proves that the yellow body trait is sex-linked because only male flies inherit the yellow body trait.

The example argument is weak for several reasons. The student's claim (underlined) is insufficient (0/1), because it provides an incomplete answer to the research question (i.e., the student never mentions the curly wing trait). The claim, however, is accurate (1/1). The student does not use genuine evidence (in bold) in the argument; he only uses a vague generalization to support his claim (0/3). The student also does not include a sufficient justification of the evidence in his argument, because he does not explain why the evidence was important by linking it to a specific principle, concept, or underlying assumption (0/2). Instead, he simply insists that the claim is accurate. The author does use scientific terms, such

## SECTION 1: GENERATE AN ARGUMENT

## 4 FRUIT FLY TRAITS
TEACHER NOTES

as *evidence*, correctly (1/1) but includes phrases (e.g., "to prove we are correct," "found the following evidence") that misrepresent the nature of science in the argument (0/1). The organization of the argument is sufficient, because the arrangement of the sentences does not distract from the development of the main idea (1/1). Finally, there are no punctuation (1/1) or grammatical errors (1/1) in the argument. The overall score for the sample argument is 5 out of the 12 points possible.

## Standards Addressed in This Activity

This activity can be used to address the following dimensions outlined in *A Framework for K–12 Science Education* (NRC 2012):

## Scientific Practices

- Developing and using models
- Using mathematics and computational thinking
- Constructing explanations
- Engaging in argument from evidence
- Obtaining, evaluating, and communicating information

## Crosscutting Concepts

- Patterns
- Cause and effect: Mechanism and explanation
- Structure and function

## Life Sciences Core Ideas

- Heredity: Inheritance and variation of traits
- Biological evolution: Unity and diversity

This activity can be used to address the following standards for literacy in science from the *Common Core State Standards for English Language Arts and Literacy* (NGA and CSSO 2010):

## Writing

- Text types and purposes
- Production and distribution of writing
- Research to build and present knowledge
- Range of writing

## Speaking and Listening

- Comprehension and collaboration
- Presentation of knowledge and ideas

## References

Campbell, N., and J. Reece. 2002. *Biology*. 6th ed. San Francisco, CA: Benjamin Cummings.

National Governors Association Center (NGA) for Best Practices, and Council of Chief State School Officers (CCSSO). 2010. *Common core state standards for English language arts and literacy*. Washington, DC: National Governors Association for Best Practices, Council of Chief State School.

National Research Council (NRC). 2012. *A framework for K–12 science education: Practices, crosscutting concepts, and core ideas*. Washington, DC: National Academies Press.

SECTION 1: GENERATE AN ARGUMENT

# DNA FAMILY RELATIONSHIP ANALYSIS (GENETICS) 5

Most of the DNA in the human genome does not encode proteins or RNA. Some of this DNA consists of regulatory sequences, which help control the processes of transcription and translation, but most of it actually consists of sequences whose functions are not yet fully understood by scientists. This DNA includes introns, stretches of noncoding DNA that are often found within the coding sequences of genes. Some of this noncoding DNA also consists of repetitive DNA, which is a nucleotide sequence (e.g., ATTGGCC) that repeats several times (e.g., ATTGGCC-ATTGGCC-ATTGGCC). These repetitive sequences are called short tandem repeats (STRs).

The exact number of times a specific sequence repeats at a specific site in genome differs from individual to individual. The size of an STR (e.g., the number of times a sequence repeats) at a specific site can therefore be used as a genetic marker. These genetic markers can then be used to determine if two people are related or not. There are several different genetic markers that scientists use to help determine family relationships (e.g., D21S11 and D7S820).

Everyone inherits two copies of these various genetic markers: one copy from the father and one from the mother. The two copies of each marker are *usually* different. Therefore, scientists can often determine which version of a particular marker was inherited from a particular parent. This information can be used to determine if two people are related or not.

Mr. and Mrs. H. had five children: three sons and two daughters. Tragically, the H.'s youngest son was abducted from them 20 years ago. This child was only six years old at the time and the police never found him or the person who took him. Recently, a young man named Jeff M., who is in his mid-20s, has contacted the H. family. He claims that he is the boy who was abducted from them. However, the H. family is skeptical and has requested a genetic test to determine if Jeff M. is related to them or not. Unfortunately, Mr. H. died in car accident several years ago. A DNA sample, as a result, can only be collected from Jeff M., Mrs. H., and the family's four children.

Your task is to use the results of an STR family relationship test that was conducted using DNA samples from these six individuals to determine if Jeff M. is the biological offspring of Mr. and Mrs. H. So the guiding question of this investigation: **Is the H. family related to Jeff M.?**

With your group, develop a claim that best answers this question. Once your group has developed your claim, prepare a whiteboard that you can use to share and justify your ideas. Your whiteboard should include all the information shown in the Figure 5.1 (p. 56).

To share your work with others, we will be using a round-robin format. This means that one member of the group will stay at your workstation to share your group's ideas while the other group

## SECTION 1: GENERATE AN ARGUMENT

### 5   DNA FAMILY RELATIONSHIP ANALYSIS

members go to the other groups one at a time in order to listen to and critique the arguments developed by your classmates.

Remember, as you critique the work of others, you need to decide if their conclusions are valid or acceptable based on the quality of their claim and how well they are able to support their ideas. In other words, you need to determine if their argument is *convincing* or not. One way to determine if their argument is convincing is to ask them some of the following questions:

*Figure 5.1. Components of the Whiteboard*

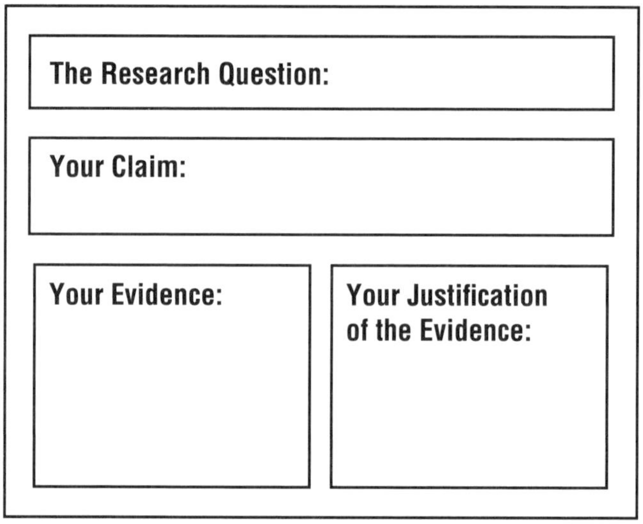

| The Research Question: |
|---|

| Your Claim: |
|---|

| Your Evidence: | Your Justification of the Evidence: |
|---|---|

- How did you analyze or interpret your data? Why did you decide to do it that way?
- How do you know that your analysis of the data is free from errors?
- Why does your evidence support your claim?
- Why did you decide to use that evidence? Why is your evidence important?
- How does your justification of the evidence fit with accepted scientific ideas?
- What are some of the other claims your group discussed before agreeing on your claim, and why did you reject them?

# SECTION 1: GENERATE AN ARGUMENT

## DNA FAMILY RELATIONSHIP ANALYSIS    5

## Results From STR Family Relationship Analysis Test

Figures 5.2–5.5 represent the results from an STR family relationship test. The profile at each DNA marker or STR region (e.g., D13S317, THO1) appears as one or two bars. The height of each bar represents the number of times that the STR is repeated in that person. For example, in Mrs. H., the STR at the D13S317 locus is repeated 3 times on one chromosome and 14 times on the other. The STR profile for Mrs. H. at the D132317 site is therefore listed as 3, 14.

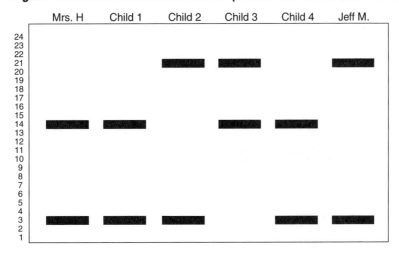

*Figure 5.2. DNA Marker: D13S317 (Found on Chromosome 13)*

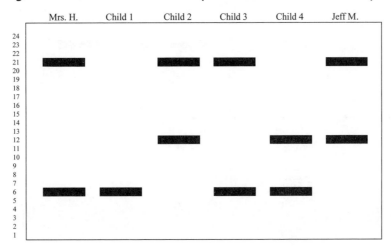

*Figure 5.3. DNA Marker: THO1 (Found on Chromosome 11)*

SCIENTIFIC ARGUMENTATION **IN BIOLOGY**: 30 CLASSROOM ACTIVITIES    **57**

## SECTION 1: GENERATE AN ARGUMENT

### 5 DNA FAMILY RELATIONSHIP ANALYSIS

*Figure 5.4. DNA Marker: D21S11 (Found on Chromosome 21)*

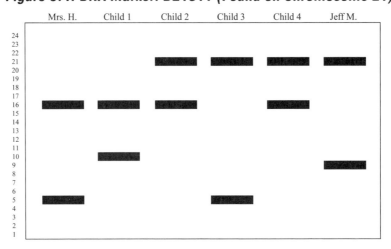

*Figure 5.5. DNA Marker: D7S820 (Found on Chromosome 7)*

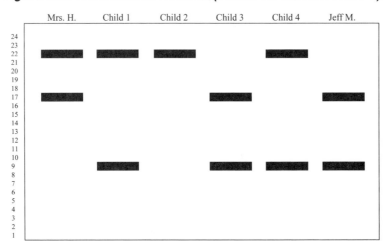

# SECTION 1: GENERATE AN ARGUMENT

## DNA FAMILY RELATIONSHIP ANALYSIS  5

A lab technician used the protocol in Figure 5.6 to create each gel.

### Figure 5.6. Protocol Used to Create Each Gel

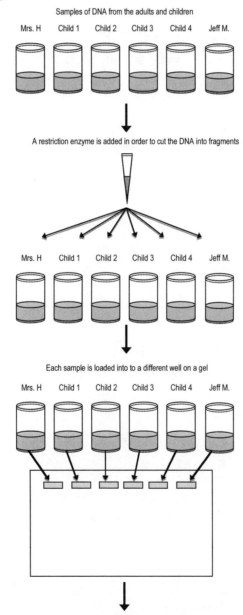

## SECTION 1: GENERATE AN ARGUMENT

Name_____ Date_____

# DNA FAMILY RELATIONSHIP ANALYSIS:
## What Is Your Argument?

In the space below, write an argument in order to persuade another biologist that your claim is valid and acceptable. As you write your argument, remember to do the following:

- State the claim you are trying to support
- Include genuine evidence (data + analysis + interpretation)
- Provide a justification of your evidence that explains why the evidence is relevant and why it provides adequate support for the claim
- Organize your argument in a way that enhances readability
- Use a broad range of words including vocabulary that we have learned
- Correct grammar, punctuation, and spelling errors

SECTION 1: GENERATE AN ARGUMENT

# DNA FAMILY RELATIONSHIP ANALYSIS 5
## TEACHER NOTES

### Purpose

The purpose of this activity is to help students understand the molecular basis of heredity and the role that DNA technology can play in solving social problems. This activity also helps students learn how to engage in practices such as constructing explanations, engaging in argument from evidence, and communicating information. This activity is also designed to give students an opportunity to learn how to write in science and develop their speaking and listening skills, which are important goals for literacy in science (see Standards Addressed in This Activity for a complete list of the practices, crosscutting concepts, core ideas, and literacy skills that are aligned with this activity).

### The Content and Related Concepts

Most of the DNA in eukaryotic genomes does not encode proteins or RNA (Campbell and Reece 2002). Although some of this DNA consists of regulatory sequences, which help control the processes of transcription and translation, most of it actually consists of sequences whose functions are not yet understood. This DNA includes introns, stretches of noncoding DNA that are often found within the coding

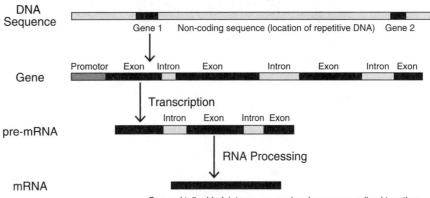

*Figure 5.7. Coding and Noncoding Sequences of DNA*

sequences of genes (see Figure 5.7). Even more of the noncoding DNA consists of repetitive DNA, which is a nucleotide sequence that is present in many copies in the genome and usually not found within a gene. There are two types of repetitive DNA. The first type of repetitive DNA is called short tandem repeats (STRs). STRs consist of a 1–10 base pair sequence (e.g., GTTAC) that repeats (e.g., GTTAC-GTTAC-GTTAC) as many 100,000 times at a specific location. The second type is called interspersed repetitive DNA. The repeated units of this type of DNA are not next to each other; instead they are scattered throughout the genome. A single unit of interspersed repetitive DNA is usually 100–1,000 base pairs long, and the dispersed copies are usually very similar but not identical to one another.

# SECTION 1: GENERATE AN ARGUMENT

## 5 DNA FAMILY RELATIONSHIP ANALYSIS
### TEACHER NOTES

The DNA sequence of every person, except for identical twins, is unique, but the number of STRs at a specific location is passed down from parents to offspring. As a result, the number of STRs at different locations in the genome can be used as a genetic marker for a DNA fingerprint. An individual's DNA contains two copies of each of these markers: one copy is inherited from the father, and one is inherited from the mother. As a result, the two versions of a marker that an individual has at a specific location can differ in length, depending on which version of the marker he or she inherited from his or her parents. These differences allow scientists to determine if two or more people are related or not. It is important to note, however, that scientists must examine several different markers before they can draw any conclusions about familiar relationships because two people can share the same marker even when they are not related. Scientists therefore typically use 16 or more markers in order to create a DNA fingerprint and to conduct a family relationship analysis, because the likelihood of two unrelated individuals sharing several different markers is quite small.

In this activity, the students are asked to determine if Jeff M. is the child of Mr. and Mrs. H. based on the results of STR analysis from four different markers: TH01, D21S11, D7S820, and D13S317. The results of the analysis are provided in Table 5.1. In this table, the version of the STR that could have been inherited from Mrs. H. is in bold while the version of the STR that could have been inherited from Mr. H. is in italics. For the D13S317 marker, Jeff M. has an STR in common with Mrs. H. (3) and an STR in common with Child 2 and Child 3 (21), which could have been inherited from Mr. H. For the TH01 marker, Jeff M. has an STR in common with Mrs. H. (21) and an STR in common with Child 2 and Child 4 (12) that could have been inherited from Mr. H.. For the D21S11 marker, Jeff M. does not have an STR in common with Mrs. H. but has an STR in common with Child 2, Child 3, and Child 4 (21) that could have been inherited from Mr. H. Finally, for the D21S11 marker, Jeff M. has an STR in common with Mrs. H. (17) and has an STR in common with Child 1, Child 3, and Child 4 (9) that could have been inherited from Mr. H. Jeff M., therefore, is not the biological child of

*Table 5.1. Results of STR Analysis*

| Marker  | Mrs. H. | Child 1 | Child 2 | Child 3 | Child 4 | Jeff M. |
|---------|---------|---------|---------|---------|---------|---------|
| D13S317 | 14, *3* | 14, *3* | **21**, *3* | **14**, *21* | **14**, *3* | **3**, *21* |
| TH01    | 21, *6* | 6, *6*  | **21**, *12* | **6**, *21* | **6**, *12* | **12**, *21* |
| D21S11  | 16, *5* | 16, *10* | **16**, *21* | **5**, *21* | **16**, *21* | **9**, *21* |
| D7S820  | 22, *17* | 22, *9* | **22**, *22* | **17**, *9* | **22**, *9* | **17**, *9* |

# SECTION 1: GENERATE AN ARGUMENT

## DNA FAMILY RELATIONSHIP ANALYSIS
### TEACHER NOTES 5

Mr. and Mrs. H., although he has several STRs in common with Mrs. H. and her children at all four genetic markers. However, a case could be made that the unique STR that Jeff M. has at marker D21S11 (9) is a mutation of the STR (10) that Child 1 inherited from Mr. H., or the STR that Child 1 has (10) is a mutation of the STR (9) inherited by Jeff M.

## Curriculum and Instructional Considerations

This activity is best used as part of a unit on genetics. However, it should only be used to give students a chance to apply their understanding of Mendelian inheritance and DNA structure in an unfamiliar context. It is therefore important for the teacher to introduce students to the concept of genes and the law of segregation as well as several different modes of inheritance such as co-dominance, incomplete dominance, multiple allele, polygenic, and sex-linked. Students should also understand the structure of DNA. Students will need a basic understanding of these ideas in order to be able to analyze and interpret the data that will be supplied to them during the activity.

The focus of the explicit discussion at the end of the activity should focus on an aspect of the nature of science or the nature of scientific inquiry. For example, a teacher could discuss how a scientific explanation must be consistent with observational evidence about nature or how scientists rely on a wide range of methods (and not just experiments) to answer research questions using what the students did as an illustrative example.

## Recommendations for Implementing the Activity

This activity takes approximately 100 minutes of instructional time to complete, but the amount of time devoted to each activity varies depending on how a teacher decides to spend time in class. For more information about how to implement the activity, see Appendix E on page 369.

Table 5.2 (p. 64) provides information about the type and amount of materials needed to implement this activity in a classroom with 28 students with groups of four and groups of three.

## Assessment

The rubric provided in Appendix B (p. 366) can be used to assess the arguments crafted by each student at the end of the activity. To illustrate how the rubric can be used to score an argument, consider the following example. This sample argument, which was written by a 10th-grade student, provides an accurate claim but the evidence and rational are rather weak.

> <u>Jeff M. is the child of Mr. and Mrs. H.</u>. I know this is right because the results of the DNA analysis proves it. **Jeff M. had markers in common with Mrs. H. and all of Mrs. H.'s children.** Mrs. H. therefore has found her long lost son. DNA does'nt lie.

The content of the example argument is weak for several reasons. The student's claim (underlined) is sufficient (1/1) but inaccurate (0/1). The student, in addition, does not use genuine evidence (in bold) to support the claim (0/3). Instead, the student relies on an unsub-

## SECTION 1: GENERATE AN ARGUMENT

### 5　DNA FAMILY RELATIONSHIP ANALYSIS
### TEACHER NOTES

*Table 5.2. Materials Needed to Implement the Activity in a Classroom of 28 Students*

| Material | Amount Needed With ... | |
| --- | --- | --- |
| | Groups of 3 | Groups of 4 |
| Whiteboards (or chart paper)+ | 10 | 7 |
| Whiteboard markers (or permanent if using chart paper)+ | 20 | 14 |
| Copy of Student Pages (pp. 55–59)* | 28 | 28 |
| Copy of Student Page (p. 60)* | 10 | 7 |
| Copy of Appendix B (p. 366)* | 28 | 28 |

+ Teachers can also have students prepare their arguments in a digital medium (such as PowerPoint or Keynote).
* Teachers can also project these materials onto a screen in order to cut down on paper use.

stantiated inference as evidence. The student also does not include a sufficient justification of the evidence in his argument because he does not explain why the evidence is important by linking it to a specific principle, concept, or underlying assumption (0/2). Instead, he simply insists that his claim is accurate. The author uses scientific terms correctly (1/1) but includes phrases (e.g., "I know this is right", "the DNA analysis proves it") that misrepresent the nature of science (0/1). The writing mechanics of the sample argument also need improvement. The organization of the argument is acceptable, because the arrangement of the sentences does not distract from the development of the main idea (1/1), although the argument is rather short. There are also some grammatical (0/1) and punctuation errors (0/1) in the argument. The overall score for the sample argument, therefore, is 3 out the 12 points possible.

## Standards Addressed in This Activity

This activity can be used to address the following dimensions outlined in *A Framework for K–12 Science Education* (NRC 2012):

### Scientific Practices

- Developing and using models
- Constructing explanations
- Engaging in argument from evidence
- Obtaining, evaluating, and communicating information

### Crosscutting Concepts

- Patterns
- Cause and effect: Mechanism and explanation

## SECTION 1: GENERATE AN ARGUMENT

### DNA FAMILY RELATIONSHIP ANALYSIS
#### TEACHER NOTES 5

## Life Sciences Core Ideas

- From molecules to organisms: Structures and processes
- Heredity: Inheritance and variation of traits

This activity can be used to address the following standards for literacy in science from the *Common Core State Standards for English Language Arts and Literacy* (NGA and CCSSO 2010):

## Writing

- Text types and purposes
- Production and distribution of writing
- Research to build and present knowledge
- Range of writing

## Speaking and Listening

- Comprehension and collaboration
- Presentation of knowledge and ideas

## References

Campbell, N., and J. Reece. 2002. *Biology*. 6th ed. San Francisco, CA: Benjamin Cummings.

National Governors Association Center (NGA) for Best Practices, and Council of Chief State School Officers (CCSSO). 2010. *Common core state standards for English language arts and literacy*. Washington, DC: National Governors Association for Best Practices, Council of Chief State School.

National Research Council (NRC). 2012. *A framework for K–12 science education: Practices, crosscutting concepts, and core ideas*. Washington, DC: National Academies Press.

# SECTION 1: GENERATE AN ARGUMENT

# EVOLUTIONARY RELATIONSHIPS IN MAMMALS (GENETICS AND EVOLUTION) 6

One of Darwin's most revolutionary ideas was that all living things are related. According to this theory, all living things found on Earth today are related to each other because all life on Earth shares a common ancestor. This ancestor, Darwin argued, once lived on Earth sometime in the distant past but is now extinct. All organisms, from alligators to algae, are connected to one another like branches on a giant tree of life. He came to this conclusion, in part, by examining homologous structures. Homologous structures are things such as limbs (see Figure 6.1) that have a similar structure even though they may have a very different function.

*Figure 6.1. Homologous Structures in Seven Different Vertebrate Limbs*

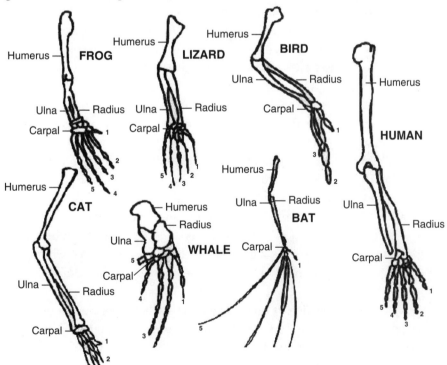

SCIENTIFIC ARGUMENTATION IN BIOLOGY: 30 CLASSROOM ACTIVITIES  **67**

# SECTION 1: GENERATE AN ARGUMENT

## 6 EVOLUTIONARY RELATIONSHIPS IN MAMMALS

To explain the similar bone structure in these animals, Darwin said that they must be descendants of the same ancestor that had a limb that consisted of a humerus, an ulna, a radius, and carpals. He reasoned that the difference in the shape of the bones was a result of gradual modifications that made the organisms better adapted to survive in a particular environment. He called this idea decent with modification. He argued that over time, natural selection could slowly select for or against subtle variations in the basic shape of the bones in the limbs of organisms but could not completely change the basic body plan. This selection process would gradually result in whale fins and bat wings that had fingers similar to the fingers of a lizard, frog, or human. These variations would give their owners an advantage in a particular environment such as the ocean in the case of the whale or the air in the case of the bat.

This idea of descent with modification also suggests that species that diverged from one another relatively recently in the history of life on Earth will share more genetic similarities than species that diverged from one another earlier. Species that share many genetic similarities are considered to be more closely related than two species that have many differences, because the random mutations that occur during DNA replication will cause differences in the DNA sequence to build up in each independent species over time. Therefore, the longer it has been since two species separated form an ancestral species, the more time there will have been for mutations to occur in each species.

Scientists can therefore use the amino acid sequence of proteins to determine the evolutionary history of a group of organisms, because proteins are determined by the DNA base sequence of a specific gene. One such protein, called hemoglobin subunit alpha, is found in virtually all animals and can be used by scientists to determine the degree of relatedness in any group of organisms. This protein enables red blood cells to transport oxygen.

However, it is often difficult to determine the degree of relatedness between different types of animals based on the amino acid sequence of a protein. The following task is an example of this problem. Figures 6.2–6.10 show nine different types of mammals.

All of these animals share certain characteristics, such as body hair and mammary glands, which make them mammals. They also all have certain physical characteristics that make them unique. For example, bats are small like a mouse, but they have wings, while elk are large like a cow but grow a huge set of antlers. These differences make it difficult to determine which types of mammals are the most closely related.

This has made many people wonder: **Which of these mammals are the most closely related?**

With your group, develop a claim that answers this question. Once your group has developed your claim, prepare a whiteboard that you can use to share and justify it. Your whiteboard should include all the information shown in Figure 6.11.

To share your work with others, we will be using a round-robin format. This means that one member of the group will stay at your workstation to share your group's ideas while the other group

# SECTION 1: GENERATE AN ARGUMENT

## EVOLUTIONARY RELATIONSHIPS IN MAMMALS   6

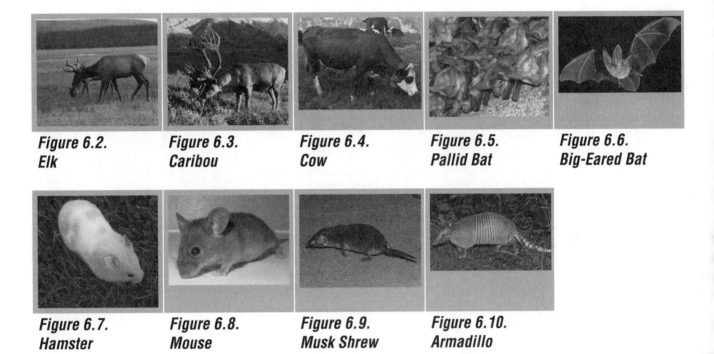

Figure 6.2. Elk

Figure 6.3. Caribou

Figure 6.4. Cow

Figure 6.5. Pallid Bat

Figure 6.6. Big-Eared Bat

Figure 6.7. Hamster

Figure 6.8. Mouse

Figure 6.9. Musk Shrew

Figure 6.10. Armadillo

members go to the other groups one at a time in order to listen to and critique the arguments developed by your classmates.

Remember, as you critique the work of others, you need to decide if their conclusions are valid or acceptable based on the quality of their claim and how well they are able to support their ideas. In other words, you need to determine if their argument is *convincing* or not. One way to determine if their argument is convincing is to ask them some of the following questions:

- How did you analyze or interpret your data? Why did you decide to do it that way?
- How do you know that your analysis of the data is free from errors?

Figure 6.11. Components of the Whiteboard

| The Research Question: |  |
|---|---|
| Your Claim: |  |
| Your Evidence: | Your Justification of the Evidence: |

SCIENTIFIC ARGUMENTATION IN BIOLOGY: 30 CLASSROOM ACTIVITIES

## SECTION 1: GENERATE AN ARGUMENT

# 6 EVOLUTIONARY RELATIONSHIPS IN MAMMALS

- Why does your evidence support your claim?
- Why did you decide to use that evidence? Why is your evidence important?
- How does your justification of the evidence fit with accepted scientific ideas?
- What are some of the other claims your group discussed before agreeing on your claim, and why did you reject them?

# SECTION 1: GENERATE AN ARGUMENT

## EVOLUTIONARY RELATIONSHIPS IN MAMMALS 6

## How to Create a Cladogram

One way to determine how groups of organisms are related to one another (which is known as phylogeny) is to compare certain features such as anatomical structures (body organs and parts) or their genetic makeup (DNA structure) in order to determine how similar these features are in each type of organism. When different organisms share a large number of features, they are described as being closely related. In order to capture the evolutionary history of a group of organisms, biologists often create a diagram of branching lines that connect those groups. These diagrams look like an upside-down mobile and are called cladograms (\kla-da-gram\). You will need to develop a cladogram in order to answer the research question.

In order to construct a cladogram (Figure 6.12), it is important to identify the characteristics of the ancestral population and those of the descendants. Characteristics shared by most or all members of related taxa are referred to as *ancestral* traits. These ancestral traits link the members of related branches to a common ancestor. On the other hand, characteristics that are found in various evolutionary branches that differ from those of the ancestors are considered *derived*. In many cases, a derived characteristic is a unique modification of a shared ancestral characteristic. Derived characteristics or traits distinguish the members of one evolutionary branch from the members of another branch.

A cladogram is constructed based on the presence of derived traits in two or more related taxa. Ideally, a cladogram should be based on branches that are defined by a unique derived trait that emerged only once and are shared by all subsequent descendants. Because living organisms are a complex combination of traits, however, sometimes it is possible to draw more than one cladogram that might reflect the

*Figure 6.12. An Example of a Cladogram*

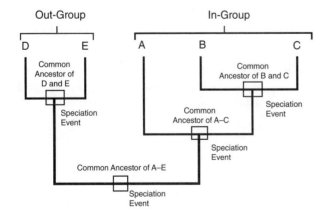

*Figure 6.13. Different Clades Represented in a Cladogram (Each rectangle represents a different clade.)*

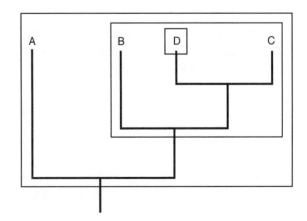

# SECTION 1: GENERATE AN ARGUMENT

## 6 EVOLUTIONARY RELATIONSHIPS IN MAMMALS

evolutionary history of a group of organisms. In this activity, you will need to create a cladogram for the nine different species of mammals. You can construct a cladogram by following these steps:

1. Identify the taxa. These taxa will be the tips of your tree and must themselves be clades. A clade is a group of organisms that includes an ancestor and all descendants of that ancestor (see Figure 6.13, p. 71). In this case, your taxa are the nine species of mammals.

2. Determine the characters and examine each taxon to determine the character states. In this investigation, you use the first 40 amino acids in the hemoglobin alpha protein as your characters. Your character states, therefore, will be the different types of amino acids at each point in the sequence.

3. Determine the order of appearance of characters. In other words, figure out the order of evolution for each character. In this investigation, you will need to determine which mutation appeared first, second, and so on in the sequence of amino acids in the hemoglobin alpha protein.

4. Group taxa together by derived or "changed" character states. Do not group the taxa together by the original character state.

5. Build your tree. In order to build your tree correctly, you must follow three rules:
   » All taxa must go on the endpoints of the tree, never at nodes.
   » All nodes must have a shared derived character, which is common to all taxa above the node.
   » All shared derived characters can appear only once on a tree.

Two fundamental principles that are used to evaluate a cladogram are parsimony and robustity. First parsimony: When there is more than one way to draw a cladogram *and* when there are no *other* data that suggest one of these is more likely than the others, we tend to choose the one in which derived traits are reinvented in different branches the fewest number of times. Second robusticity: We prefer trees that maintain their basic form, even when different options are applied to the sequence of changes in one or more of their branches. However, when more data are available about the history or the origin of a particular feature, these data are more important tools in determining which of the alternative trees is better.

# SECTION 1: GENERATE AN ARGUMENT

## EVOLUTIONARY RELATIONSHIPS IN MAMMALS — 6

### Table 6.1. Amino Acid Sequence for Hemoglobin Subunit Alpha Protein 1–20

| Organism | 1 | 2 | 3 | 4 | 5 | 6 | 7 | 8 | 9 | 10 | 11 | 12 | 13 | 14 | 15 | 16 | 17 | 18 | 19 | 20 |
|---|---|---|---|---|---|---|---|---|---|---|---|---|---|---|---|---|---|---|---|---|
| Elk | M | V | L | S | A | T | D | K | S | N | V | K | A | A | W | G | K | V | G | G |
| Caribou | - | V | L | S | A | A | D | K | S | N | V | K | A | A | W | G | K | V | G | G |
| Cow | M | V | L | S | A | A | D | K | S | N | V | K | A | A | W | G | K | V | G | G |
| Pallid Bat | M | V | L | S | P | A | D | K | G | N | V | K | A | A | W | D | K | V | G | G |
| Big-Eared Bat | - | V | L | S | A | A | D | K | G | N | V | K | A | A | W | D | K | V | G | G |
| Golden Hamster | - | V | L | S | A | K | D | K | T | N | I | S | E | A | W | G | K | - | G | G |
| Mouse | M | V | L | S | G | E | D | K | S | N | I | K | A | A | W | G | K | - | G | G |
| Musk Shrew | - | V | L | S | A | N | D | K | A | N | V | K | A | A | W | D | K | V | G | G |
| Nine-Banded Armadillo | - | V | L | S | A | A | D | K | T | H | V | K | A | F | W | G | K | V | G | G |

### Table 6.2. Amino Acid Sequence for Hemoglobin Subunit Alpha Protein 20–40

| Organism | 21 | 22 | 23 | 24 | 25 | 26 | 27 | 28 | 29 | 30 | 31 | 32 | 33 | 34 | 35 | 36 | 37 | 38 | 39 | 40 |
|---|---|---|---|---|---|---|---|---|---|---|---|---|---|---|---|---|---|---|---|---|
| Elk | N | A | P | A | Y | G | A | E | A | L | E | R | M | F | L | S | F | P | T | T |
| Caribou | N | A | P | A | Y | G | A | E | A | L | E | R | M | F | L | S | F | P | T | T |
| Cow | H | A | A | E | Y | G | A | E | A | L | E | R | M | F | L | S | F | P | T | T |
| Pallid Bat | H | A | G | D | Y | G | A | E | A | L | E | R | M | F | L | S | F | P | T | T |
| Big-Eared Bat | Q | A | G | E | Y | G | A | E | A | L | E | R | M | F | L | S | F | P | T | T |
| Golden Hamster | H | A | G | E | Y | G | A | E | A | L | E | R | M | F | F | V | Y | P | T | T |
| Mouse | H | A | G | E | Y | G | A | E | A | L | E | R | M | F | A | S | F | P | T | T |
| Musk Shrew | Q | A | A | N | Y | G | A | E | A | L | E | R | T | F | A | S | F | P | T | T |
| Nine-Banded Armadillo | H | A | A | E | F | G | A | E | A | L | E | R | M | F | A | S | F | P | P | T |

## SECTION 1: GENERATE AN ARGUMENT

Name_____ Date_____

# EVOLUTIONARY RELATIONSHIPS IN MAMMALS:
## What Is Your Argument?

In the space below, write an argument in order to persuade another biologist that your claim is valid and acceptable. As you write your argument, remember to do the following:

- State the claim you are trying to support
- Include genuine evidence (data + analysis + interpretation)
- Provide a justification of your evidence that explains why the evidence is relevant, and why it provides adequate support for the claim
- Organize your argument in a way that enhances readability
- Use a broad range of words including vocabulary that we have learned
- Correct grammar, punctuation, and spelling errors

SECTION 1: GENERATE AN ARGUMENT

# EVOLUTIONARY RELATIONSHIPS IN MAMMALS 6
## TEACHER NOTES

## Purpose

The purpose of this activity is to help students understand how modern phylogenetic systematics (i.e., the classification based on evolutionary history) is based on cladistics analysis. This activity also gives students an opportunity to create a cladogram, which is a diagram that depicts a hypothetical branching sequence of lineages leading to specific species or genera. This activity also helps students learn how to engage in practices such developing and using models, constructing explanations, arguing from evidence, and communicating information. This activity is also designed to give students an opportunity to learn how to write in science and develop their speaking and listening skills, which are important goals for literacy in science (see Standards Addressed in This Activity for a complete list of the practices, crosscutting concepts, core ideas, and literacy skills that are aligned with this activity).

## The Content and Related Concepts

Phylogenic systematics is the study of biological diversity in an evolutionary context. It is the way biologists attempt to reconstruct the pattern of events that led to the distribution and diversity of life on Earth. Modern phylogenetic systematics is based on cladistics (Campbell and Reece 2002). Cladistics refers to a particular method of examining evolutionary relationships among organisms. Like other methods in science, it is based on a set of assumptions, and it has several limitations, but it is the best method available for phylogenetic analysis because it results in an explicit and testable hypothesis of the evolutionary relationships among organisms.

The basic idea behind cladistics is that members of a group share a common evolutionary history and are more closely related to other members of the same group than to other organisms. These groups share unique features that were not present in their ancestors. These unique characteristics are often called shared derived characters. The members of a group, therefore, consist of an evolutionary ancestor and all the decedents who share the same derived character. A group of organisms that are defined by a shared derived character is called a clade.

It is important to remember that cladistics is not based on the number of characteristics a group of organisms share; in fact, two organisms may share several characteristics and not be considered members of the same group. For example, consider a shark, a dolphin, and a bat. The shark and the dolphin both live in the water and have fins, so one might think that these two organisms belong together in a group. This grouping, however, would not reflect the evolutionary history of these animals because

# SECTION 1: GENERATE AN ARGUMENT

## 6 EVOLUTIONARY RELATIONSHIPS IN MAMMALS
### TEACHER NOTES

the dolphin and bat are actually more closely related; they are warm-blooded and have lungs, which are shared derived characteristics not found in the shark. It is not just the presence of shared characteristics that is important; it is the presence of shared derived characteristics that make a group unique from other groups.

There are three basic assumptions in cladistics. The first assumption is that life arose on Earth only once, and therefore all organisms are related to each other. The second assumption is perhaps the most controversial: New kinds of organisms arise when existing species or populations divide into two groups. This assumption is controversial because some biologists think that populations can divide into more than two groups. The third and final assumption is the most important assumption in cladistics: Characteristics of organisms change over time. It is only when characteristics change that we are able to identify different lineages or groups. Biologists call the original state of a characteristic the plesiomorphic character and the new or "changed" state the apomorphic character (Campbell and Reece 2002).

The following sequence of steps is used to construct a cladogram:

1. Choose a group of species. In this activity, the students are given nine different species of mammal.

2. Determine the characters, or features of the organisms, that will be used to make the cladogram and examine each species to determine the character states (i.e., decide whether each species does or does not have a particular shared derived character). In this activity, the students use the first 40 amino acids in the protein hemoglobin alpha.

3. Determine whether the character state in each species is plesiomorphic (original) or apomorphic (derived). In the case of this activity, shared derived characters refer to a difference in amino acids at specific location in the original amino acid sequence.

4. Group the species by the number of shared derived characteristics. That is, in this activity, all the species that share a difference a specific location in the amino acid sequence.

5. Draw the cladogram so each species is labeled on the endpoints of the figure and not at the nodes. The cladogram nodes should also list derived character (synapomorphies), which is common to all species above the node, and any derived character should only appear on the cladogram once.

It is important to remember that cladograms should not be viewed as either right or wrong; rather, they should be judged based on how well they account for the available data.

## Curriculum and Instructional Considerations

This activity is best used as part of a unit on evolution or biological classification. Teachers should use it to give students a chance to apply their understanding of phylogenetics and their ability to construct a cladogram in an unfamiliar context. When used as an application activity, it

## SECTION 1: GENERATE AN ARGUMENT

### EVOLUTIONARY RELATIONSHIPS IN MAMMALS
#### TEACHER NOTES 6

is important for the teacher to teach students about various biological classification systems, the principle of descent with modification, phylogenetics, and cladistics before students are asked to complete the task. The students will need to have this foundation of important biological ideas in order to make sense of the supplied data and then construct high-quality arguments. The focus of the explicit discussion at the end of the activity should focus on an aspect of the nature of science or the nature of scientific inquiry. For example, a teacher could discuss how all scientific knowledge is, in principle, subject to change as new evidence becomes available or how scientific explanation must be consistent with observational evidence about nature, and must make accurate predictions when appropriate about systems being studied using what the students did as an illustrative example.

## Recommendations for Implementing the Activity

This activity takes approximately 100 minutes of instructional time to complete, but the amount of time devoted to each activity varies depending on how a teacher decides to spend time in class. For more information about how to implement the activity, see Appendix E on page 369.

Table 6.3 provides information about the type and amount of materials needed to implement this activity in a classroom with 28 students with groups of four and groups of three.

## Assessment

The rubric provided in Appendix B (p. 366) can be used to assess the arguments crafted by each student at the end of the activity. To illustrate how the rubric can be used to score

*Table 6.3. Materials Needed to Implement the Activity in a Classroom of 28 Students*

| Material | Amount Needed With… | |
| --- | --- | --- |
|  | Groups of 3 | Groups of 4 |
| Whiteboards (or chart paper)+ | 10 | 7 |
| Whiteboard markers (or permanent if using chart paper)+ | 20 | 14 |
| Copy of Student Pages (pp. 67–73)* | 28 | 28 |
| Copy of Student Page (p. 74)* | 10 | 7 |
| Copy of Appendix B (p. 366)* | 28 | 28 |

+ Teachers can also have students prepare their arguments in a digital medium (such as PowerPoint or Keynote).
* Teachers can also project these materials onto a screen in order to cut down on paper use.

# SECTION 1: GENERATE AN ARGUMENT

## 6 EVOLUTIONARY RELATIONSHIPS IN MAMMALS
### TEACHER NOTES

an argument, consider the following example. This sample argument, which was written by a 10th-grade student, is an example of an argument that is weak in terms of both content and writing mechanics.

> <u>The bats and the mice are most closely related.</u> I know that this is true because the cladogram I made by looking at the differences in the amino-acid sequence of hemoglobin-a. **The armadillo and pallid bat humans have the most difference in they're sequences so they are on the ends and pallid bat and the big-eared bat have the most in common so they are next to each other.** These differences are important because more differences in the amino acid sequence means two species shared a common ancestor a long time ago. Therefore, two species with a similar amino acid sequence and more closely related than two species that had a lot of differences. **The hamster shrew and mouse have almost identical sequences and the sequences of the elk, caribou, and cow is very similar to each other and the armadillo is just odd.**

The content of the example argument is weak for several reasons. The student's claim (underlined) is insufficient (0/1). The claim is also inaccurate, because it reflects a poor interpretation of the data set (0/1). The student does not use genuine evidence (in bold) in the argument, because he basically provides several unsubstantiated inferences rather than attempting to quantify the difference in the amino acid sequences by referring to specific elements of the cladogram that he created (0/3). The student provides a decent justification of the evidence in his argument because he explains why differences in the amino acid sequence are important (1/1) but does not attempt to link the evidence to an important biological concept (e.g., common decent) or a principle (e.g., phylogenetic) (0/1). The organization of the argument is adequate (1/1), but there are a few punctuation (0/1) and grammatical errors (0/1). The overall score for the sample argument come from the 2 mechanics points, out the 12 points possible.

## Standards Addressed in This Activity

This activity can be used to address the following dimensions outlined in *A Framework for K–12 Science Education* (NRC 2012):

### Scientific Practices

- Developing and using models
- Constructing explanations
- Engaging in argument from evidence
- Obtaining, evaluating, and communicating information

### Crosscutting Concepts

- Patterns
- Cause and effect: Mechanism and explanation
- Structure and function

## SECTION 1: GENERATE AN ARGUMENT

### EVOLUTIONARY RELATIONSHIPS IN MAMMALS
#### TEACHER NOTES 6

## Life Sciences Core Ideas

- From molecules to organisms: Structures and processes
- Biological evolution: Unity and diversity

This activity can be used to address the following standards for literacy in science from the *Common Core State Standards for English Language Arts and Literacy* (NGA and CCSSO 2010):

## Writing

- Text types and purposes
- Production and distribution of writing
- Research to build and present knowledge
- Range of writing

## Speaking and Listening

- Comprehension and collaboration
- Presentation of knowledge and ideas

## References

Campbell, N., and J. Reece. 2002. *Biology*. 6th ed. San Francisco, CA: Benjamin Cummings.

National Governors Association Center (NGA) for Best Practices, and Council of Chief State School Officers (CCSSO). 2010. *Common core state standards for English language arts and literacy*. Washington, DC: National Governors Association for Best Practices, Council of Chief State School.

National Research Council (NRC). 2012. *A framework for K–12 science education: Practices, crosscutting concepts, and core ideas*. Washington, DC: National Academies Press.

SECTION 1: GENERATE AN ARGUMENT

# DECLINE IN SALTWATER FISH POPULATIONS (ECOLOGY AND HUMAN IMPACT ON THE ENVIRONMENT)

7

Freshwater and saltwater fisheries are important to both our culture and our economy. Fishing provides recreational opportunities and important food sources for society. There have been many debates about what kinds of policies should be implemented to safeguard our fisheries. According to the Florida Fish and Wildlife Conservation Commission, Florida's marine fisheries provide over 2.5 million recreational anglers with sport fishing opportunities and over 15,000 commercial fishers with employment.

Previously, it was a general belief that the oceans held an endless supply of fish. There are now many groups of people who think otherwise. Because these groups of people have different interests in the opportunities for fishing, they also have different ideas about the policies that should be enforced to protect the fish in the ocean. By protecting the fish, they are protecting their interests in fishing, including recreational fishing, economic impact (i.e., commercial fishing), and traditional lifestyle. Policies are generally designed to provide limitations by restricting opportunities. Restrictions can mean:

- reducing opportunities for businesses to grow by limiting the number of fishermen, the number of fish allowed to be caught, the size of fish to be caught, and the time of year that fish can be caught;
- preventing groups of people from practicing fishing as a part of their traditional lifestyles;
- impacting the general economy by reducing availability of (or by flooding the market with) fish;
- impacting the general economy by changing the costs for fishing and fishing supplies; and
- preventing groups of people from fishing for recreation.

Fisheries around the world claim that even with current policies, the fish populations are declining. For example, in 2005, the State of the World of Fisheries and Aquaculture (SOFIA) released a statement that 3% of marine stocks were underexploited, 21% were moderately exploited, 52% were fully exploited, 16% were overexploited and the remaining 7% were recovering from being overexploited (due to strict policies) (Kourous 2005). The authors of this report blame the fish population decline on growing human populations and insufficient monitoring policies that would allow for this increased demand to be met without harming fish populations (i.e., limiting the numbers, the sizes, and the times that fish may be caught).

In 2006, the *Washington Post* published an article describing a report from ecologists and economists claiming that at least 90% of fish species were below their historic maximum catch levels, due in part to the increase in commercial fishing and the inability of fish populations to resist environmental stresses created by the specific practices of commercial fishing (Eilperin).

## SECTION 1: GENERATE AN ARGUMENT

## 7  DECLINE IN SALTWATER FISH POPULATIONS

In 2007 and 2008, the number of smolt (salmon migrating to the ocean) increased in the Sacramento River and in Alaska; however, the number of salmon returning to spawn substantially decreased, resulting in a ban on both commercial and recreational fishing of Chinook salmon (king salmon) for two years in California and in most of Oregon. The decline of salmon is blamed on the dams that have been built and the pesticides from farms bordering the rivers. Environmentalists assert that the salmon are unable to return upriver to spawn, and those that spawn in the lower river areas develop abnormalities caused by the pollutants that seep into the waters from farms. In an attempt to address the concern, biologists have been spawning salmon in hatcheries and releasing them into the oceans and rivers. Newspaper reports claim that 90% of the Chinook salmon caught by fisherman in 2008 were from hatcheries rather than naturally spawned. Although several reports continue to express warnings about the decline of saltwater fisheries, in 2010 newspapers and television reports claimed that the salmon numbers increased in Vancouver to numbers greater than any seen in over 100 years.

In Florida, regulations for the past 20 years have included a strict management rule, known as a bag limit, of only one red drum fish permitted to be caught per day by recreational fishermen. This regulation has recently been changed to allow for a two-fish bag limit (two fish per person per day). The Gulf red snapper has reportedly increased, yet the recreational harvest season was reduced and the commercial quota was increased in 2012. The spotted seatrout have been strictly regulated by allowing anglers to only fish for them in certain parts of Florida during limited months of the year. The previous harvest prohibition for roundscale spearfish has recently been removed and is now a 250 fish season (i.e., the season closes once 250 of them are caught).

These observations raise an interesting question: **Is our saltwater fish population declining? If so, what policies would be most effective in slowing that decline?**

Because this is a complicated question that may have different answers for different regions, in this discussion we will be considering the fish populations around the Florida coast.

You can use the following materials to generate your argument:

- Data tables that have been provided
- Information regarding policies and regulations that have been suggested or enforced
- Florida Fish and Wildlife Conservation Commission (*http://myfwc.com*)
- National Oceanic and Atmospheric Administration (NOAA) education resources (*www.education.noaa.gov/Marine_Life*)

With your group, determine if any of the fish populations are changing and if there is any particular group of people who should be regulated in fishing these populations. Be prepared to discuss what policies would be most appropriate for various groups of people who rely on the fish populations. Use the data that is provided to make inferences about the fluctuations in fish populations and the use of the fish for recreation, economic, and cultural purposes. You can use any resources online as well as any

## SECTION 1: GENERATE AN ARGUMENT

### DECLINE IN SALTWATER FISH POPULATIONS   7

classroom supplies available to you to test your ideas. Make sure that you generate the evidence you will need to support your explanation as you work. You can record your method and any observation you make in the space below.

With your group, develop a claim that best answers this question. Once your group has developed your claim, prepare a whiteboard that you can use to share and justify your ideas. Your whiteboard should include all the information shown in Figure 7.1.

*Figure 7.1. Components of the Whiteboard*

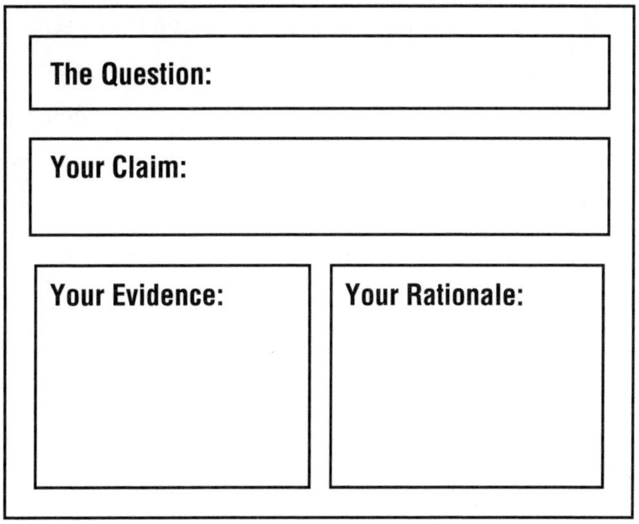

| The Question: |  |
|---|---|
| Your Claim: | |
| Your Evidence: | Your Rationale: |

To share your work with others, we will be using a round-robin format. This means that one member of the group will stay at your workstation to share your groups' ideas while the other group members will go to the other group one at a time in order to listen to and critique the arguments developed by your classmates.

Remember, as you critique the work of others, you have to decide if their conclusions are valid or acceptable based quality of their claim and how well they are able to support their ideas. In other words, you need to determine if their argument is *convincing* or not. One way to determine if their argument is convincing is to ask them some of the following questions:

- How did you analyze or interpret your data? Why did you decide to do it that way?
- How do you know that your analysis of the data is free from errors?
- Why does your evidence support your claim?
- Why did you decide to use that evidence? Why is your evidence important?
- How does your rationale fit with accepted scientific ideas?
- What are some of the other claims your group discussed before agreeing on your claim, and why did you reject them?

## References

Eilperin, J. *The Washington Post*. 2006. World's Fish Supply Running Out, Researchers Warn. November 3.

Kourous, G., 2005. Depleted fish stocks require recovery efforts. Food and Agriculture Organization of the United Nations. *http://www.fao.org/newsroom/en/news/2005/100095/index.html*.

## SECTION 1: GENERATE AN ARGUMENT

Name_____  Date_____

# DECLINE IN SALTWATER FISH POPULATIONS:
## What Is Your Argument?

In the space below, write a one- to three-paragraph argument to *support* the explanation that you think is the most valid or acceptable. As you write your argument, remember to do the following:

- State the explanation you are trying to support
- Include appropriate and relevant evidence
- Make your rationale for including the evidence explicit
- Organize your paper in a way that enhances readability
- Use a broad range of words including vocabulary that we have learned
- Make sure your writing has an easy flow and rhythm
- Correct grammar, punctuation, and spelling errors

*Table 7.1. Annual Standardized Commercial Catch Rates of Select Fish Along the Florida Atlantic Coast from 1999 to 2009 (Average Pounds per Trip)*

| Atlantic Coast, Commercial Landings Rates (Average Pounds per Trip) | | | | | | | | | | | | | | | | | | | | | | |
|---|---|---|---|---|---|---|---|---|---|---|---|---|---|---|---|---|---|---|---|---|---|---|
| | 1999 | | 2000 | | 2001 | | 2002 | | 2003 | | 2004 | | 2005 | | 2006 | | 2007 | | 2008 | | 2009 | |
| Catch | Avg lbs. | Trips | Avg lbs. | Trips | Avg lbs. | Trips | Avg lbs. | Trips | Avg lbs. | Trips | Avg lbs. | Trips | Avg lbs. | Trips | Avg lbs. | Trips | Avg lbs. | Trips | Avg lbs. | Trips | Avg lbs. | Trips |
| Atlantic Croaker | 4591 | 17 | 3796 | 18 | 2982 | 15 | 2892 | 16 | 3018 | 23 | 2684 | 22 | 2634 | 21 | 2807 | 20 | 3205 | 19 | 2376 | 18 | 2063 | 15 |
| Bluefish | 2638 | 88 | 1778 | 60 | 2047 | 64 | 2168 | 61 | 2365 | 75 | 2168 | 70 | 2758 | 60 | 2364 | 58 | 2611 | 70 | 2699 | 60 | 3913 | 67 |
| Flounders | 4591 | 18 | 3796 | 20 | 2982 | 16 | 2892 | 17 | 3013 | 17 | 2684 | 24 | 2634 | 22 | 2807 | 21 | 3205 | 20 | 2376 | 19 | 2063 | 17 |
| Striped mullet | 6888 | 125 | 6463 | 127 | 5758 | 143 | 6336 | 129 | 6608 | 150 | 5065 | 151 | 4927 | 148 | 5436 | 155 | 5024 | 150 | 6093 | 145 | 5676 | 130 |
| Sheepshead | 4980 | 32 | 4999 | 41 | 4698 | 37 | 4749 | 35 | 5134 | 32 | 4446 | 31 | 4446 | 36 | 4677 | 33 | 4757 | 34 | 5530 | 32 | 5206 | 32 |
| Red Drum | N/A | N/A | N/A | N/A | N/A | N/A | N/A | N/A | N/A | N/A | N/A | N/A | N/A | N/A | N/A | N/A | N/A | N/A | N/A | N/A | N/A | N/A |

*Table 7.2. Annual Standardized Commercial Catch Rates of Select Fish Along the Florida Gulf Coast Between 1999 and 2009 (Average Pounds per Trip)*

| Gulf Coast, Commercial Landings Rates (Pounds per Trip) | | | | | | | | | | | | | | | | | | | | | | |
|---|---|---|---|---|---|---|---|---|---|---|---|---|---|---|---|---|---|---|---|---|---|---|
| | 1999 | | 2000 | | 2001 | | 2002 | | 2003 | | 2004 | | 2005 | | 2006 | | 2007 | | 2008 | | 2009 | |
| Catch | Avg lbs. | Trips | Avg lbs. | Trips | Avg lbs. | Trips | Avg lbs. | Trips | Avg lbs. | Trips | Avg lbs. | Trips | Avg lbs. | Trips | Avg lbs. | Trips | Avg lbs. | Trips | Avg lbs. | Trips | Avg lbs. | Trips |
| Atlantic Croaker | 189 | 24 | 153 | 40 | 117 | 25 | 70 | 26 | 35 | 28 | 45 | 65 | 58 | 54 | 79 | 25 | 76 | 28 | 98 | 53 | 72 | 52 |
| Bluefish | 1451 | 34 | 1116 | 39 | 917 | 35 | 1126 | 40 | 878 | 38 | 640 | 37 | 906 | 36 | 801 | 35 | 731 | 35 | 987 | 36 | 1099 | 31 |
| Flounders | 3962 | 8 | 3048 | 7 | 3241 | 9 | 2954 | 9 | 2464 | 10 | 2503 | 10 | 2365 | 11 | 1729 | 10 | 1515 | 11 | 1580 | 9 | 2133 | 11 |
| Striped mullet | 22889 | 300 | 21465 | 305 | 20675 | 375 | 19329 | 330 | 18438 | 310 | 18868 | 310 | 16221 | 315 | 18722 | 330 | 15402 | 350 | 16714 | 375 | 17799 | 400 |
| Sheepshead | 8123 | 22 | 7675 | 23 | 6956 | 24 | 6079 | 23 | 5899 | 25 | 5816 | 25 | 5104 | 22 | 4645 | 21 | 3881 | 22 | 4376 | 19 | 4666 | 22 |
| Red Drum | N/A | N/A | N/A | N/A | N/A | N/A | N/A | N/A | N/A | N/A | N/A | N/A | N/A | N/A | N/A | N/A | N/A | N/A | N/A | N/A | N/A | N/A |

## SECTION 1: GENERATE AN ARGUMENT

## 7 DECLINE IN SALTWATER FISH POPULATIONS

*Table 7.3. Annual Standardized Recreational Catch Rates of Select Fish Along the Florida Atlantic Coast Between 1999 and 2009 (Average Number of Fish per Trip)*

| | Atlantic Coast, Recreational Landings Rates (Average Fish per Trip) | | | | | | | | | | | | | | | | | | | | | |
|---|---|---|---|---|---|---|---|---|---|---|---|---|---|---|---|---|---|---|---|---|---|---|
| | 1999 | | 2000 | | 2001 | | 2002 | | 2003 | | 2004 | | 2005 | | 2006 | | 2007 | | 2008 | | 2009 | |
| Catch | Avg Fish per Trip | Trips | Avg Fish per Trip | Trips | Avg Fish per Trip | Trips | Avg Fish per Trip | Trips | Avg Fish per Trip | Trips | Avg Fish per Trip | Trips | Avg Fish per Trip | Trips | Avg Fish per Trip | Trips | Avg Fish per Trip | Trips | Avg Fish per Trip | Trips | Avg Fish per Trip | Trips |
| Atlantic Croaker | 4.5 | 273 | 4 | 254 | 4.25 | 232 | 3 | 208 | 3 | 181 | 4 | 218 | 3.25 | 175 | 3.25 | 161 | 3.25 | 207 | 3.50 | 215 | 4 | 275 |
| Bluefish | 2 | 650 | 2.5 | 678 | 2 | 662 | 2 | 1031 | 1.75 | 826 | 2 | 426 | 2.25 | 471 | 2.5 | 603 | 2.25 | 614 | 2 | 495 | 2.5 | 569 |
| Flounders | 1 | 672 | 0.75 | 463 | 0.75 | 430 | 0.75 | 466 | 1 | 505 | 1 | 358 | .75 | 316 | .75 | 367 | 1 | 405 | 0.75 | 407 | 0.75 | 350 |
| Striped Mullet | 11 | 50 | 10 | 42 | 5 | 45 | 6 | 58 | 7 | 38 | 10 | 100 | 9 | 86 | 10 | 112 | 10 | 59 | 9 | 93 | 9 | 94 |
| Sheepshead | 2 | 769 | 1.5 | 542 | 1.5 | 72 | 1.5 | 577 | 1.5 | 483 | 1.5 | 409 | 1.5 | 483 | 1.5 | 493 | 1.5 | 391 | 2 | 453 | 1.5 | 387 |
| Red Drum | 0.6 | 1913 | 0.5 | 2235 | 0.6 | 2577 | 0.4 | 2369 | 0.6 | 1874 | 0.7 | 2182 | 0.7 | 2261 | 0.6 | 2166 | 0.5 | 1920 | 0.5 | 1848 | 0.6 | 1562 |

*Table 7.4. Annual Standardized Recreational Catch Rates of Select Fish Along the Florida Gulf Coast Between 1999 and 2009 (Average Number of Fish per Trip)*

| | Gulf Coast, Recreational Landings Rates (Average Fish per Trip) | | | | | | | | | | | | | | | | | | | | | |
|---|---|---|---|---|---|---|---|---|---|---|---|---|---|---|---|---|---|---|---|---|---|---|
| | 1999 | | 2000 | | 2001 | | 2002 | | 2003 | | 2004 | | 2005 | | 2006 | | 2007 | | 2008 | | 2009 | |
| Catch | Avg Fish per Trip | Trips | Avg Fish per Trip | Trips | Avg Fish per Trip | Trips | Avg Fish per Trip | Trips | Avg Fish per Trip | Trips | Avg Fish per Trip | Trips | Avg Fish per Trip | Trips | Avg Fish per Trip | Trips | Avg Fish per Trip | Trips | Avg Fish per Trip | Trips | Avg Fish per Trip | Trips |
| Atlantic Croaker | 3.5 | 158 | 3 | 134 | 3 | 105 | 3.25 | 114 | 4 | 81 | 3 | 125 | 2.5 | 58 | 4 | 61 | 3 | 116 | 3 | 97 | 3 | 132 |
| Bluefish | 2 | 210 | 2 | 160 | 2 | 250 | 2 | 323 | 2 | 248 | 2 | 336 | 2 | 325 | 2.5 | 601 | 2 | 580 | 2 | 430 | 2 | 397 |
| Flounders | .5 | 99 | 0.5 | 98 | 0.5 | 82 | 0.5 | 84 | 0.5 | 129 | 0.5 | 130 | 0.5 | 127 | 0.5 | 71 | 0.8 | 88 | 0.5 | 98 | 0.5 | 92 |
| Striped Mullet | 7.5 | 238 | 4 | 166 | 6 | 205 | 5 | 179 | 5 | 159 | 5 | 216 | 5 | 251 | 6 | 180 | 5 | 120 | 7 | 140 | 6 | 130 |
| Sheepshead | 2.4 | 1245 | 2.3 | 756 | 2.1 | 802 | 2.2 | 890 | 1.9 | 1011 | 2.2 | 1020 | 2 | 1063 | 2 | 782 | 1.9 | 752 | 1.7 | 729 | 2.1 | 815 |
| Red Drum | 0.8 | 2855 | 0.7 | 2238 | 0.7 | 2831 | 0.6 | 3122 | 0.7 | 3273 | 0.75 | 3772 | 0.8 | 3764 | 1.0 | 3378 | 0.8 | 3749 | 0.8 | 3636 | 0.7 | 3553 |

# SECTION 1: GENERATE AN ARGUMENT

## DECLINE IN SALTWATER FISH POPULATIONS    7

*Table 7.5. Annual Observations of Young of the Year (YOY) of Select Fish Along the Florida Atlantic Coast Between 1999 and 2009 (Number Counted) From Fishery-Independent Monitoring Sets That Captured Fish*

| Atlantic Coast YOY | | | | | | | | | | | |
|---|---|---|---|---|---|---|---|---|---|---|---|
| Catch | 1999 | 2000 | 2001 | 2002 | 2003 | 2004 | 2005 | 2006 | 2007 | 2008 | 2009 |
| Atlantic Croaker | 260 | 260 | 256 | 261 | 262 | 263 | 261 | 260 | 260 | 260 | 260 |
| Bluefish | N/A | N/A | N/A | N/A | N/A | N/A | N/A | N/A | N/A | N/A | N/A |
| Flounders | 260 | 260 | 256 | 261 | 262 | 263 | 261 | 260 | 260 | 260 | 260 |
| Striped Mullet | 260 | 260 | 256 | 261 | 262 | 263 | 261 | 260 | 260 | 260 | 260 |
| Sheepshead | 260 | 260 | 256 | 261 | 262 | 263 | 261 | 260 | 260 | 260 | 260 |
| Red Drum | 260 | 260 | 256 | 261 | 262 | 263 | 261 | 260 | 260 | 260 | 260 |

*Table 7.6. Annual Observations of Post-Young of the Year (YOY) of Select Fish Along the Florida Atlantic Coast Between 1999 and 2009 (Number Counted) From Fishery-Independent Monitoring Sets That Captured Fish*

| Atlantic Coast Post-YOY | | | | | | | | | | | |
|---|---|---|---|---|---|---|---|---|---|---|---|
| Catch | 1999 | 2000 | 2001 | 2002 | 2003 | 2004 | 2005 | 2006 | 2007 | 2008 | 2009 |
| Atlantic Croaker | 319 | 323 | 321 | 325 | 325 | 323 | 317 | 324 | 324 | 309 | 321 |
| Bluefish | N/A | N/A | N/A | N/A | N/A | N/A | N/A | N/A | N/A | N/A | N/A |
| Flounders | 324 | 324 | 321 | 325 | 325 | 326 | 322 | 324 | 326 | 324 | 324 |
| Striped Mullet | 324 | 323 | 321 | 325 | 325 | 326 | 321 | 324 | 326 | 322 | 321 |
| Sheepshead | 324 | 323 | 321 | 325 | 325 | 326 | 321 | 324 | 326 | 322 | 321 |
| Red Drum | 324 | 323 | 321 | 325 | 325 | 326 | 321 | 324 | 326 | 322 | 321 |

## SECTION 1: GENERATE AN ARGUMENT

### 7  DECLINE IN SALTWATER FISH POPULATIONS

*Table 7.7. Annual Observations of Young of the Year (YOY) of Select Fish Along the Florida Gulf Coast Between 1999 and 2009 (Number Counted) From Fishery-Independent Monitoring Sets That Captured Fish*

| Gulf Coast YOY | | | | | | | | | | | |
|---|---|---|---|---|---|---|---|---|---|---|---|
| Catch | 1999 | 2000 | 2001 | 2002 | 2003 | 2004 | 2005 | 2006 | 2007 | 2008 | 2009 |
| Atlantic Croaker | 1150 | 1152 | 1152 | 1152 | 1156 | 1176 | 1303 | 1303 | 1304 | 1316 | 1314 |
| Bluefish | N/A | N/A | N/A | N/A | N/A | N/A | N/A | N/A | N/A | N/A | N/A |
| Flounders | 1150 | 1152 | 1152 | 1152 | 1156 | 1176 | 1303 | 1303 | 1304 | 1316 | 1314 |
| Striped Mullet | 1150 | 1152 | 1152 | 1152 | 1156 | 1176 | 1303 | 1303 | 1304 | 1316 | 1314 |
| Sheepshead | 1147 | 1150 | 1152 | 1152 | 1155 | 1176 | 1302 | 1300 | 1303 | 1316 | 1314 |
| Red Drum | 1126 | 1104 | 1126 | 1122 | 1126 | 1133 | 1267 | 1287 | 1286 | 1297 | 1313 |

*Table 7.8. Annual Observations of Post-Young of the Year (YOY) of Select Fish Along the Florida Gulf Coast Between 1999 and 2009 (Number Counted) From Fishery-Independent Monitoring Sets That Captured Fish*

| Gulf Coast Post-YOY | | | | | | | | | | | |
|---|---|---|---|---|---|---|---|---|---|---|---|
| Catch | 1999 | 2000 | 2001 | 2002 | 2003 | 2004 | 2005 | 2006 | 2007 | 2008 | 2009 |
| Atlantic Croaker | 444 | 439 | 444 | 444 | 444 | 444 | 444 | 444 | 444 | 444 | 444 |
| Bluefish | N/A | N/A | N/A | N/A | N/A | N/A | N/A | N/A | N/A | N/A | N/A |
| Flounders | 444 | 439 | 444 | 444 | 444 | 444 | 444 | 444 | 444 | 444 | 444 |
| Striped Mullet | 442 | 439 | 444 | 444 | 444 | 444 | 444 | 444 | 444 | 444 | 444 |
| Sheepshead | 442 | 439 | 444 | 444 | 444 | 444 | 444 | 444 | 444 | 444 | 444 |
| Red Drum | 435 | 434 | 431 | 424 | 437 | 430 | 436 | 440 | 441 | 431 | 444 |

# SECTION 1: GENERATE AN ARGUMENT

## DECLINE IN SALTWATER FISH POPULATIONS  7

*Table 7.9. Information About Selected Fish Populations Found Around the Florida Coast*

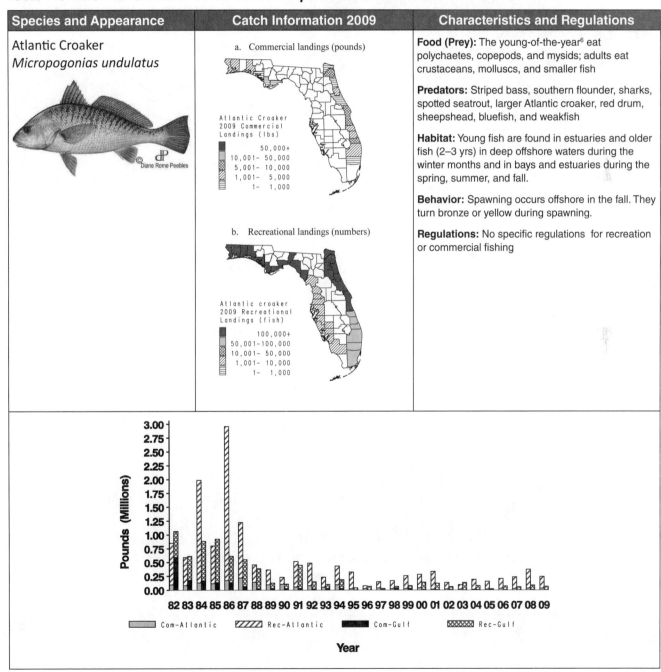

*(continued)*

## SECTION 1: GENERATE AN ARGUMENT

### 7  DECLINE IN SALTWATER FISH POPULATIONS

*Table 7.9. Information About Selected Fish Populations Found Around the Florida Coast* (continued)

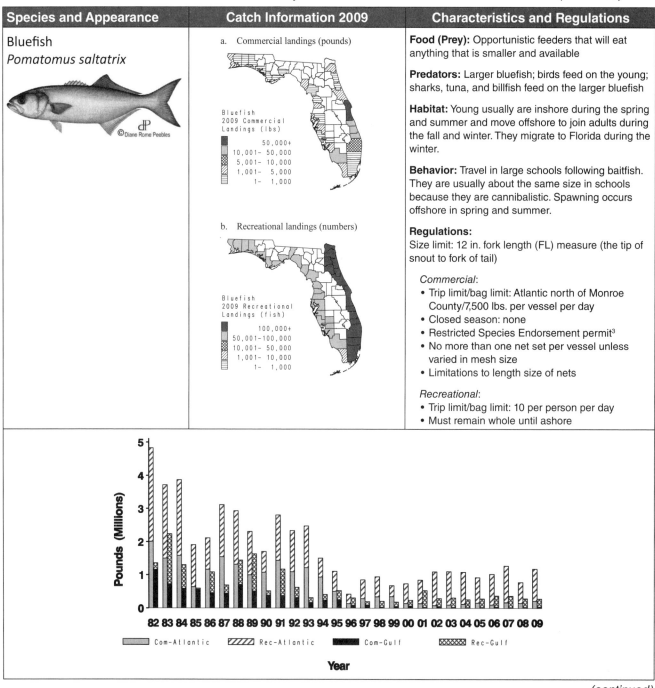

| Species and Appearance | Catch Information 2009 | Characteristics and Regulations |
|---|---|---|
| Bluefish<br>*Pomatomus saltatrix* | a. Commercial landings (pounds)<br><br>Bluefish 2009 Commercial Landings (lbs)<br>50,000+<br>10,001– 50,000<br>5,001– 10,000<br>1,001– 5,000<br>1– 1,000<br><br>b. Recreational landings (numbers)<br><br>Bluefish 2009 Recreational Landings (fish)<br>100,000+<br>50,001–100,000<br>10,001– 50,000<br>1,001– 10,000<br>1– 1,000 | **Food (Prey):** Opportunistic feeders that will eat anything that is smaller and available<br>**Predators:** Larger bluefish; birds feed on the young; sharks, tuna, and billfish feed on the larger bluefish<br>**Habitat:** Young usually are inshore during the spring and summer and move offshore to join adults during the fall and winter. They migrate to Florida during the winter.<br>**Behavior:** Travel in large schools following baitfish. They are usually about the same size in schools because they are cannibalistic. Spawning occurs offshore in spring and summer.<br>**Regulations:**<br>Size limit: 12 in. fork length (FL) measure (the tip of snout to fork of tail)<br>*Commercial*:<br>• Trip limit/bag limit: Atlantic north of Monroe County/7,500 lbs. per vessel per day<br>• Closed season: none<br>• Restricted Species Endorsement permit[3]<br>• No more than one net set per vessel unless varied in mesh size<br>• Limitations to length size of nets<br>*Recreational*:<br>• Trip limit/bag limit: 10 per person per day<br>• Must remain whole until ashore |

(continued)

## SECTION 1: GENERATE AN ARGUMENT

### DECLINE IN SALTWATER FISH POPULATIONS  7

*Table 7.9. Information About Selected Fish Populations Found Around the Florida Coast* (continued)

| Species and Appearance | Catch Information 2009 | Characteristics and Regulations |
|---|---|---|
| Gulf Flounder<br>*Paralichthys albigutta* | a. Commercial landings (pounds)<br><br>Flounders 2009 Commercial Landings (lbs)<br>50,000+<br>10,001– 50,000<br>5,001– 10,000<br>1,001– 5,000<br>1– 1,000<br><br>b. Recreational landings (numbers)<br><br>Flounders 2009 Recreational Landings (fish)<br>100,000+<br>50,001–100,000<br>10,001– 50,000<br>1,001– 10,000<br>1– 1,000 | **Food (Prey):** The young are bottom feeders eating shrimp, crabs, and small fish; adults feed on schooling menhaden, bay anchovy, pinfish, grunts, pigfish, Atlantic croaker, and mullets.<br>**Predators:** Sharks, oyster toadfish, and striped bass<br>**Habitat:** Found inshore on sandy or muddy bottoms and in tidal creeks; occasionally on near-shore rocky reefs<br>**Behavior:** They lie on the bottom often partially covered by sand or mud waiting for prey to come near, and then they strike suddenly. They hatch with a typical fish form, but the right eye migrates over to the left side early in life. They spawn offshore.<br>**Regulations:**<br>Size limitations: 12 in. in total length (TL) (the measure from tip of snout to fork of tail)<br>*Commercial*:<br>• Trip limit/bag limit: incidental bycatch (accidental or unintentional catching while fishing for other species or organisms)<br>• Closed season: none<br>• Restricted Species Endorsement permit; beach and cast net allowed; hook and line allowed; spearing only in Volusia County but not with more than three prongs; must remain in whole condition until landed ashore; no multiple hooks permitted in catching<br>*Recreational*:<br>• Trip limit/bag limit: 10 per person per day<br>• Spearing is OK; snatching prohibited; must remain whole until landed ashore |

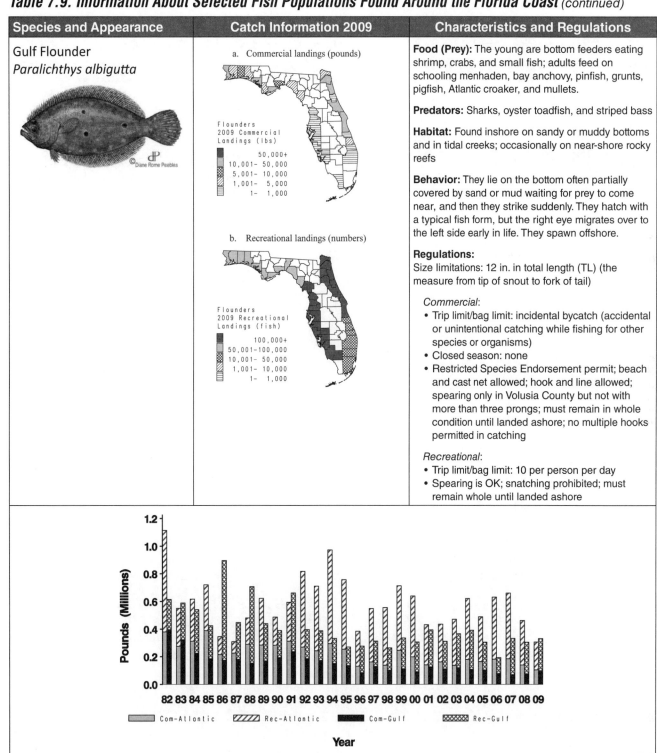

(continued)

## SECTION 1: GENERATE AN ARGUMENT

### 7  DECLINE IN SALTWATER FISH POPULATIONS

*Table 7.9. Information About Selected Fish Populations Found Around the Florida Coast* (continued)

| Species and Appearance | Catch Information 2009 | Characteristics and information |
|---|---|---|
| Red Drum<br>*Sciaenops ocellatus* | a. Commercial landings (pounds)<br><br>This fish is not a commercial fish. Therefore there is no data for geographical landings<br><br>Recreational landings (numbers)<br><br>Red Drum 2009 Recreational Landings (fish)<br>100,000+<br>50,001–100,000<br>10,001– 50,000<br>1,001– 10,000<br>1– 1,000 | **Food (Prey):** The young eat copepods, mysid shrimp, and amphipods; adults eat menhaden and anchovies in the winter and spring; adults eat crabs and shrimp in the summer and fall.<br><br>**Predators:** Larger fish, birds, bottlenose dolphin<br><br>**Habitat:** In the winter, they are found in sea grass, over muddy or sand bottoms, or near oyster bars or spring fed creeks.<br><br>**Behavior:** Young remain inshore until they reach roughly 30 inches (four years) and then they migrate to near-shore populations. They produce tens of millions of eggs from August through December in passes, inlets, and lagoon estuaries.<br><br>**Regulations:**<br>*Commercial*: prohibited completely<br><br>*Recreational*:<br>• Size limitations: not less than 18 in. or more than 27 in.<br>• Trip limit/bag limit: two per person per day in the northern regions and one per person per day in the southern regions<br>• Must remain in whole condition until landed ashore; measured as total length from the mouth to the tip of tail; single hook gear only; gigging, spearing, and snatching are prohibited |

[Bar chart: Pounds (Millions) vs. Year (82–09), showing Com-Atlantic, Rec-Atlantic, Com-Gulf, Rec-Gulf landings]

*(continued)*

# SECTION 1: GENERATE AN ARGUMENT

## DECLINE IN SALTWATER FISH POPULATIONS  7

*Table 7.9. Information About Selected Fish Populations Found Around the Florida Coast* (continued)

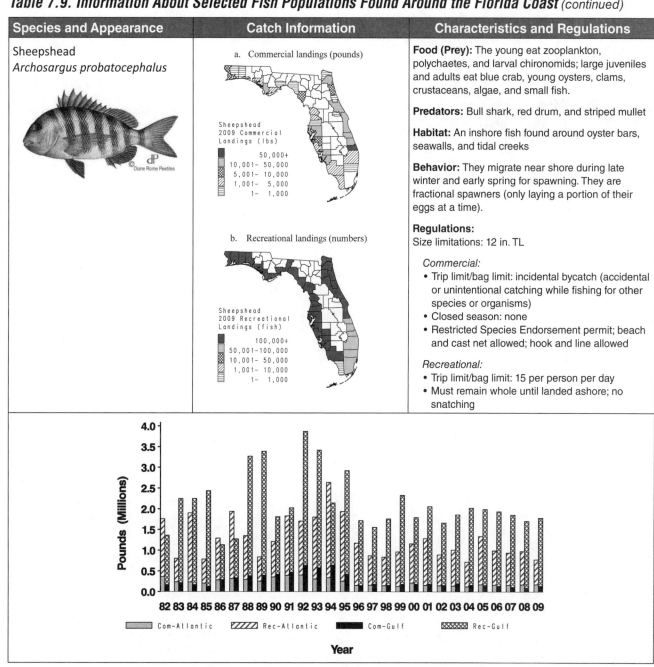

(continued)

## SECTION 1: GENERATE AN ARGUMENT

### 7 DECLINE IN SALTWATER FISH POPULATIONS

*Table 7.9. Information About Selected Fish Populations Found Around the Florida Coast* (continued)

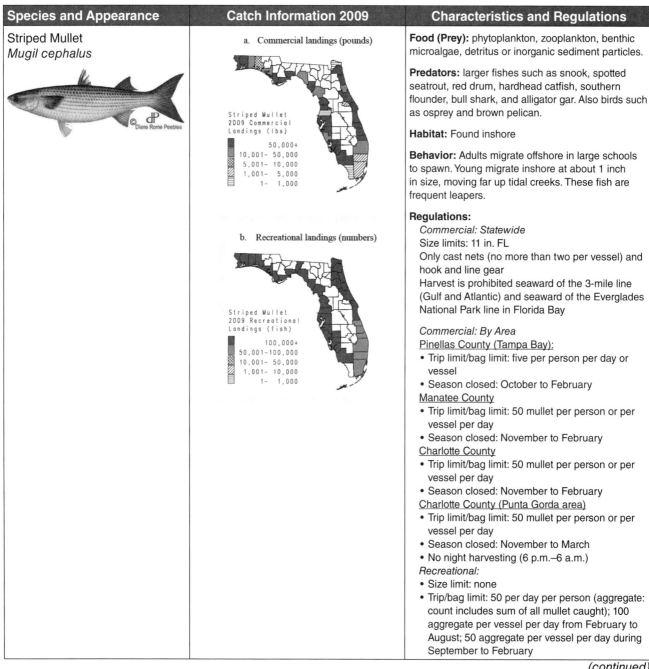

| Species and Appearance | Catch Information 2009 | Characteristics and Regulations |
|---|---|---|
| Striped Mullet<br>*Mugil cephalus* | a. Commercial landings (pounds)<br><br>Striped Mullet 2009 Commercial Landings (lbs)<br>50,000+<br>10,001– 50,000<br>5,001– 10,000<br>1,001– 5,000<br>1– 1,000<br><br>b. Recreational landings (numbers)<br><br>Striped Mullet 2009 Recreational Landings (fish)<br>100,000+<br>50,001–100,000<br>10,001– 50,000<br>1,001– 10,000<br>1– 1,000 | **Food (Prey):** phytoplankton, zooplankton, benthic microalgae, detritus or inorganic sediment particles.<br>**Predators:** larger fishes such as snook, spotted seatrout, red drum, hardhead catfish, southern flounder, bull shark, and alligator gar. Also birds such as osprey and brown pelican.<br>**Habitat:** Found inshore<br>**Behavior:** Adults migrate offshore in large schools to spawn. Young migrate inshore at about 1 inch in size, moving far up tidal creeks. These fish are frequent leapers.<br>**Regulations:**<br>*Commercial: Statewide*<br>Size limits: 11 in. FL<br>Only cast nets (no more than two per vessel) and hook and line gear<br>Harvest is prohibited seaward of the 3-mile line (Gulf and Atlantic) and seaward of the Everglades National Park line in Florida Bay<br>*Commercial: By Area*<br><u>Pinellas County (Tampa Bay)</u>:<br>• Trip limit/bag limit: five per person per day or vessel<br>• Season closed: October to February<br><u>Manatee County</u><br>• Trip limit/bag limit: 50 mullet per person or per vessel per day<br>• Season closed: November to February<br><u>Charlotte County</u><br>• Trip limit/bag limit: 50 mullet per person or per vessel per day<br>• Season closed: November to February<br><u>Charlotte County (Punta Gorda area)</u><br>• Trip limit/bag limit: 50 mullet per person or per vessel per day<br>• Season closed: November to March<br>• No night harvesting (6 p.m.–6 a.m.)<br>*Recreational:*<br>• Size limit: none<br>• Trip/bag limit: 50 per day per person (aggregate: count includes sum of all mullet caught); 100 aggregate per vessel per day from February to August; 50 aggregate per vessel per day during September to February |

*(continued)*

# SECTION 1: GENERATE AN ARGUMENT

## DECLINE IN SALTWATER FISH POPULATIONS    7

*Table 7.9. Information About Selected Fish Populations Found Around the Florida Coast* (continued)

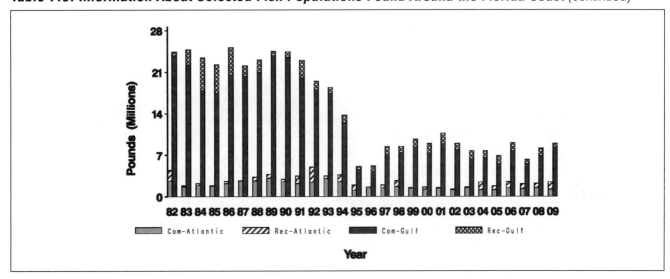

## Reference

Florida Fish and Wildlife Conservation Commission. Saltwater fish. Florida Fish and Wildlife Conservation Commission. *http://myfwc.com/wildlifehabitats/profiles/fish/saltwater/*.

SECTION 1: GENERATE AN ARGUMENT

# 7 DECLINE IN SALTWATER FISH POPULATIONS
## TEACHER NOTES

## Purpose

The purpose of this activity is to help students understand the factors influencing the survival of organisms in an environment, focusing on the interdependent relationships in an ecosystem. It calls attention to the limitations of our resources and to the human impact on the environment. This activity will help students develop the skills needed for scientific inquiry such as interpreting data to make inferences about fluctuations in fish populations and analyzing the effects of human use and habitat changes on fish populations. This activity, with teacher input, can also be used to help students understand the nature of scientific inquiry and the nature of scientific explanations as they consider the reliability and validity of data to support their claims and as they sort through the data to identify specific relationships between factors that exist in a complex system of interactions. Students can learn to use causal patterns as they reason about patterns in relationships and forms of cause and effect with teacher guidance.

## The Content and Related Concepts

It has been a common misconception that the oceans provide an endless supply of fish. People around the world have historically relied on the freedom to use this resource to supply their needs. In the late 1960s to early 1970s, people began to be concerned with the idea that the oceans were indeed not unlimited, and there was a call to regulate marine fisheries. A fishery is a waterway or a portion of a sea where aquatic species can be harvested.

There has been some debate about whether fisheries should be regulated and whether they should be separated by commercial and recreational interests. Some fear that if commercial interests were left to their own devices, they would catch all of the fish until the fisheries were depleted. Others argue that if recreational interests are not regulated, individuals or private interests would create situations in which the fishing would not be sustainable (i.e., catching young fish and interfering with the cycle of population growth).

In 1976, Congress passed the Magnuson-Stevens Fishery Conservation and Management Act (MSA) which addressed policies on fisheries up to 200 miles offshore with the charge to conserve and manage America's fishery resources as well as to promote commercial and recreational fishing. Eight Regional Fishery Management Councils were formed to address the needs and stresses of different areas. This was in part to protect the indigenous practices, such as those in Alaska and Hawaii, and the small and private fisherman in communities such as New England where economic livelihood depends on the fishing industry. Regulations included shortening fish-

# SECTION 1: GENERATE AN ARGUMENT

## DECLINE IN SALTWATER FISH POPULATIONS
### TEACHER NOTES 7

ing seasons, restricting vessel sizes and types, restricting the number of days allowed at sea, restricting the type of gear, restricting who could fish (i.e., giving preference to indigenous peoples or recreational activity), restricting size of catch, and restricting the kind of fish (organism) harvested (Environmental Defense Fund 2011). Other initiatives have included raising fish in hatcheries and releasing the young in streams to head down to the ocean fisheries or releasing them in the ocean fisheries. Even with these regulations and initiatives, the concerns for a total collapse of global fisheries including red snapper, cod, tuna, and salmon were published in several studies, news articles, and documentaries (Environmental Defense Fund 2011). These concerns were followed by calls for even stricter regulations and for new practices in fishing.

While much of the concern for the collapse of ocean fisheries has been attributed to fishing and catching practices, there has also been attention brought to other factors impacting the fisheries. For example, global warming has been blamed for some of the reduction in fisheries. Some people claim that the ocean temperatures are increasing, and others claim that the salt content is changing due to global warming. These kinds of changes could impact the ecosystems, preventing some species from surviving while promoting survival of others. In addition, farming and industrial practices have been blamed for negative impacts on fisheries, adding to the stresses of the fisheries through pollution: soil and chemical runoff, oil spills, dumping of waste, and so on. These impacts have been blamed for not only reduc-ing the number of fish but also causing abnormalities and poisoning fish beyond acceptable levels for human consumption.

## Curriculum and Instructional Considerations

### Middle School

Students in the middle grades continue to learn about the comparisons of species and their relationships to each other in an ecosystem through food webs and food chains. There are many resources that provide examples of food webs for various ecosystems; however, these often lack the inclusion of abiotic factors (aside from perhaps the Sun). Because of this, students are likely to eliminate interactions between abiotic and biotic factors in their concepts of ecosystems, and they are not likely to understand indirect impacts in an ecosystem. In addition, the discussions usually present the relationships of a food web as a "who ate whom," and therefore the direction of the energy in a food web is often misunderstood, and cyclic causal patterns are not recognized (Grotzer and Basca, forthcoming). In learning about various types of ecosystems, students often develop a sense of closed systems in which ecosystems do not interact with each other. This affects the students' ability to identify impacts that are nonobvious and/or time delayed, and it is likely to affect their ability to understand two-way patterns of causality (Grotzer and Perkins 2003). Using this activity in the middle school will help students to develop a stronger understanding of the complex relationships between living organ-

## SECTION 1: GENERATE AN ARGUMENT

### 7 DECLINE IN SALTWATER FISH POPULATIONS
TEACHER NOTES

isms and abiotic factors in an ecosystem while developing their critical-thinking skills using different cause-and-effect patterns.

### High School

In the high school curriculum, systems are discussed within various applications: machines, cells, human anatomy, and ecosystems. Students will be focusing on learning about the ecosystem as a complex system of interactions but still may not recognize the abiotic role within the system. They will be reasoning about population level effects. This will be difficult unless they move beyond reasoning from one's own perspective. This involves reasoning about multiple interacting organisms, which presents a problem of cognitive load (Grotzer et al. 2009).

Students will learn about limiting factors in an environment, and they learn about the flow of energy that occurs through an ecosystem on both macro and micro levels. The curriculum will focus on recognizing relationships between factors within the ecosystem, with greater emphasis on anthropogenic changes (human impacts). Misconceptions about the flow of energy in an ecosystem are common as students attempt to make sense of the relationship between system causal patterns and energy flow in a system. Students are likely to have the following misconceptions: An animal that is high on the food web preys on all populations below it; if the size of one population in a food web is changed, all other populations in the web are changed in the same way; the top of the web has the most energy or that energy accumulates at the top; and, populations on the top increase as the organisms below decrease (Annenberg Foundation n.d.; Grotzer et al. 2009). These misconceptions remain because of students' inability to understand the interdependence of causal relationships that include various patterns and their disconnection between the abiotic and biotic factors (Honey and Grotzer 2009).

Also, by high school, students are likely to have had experiences in labs that have included observable changes over short periods of time between isolated variables. While these experiences are important for scientific thinking and doing, the observable changes over short periods of times may add to the unfamiliarity of using data to examine reasoning for populations through causal patterns. In addition, the students will likely be learning about variations that allow different organism to survive in an environment. Students may develop a misconception that ecosystems which experience a change will result in individual organisms that mutate and adapt immediately. Implementing this activity in the high school curriculum will help high school students to understand different scientific processes and develop an understanding for the interactions and interdependence of the organisms and abiotic factors in an ecosystem.

### Recommendations for Implementing the Activity

Students will find this activity interesting as it has real-life applications and connects to social and historical perspectives. Current newspapers publish stories about the debate,

## SECTION 1: GENERATE AN ARGUMENT

### DECLINE IN SALTWATER FISH POPULATIONS
#### TEACHER NOTES 7

and students are likely to find the social and traditional impacts on society relevant. Resistance to this activity is likely to come from the overwhelming amount of data that is involved in understanding the complexity of an ecosystem. Teachers using this activity will likely need to help students with this by guiding them through portions of the data, depending on the students' age and the placement of the activity in the curriculum. It is suggested that this activity follow a lesson on ecosystems or food chains or food webs.

This activity takes approximately 100 minutes of instructional time to complete, but the amount of time devoted to each stage of the activity varies depending on how a teacher decides to spend time in class. For more information about how to implement the activity, see Appendix E on page 369.

Table 7.10 provides information about the type and amount of materials needed to implement this activity in a classroom with 28 students with groups of four and groups of three.

## Assessment

The rubric provided in Appendix B (p. 366) can be used to assess the arguments and counterarguments crafted by each student at the end of the activity. The rubric includes categories for the adequacy and conceptual quality of the claim, the appropriate use of evidence, the sufficiency of the rationale, and the overall quality of the writing. We strongly recommend that teachers use the Comments or Suggestions section to give students detailed feedback so they will understand what they did wrong, why it is wrong, and ways they can improve their performance next time. To illustrate how to score the arguments and counterarguments, consider the following example written by a seventh-grade student:

There should be some restrictions placed on fishing for both commercial

*Table 7.10. Materials Needed to Implement the Activity in a Classroom of 28 Students*

| Material | Amount Needed With... | |
| --- | --- | --- |
| | Groups of 3 | Groups of 4 |
| Whiteboards (or chart paper)+ | 10 | 7 |
| Whiteboard markers (or permanent if using chart paper)+ | 20 | 14 |
| Copy of Student Background Info Pages (pp. 81–83)* | 28 | 28 |
| Copy of Student Directions Page (p. 84)* | 10 | 7 |
| Copy of Student Data Pages (pp. 85–95)* | 28 | 28 |
| Copy of Appendix B (p. 366)* | 28 | 28 |

+ Teachers can also have students prepare their arguments in a digital medium (such as PowerPoint or Keynote)
* Teachers can also project these materials onto a screen in order to cut down on paper use

**SCIENTIFIC ARGUMENTATION IN BIOLOGY: 30 CLASSROOM ACTIVITIES**

## SECTION 1: GENERATE AN ARGUMENT

### 7 DECLINE IN SALTWATER FISH POPULATIONS
**TEACHER NOTES**

and recreational users for all fish. **The data shows that all fish populations are both increasing and decreasing at different times.** We figured this out by graphing the number of fish caught per year by recreational and commercial landings. Even though most of the trends are mostly increasing in populations, that is probably because rules about fishing have been getting enforced. Since fish populations do have an increase and a decrease, that means that they are probably not endless supplies. If we want to keep having fish we need to protect the populations from going extinct. Some of the reasons that the populations might be decreasing are not just human cause (anthropogenic). We know this because when we graphed the fish populations and compared them. It seemed like the Atlantic croaker was decreasing when other fish populations like the Red drum were increasing. Since the Red drum eats the Atlantic croaker, and the Atlantic croaker is eaten by many other fish, when those populations are increasing, the Atlantic croaker population is probably going to decrease. This is because of the food chains that have a reaction in one place when something in the chain changes. And the rules for the Red drum to protect it helped the Red drum to increase but that made the Atlantic croaker decrease. With this information, we think that there should definitely be some rules about who can fish and how much they can fish. We think that if fishing is recreational then the rules should make it so that people can't catch the fish super easy. That would help slow down the fishing catches for recreation. And if the prey fish (like the Atlantic croaker) starts to decrease a lot then there should be rules keeping them from being caught and there should be rules removed from their predators (like the Red drum and the sheepshead) from being caught. This way there will be a better balance in the ocean so that everyone can exist and the fish populations won't go extinct. If we let one of the fish populations go extinct we would be in a lot of trouble since we are all a part of the same food web.

The content of the example argument is adequate. The student's claim (underlined) is sufficient (1/1) and accurate (1/1). The student, however, does not use genuine evidence (in bold) to support the claim (0/3). Instead, the student relies on an unsubstantiated inference as evidence. The student provides an in-depth justification of the evidence in her argument by explaining why the "evidence" was important (2/2). The author also uses scientific terms correctly (1/1) and uses phrases that are consistent with the nature of science (1/1). However, the writing mechanics of the sample argument could use some improvement. The organization of the argument needs to be modified because the arrangement of the

## SECTION 1: GENERATE AN ARGUMENT

### DECLINE IN SALTWATER FISH POPULATIONS
### TEACHER NOTES 7

sentences distracts from the development of the main idea (0/1). There are also some grammatical errors (0/1) in the argument, although the punctuation is correct (1/1). The overall score for the sample argument, therefore, is 7 out of the 12 points possible.

It should be noted that this is a very complex topic as there are many variables and interactions that are not well understood and are not easily isolated. Allowing students to discuss the data and analyze it will introduce these complexities and help them to build their skills in scientific argumentation. Teachers should also be sure to pay attention to how the students are evaluating claims during the generation of a tentative argument phase and the argumentation phases of the lesson. Teachers should remind students to rely on both empirical (fit with data) and theoretical (fit with the historical body of scientific knowledge) criteria to evaluate or support ideas rather than plausibility, past experiences, or an authority figure.

## Standards Addressed in This Activity

This activity can be used to address the following dimensions outlined in *A Framework for K–12 Science Education* (NRC 2012):

### Scientific Practices

- Constructing explanations
- Engaging in argument from evidence
- Obtaining, evaluating, and communicating information

### Crosscutting Concepts

- Cause and effect: Mechanism and explanation
- Structure and function

### Life Sciences Core Ideas

- Interdependent Relationships in Ecosystems
- Flow of Matter and Energy in Ecosystems
- Ecosystems: Interactions, Energy, and Dynamics

This activity can be used to address the following standards for literacy in science from the *Common Core State Standards for English Language Arts and Literacy* (NGA and CCSSO 2010):

### Writing

- Text types and purposes
- Production and distribution of writing
- Research to build and present knowledge
- Range of writing

### Speaking and Listening

- Comprehension and collaboration
- Presentation of knowledge and ideas

## References

Annenberg Foundation. n.d. The habitable planet: ecosystems. President and Fellows of Harvard College. *http://www.pz.harvard.edu/ucp/curriculum/ecosystems/*.

# SECTION 1: GENERATE AN ARGUMENT

## 7 DECLINE IN SALTWATER FISH POPULATIONS
### TEACHER NOTES

Environmental Defense Fund. 2011. Brief history of fisheries management. *www.edf.org/page.cfm?tagID=1544*.

Florida Fish and Wildlife Conservation Commission. n.d. *http://myfwc.com*.

Grotzer, T. A., and B. B. Basca. Forthcoming. Helping students to grasp the underlying causal structures when learning about ecosystems: How does it impact understanding? *Journal of Biological Education*.

Grotzer, T., C. Dede, S. Metcalfe, and J. Clarke. 2009. Addressing the challenges in understanding ecosystems: Classroom studies. Paper presented at the annual conference for National Association for Research in Science Teaching (NARST), Garden City, CA.

Grotzer, T. A., and D. Perkins. 2003. Understandings of Consequence Project: Causal patterns in ecosystems. President and Fellows of Harvard College. *www.pz.harvard.edu/ucp/curriculum/ecosystems*.

Honey, R., and T. A. Grotzer. 2009. *Cultural diversity in the classroom: Salish/Kootenai students' perceptions of ecosystems relationships*. Presented at the National Association of Research in Science Teaching (NARST) Conference, Garden Grove, California.

Kourous, G., 2005. Depleted fish stocks require recovery efforts. Food and Agriculture Organization of the United Nations. *www.fao.org/newsroom/en/news/2005/100095/index.html*.

Leach, J., R. Driver, P. Scott, and C. Wood-Robinson. 1996. Children's ideas about ecology 3: Ideas found in children aged 5–16 about the interdependency of organisms. *International Journal of Science Education* 18: 19–34.

National Governors Association Center (NGA) for Best Practices, and Council of Chief State School Officers (CCSSO). 2010. *Common core state standards for English language arts and literacy*. Washington, DC: National Governors Association for Best Practices, Council of Chief State School.

National Research Council (NRC). 2012. *A framework for K–12 science education: Practices, crosscutting concepts, and core ideas*. Washington, DC: National Academies Press.

Webb, P., and G. Boltt. 1990. Food chain to food web: A natural progression? *Journal of Biological Education* 24 (3): 187–190.

## SECTION 1: GENERATE AN ARGUMENT

# HISTORY OF LIFE ON EARTH (TRENDS IN EVOLUTION) 8

Scientists have identified approximately 1.5 million different species of organisms on Earth. These species have then been placed into groups called genus. Each genus can also be placed into a larger group called a family. Families, in turn, can be grouped together to create an order. Figure 8.1 provides an overview of the number of different families that are found within some common types of organisms.

*Figure 8.1. The Number of Families Within Some Common Types of Organisms*

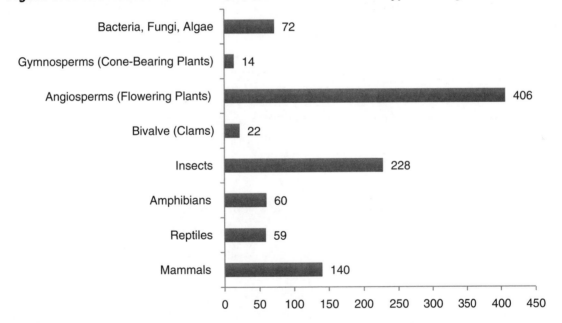

| Organism | Number of Families |
|---|---|
| Bacteria, Fungi, Algae | 72 |
| Gymnosperms (Cone-Bearing Plants) | 14 |
| Angiosperms (Flowering Plants) | 406 |
| Bivalve (Clams) | 22 |
| Insects | 228 |
| Amphibians | 60 |
| Reptiles | 59 |
| Mammals | 140 |

All this biodiversity on Earth has made many scientists wonder: **How has biodiversity on Earth changed over time?**

With your group, use the supplied data set to develop a claim that best answers this research question. Once your group has developed your claim, prepare a whiteboard that you can use to share and justify your claim. Your whiteboard should include all the information shown in Figure 8.2 (p. 104).

## SECTION 1: GENERATE AN ARGUMENT

### 8 HISTORY OF LIFE ON EARTH

To share your work with others, we will be using a round-robin format. This means that one member of the group will stay at your workstation to share your group's ideas while the other group members go to the other groups one at a time in order to listen to and critique the arguments developed by your classmates.

Remember, as you critique the work of others, you need to decide if their conclusions are valid or acceptable based on the quality of their claim and how well they are able to support their ideas. In other words, you need to determine if their argument is *convincing* or not. One way to determine if their argument is convincing is to ask them some of the following questions:

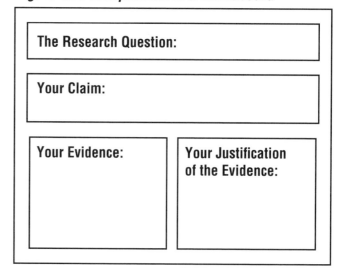

*Figure 8.2. Components of a Whiteboard*

- How did you analyze or interpret your data? Why did you decide to do it that way?
- How do you know that your analysis of the data is free from errors?
- Why does your evidence support your claim?
- Why did you decide to use that evidence? Why is your evidence important?
- How does your justification of the evidence fit with accepted scientific ideas?
- What are some of the other claims your group discussed before agreeing on your claim, and why did you reject them?

# SECTION 1: GENERATE AN ARGUMENT

## HISTORY OF LIFE ON EARTH  8

*Table 8.1. Information About the Number of Different Families[1] That Have Been Identified in the Fossil Record (Benton 1993, 1995)*

| Time Before Present[2] (in Millions of Years) | Number of Different Families That Have Been Found in the Fossil Record | | | | | | | |
|---|---|---|---|---|---|---|---|---|
| | Mammals | Reptiles | Amphibians | Insects | Bivalves (Clams) | Angiosperms (Flowering Plants) | Gymnosperms (Cone-Bearing Plants) | Bacteria, Fungi, & Algae |
| 0.01 | 124 | 38 | 23 | 550 | 17 | 153 | 11 | 98 |
| 0.8 | 138 | 41 | 21 | 551 | 17 | 153 | 11 | 99 |
| 26 | 138 | 41 | 20 | 534 | 15 | 123 | 11 | 108 |
| 53 | 110 | 47 | 18 | 323 | 17 | 101 | 11 | 107 |
| 78 | 27 | 65 | 9 | 237 | 15 | 25 | 13 | 100 |
| 104 | 17 | 46 | 9 | 308 | 15 | 14 | 17 | 99 |
| 128 | 8 | 38 | 7 | 125 | 12 | 3 | 18 | 89 |
| 153 | 10 | 47 | 4 | 87 | 15 | 2 | 19 | 75 |
| 175 | 4 | 14 | 2 | 104 | 13 | 2 | 16 | 58 |
| 199 | 5 | 23 | 2 | 56 | 12 | 2 | 14 | 45 |
| 229 | 0 | 31 | 8 | 90 | 18 | 1 | 13 | 42 |
| 252 | 0 | 31 | 9 | 18 | 51 | 0 | 9 | 38 |
| 275 | 0 | 8 | 27 | 14 | 63 | 0 | 10 | 39 |
| 299 | 0 | 5 | 22 | 19 | 56 | 0 | 9 | 37 |
| 327 | 0 | 0 | 10 | 7 | 57 | 0 | 6 | 34 |
| 356 | 0 | 0 | 0 | 1 | 60 | 0 | 5 | 35 |
| 372 | 0 | 0 | 0 | 1 | 63 | 0 | 0 | 36 |
| 402 | 0 | 0 | 0 | 0 | 77 | 0 | 0 | 29 |
| 427 | 0 | 0 | 0 | 0 | 71 | 0 | 0 | 24 |
| 453 | 0 | 0 | 0 | 0 | 67 | 0 | 0 | 23 |
| 472 | 0 | 0 | 0 | 0 | 51 | 0 | 0 | 20 |
| 501 | 0 | 0 | 0 | 0 | 27 | 0 | 0 | 21 |
| 527 | 0 | 0 | 0 | 0 | 18 | 0 | 0 | 20 |
| 555 | 0 | 0 | 0 | 0 | 18 | 0 | 0 | 19 |

1. In biology, *family* refers to a taxonomic rank that falls between order and genus. The levels of classification include kingdom, phylum, class, order, family, genus, and species. For example, the Bonobo (*P. paniscus*) is a part of the genus *Pan*, the family Hominidae, the order Primates, the class Mammalia, the phylum Chordata, and the Kingdom Animalia. Many different species make up a particular family.
2. These dates represent the midpoint in different geologic stages. For example, 0.01 mya is the midpoint of the Holocene stage and 0.8 mya is the midpoint of the Pleistocene stage.

## References

Benton, M. J. 1993. *The fossil record 2*. London: Chapman & Hall.

Benton, M. J. 1995. Diversification and extinction in the history of life. *Science* 268: 52–58.

## SECTION 1: GENERATE AN ARGUMENT

Name_____ Date_____

# HISTORY OF LIFE ON EARTH:
## What Is Your Argument?

In the space below, write an argument in order to persuade another biologist that your claim is valid and acceptable. As you write your argument, remember to do the following:

- State the claim you are trying to support
- Include genuine evidence (data + analysis + interpretation)
- Provide a justification of your evidence that explains why the evidence is relevant and why it provides adequate support for the claim
- Organize your argument in a way that enhances readability
- Use a broad range of words including vocabulary that we have learned
- Correct grammar, punctuation, and spelling errors

SECTION 1: GENERATE AN ARGUMENT

# HISTORY OF LIFE ON EARTH
## TEACHER NOTES 8

## Purpose

The purpose of this activity is to help students understand the diversification and extinction of species within families during the history of life on Earth. This activity also helps students learn how to engage in practices such as using mathematics and computational thinking, constructing explanations, arguing from evidence, and communicating information. This activity is also designed to give students an opportunity to learn how to write in science and develop their speaking and listening skills, which are important goals for literacy in science (see Standards Addressed in This Activity for a complete list of the practices, crosscutting concepts, core ideas, and literacy skills that are aligned with this activity).

## The Content and Related Concepts

The fossil record provides the historical archives that biologists use to study the history of life on Earth. The fossil record is a substantial but incomplete chronicle of life on Earth; species that existed for a long period of time, that were abundant and widespread, and that had hard shells or skeletons are more likely to be preserved than species with soft bodies that lived in specific locations. Sedimentary strata can be used to determine the relative age of fossils in successive geological periods, and radiometric dating can be used to identify their absolute age. The history of life involves enormous change. Major life forms have appeared, flourished, and gone extinct. At some points in Earth's history, many different species went extinct in a short amount of time. These are called mass extinctions. Although extinction is common, life on Earth has become more diverse over time. The following are some of the major events in the history of life on Earth (Benton 1995; Campbell and Reece 2006):

- The evolutionary history of life began between 3.5 and 4.0 billion years ago.
- Prokaryotes dominated evolutionary history from 3.5 to 2.0 billion years ago.
- Oxygen began to accumulate in the atmosphere 2.7 billion years ago.
- Eukaryotic life evolved 2.1 billion years ago.
- Multicellular eukaryotes evolved 1.2 billion years ago.
- Animals evolved about 600 million years ago.
- Ordovician mass extinction occurred about 445 million years ago, and 57% of all genera were lost.
- Insects evolved about 380 million years ago.
- Devonian mass extinction occurred about 370 million years ago, and 50% of all genera were lost.

## SECTION 1: GENERATE AN ARGUMENT

### 8 HISTORY OF LIFE ON EARTH
TEACHER NOTES

- Amphibians evolved about 330 million years ago.
- Reptiles evolved about 300 million years ago.
- Permian mass extinction occurred about 250 million years ago, and 83% of all genera were lost.
- Flowering plants evolved about 230 million years ago.
- Mammals evolved about 210 million years ago.
- Triassic mass extinction occurred about 200 million years ago, and 48% of all genera were lost.
- Cretaceous mass extinction occurred about 65 million years ago, and 50% of all genera were lost.

In this activity, the students are asked to determine if number and types of species found on Earth have changed over time using a data set that includes the number of different families found in the fossil record at different points in time. This data set is a simplified version of a more comprehensive one compiled by Benton (1993, 1995) and published online at *www.fossilrecord.net*. The activity's data set clearly illustrates how the number of types of families on Earth has, in general, increased over time. However, mass extinctions that decreased the overall number and type of families found on Earth are also evident in the data set. An example of one such mass extinction is the one that marked the end of the Triassic period (about 200 million years ago).

## Curriculum and Instructional Considerations

This activity is best used at the beginning of a unit on biological evolution, because it is a good way to address many of the misconceptions students have about the history of life on Earth. For example, some students think that the number of species on Earth has remained constant over time or that there are fewer species on Earth now then there was in the past due to catastrophic events, such as a great flood. These views are often rooted in creationist-based explanations for the origin and diversity of life on Earth. Other students will think that the number of species on Earth has steadily increased over time and do not realize how several mass extinctions have resulted in a substantial decrease in the amount of biodiversity on Earth.

The focus of the explicit discussion at the end of the activity should focus on trends in the history of life on Earth and an aspect of the nature of science. For example, a teacher could discuss how science is different from other ways of knowing (such as religion) and how a scientific explanation must be consistent with observational evidence about nature or how scientists rely on a wide range of methods (and not just experiments) to answer research questions. Teachers can use what the students did during this this activity as an illustrative example. This is also a good time to discuss the difference between inferences and observations or theories and laws.

# SECTION 1: GENERATE AN ARGUMENT

## HISTORY OF LIFE ON EARTH
### TEACHER NOTES 8

## Recommendations for Implementing the Activity

This activity takes approximately 100 minutes of instructional time to complete, but the amount of time devoted to each stage of the activity varies depending on how a teacher decides to spend time in class. For more information about how to implement the activity, see Appendix E on page 369.

Table 8.2 provides information about the type and amount of materials needed to implement this activity in a classroom with 28 students with groups of four and groups of three.

## Assessment

The rubric provided in Appendix B (p. 366) can be used to assess the arguments crafted by each student at the end of the activity. To illustrate how the rubric can be used to score an argument written by a student, consider the following example. This sample argument, which was written by an 11th-grade student, is an example of a high-quality argument.

> The number of and type of families found on Earth have increased over time. **The figure to the right, which I made using the data supplied to us, shows the total number of families found in the fossil record over time. This figure clearly shows how the total number of families increases from 37 at 555 million years ago to 1014 at .01 million years ago. If the number of species remained constant over time we would expect to see no change or a decrease in the number of families**

*Table 8.2. Materials Needed to Implement the Activity in a Classroom of 28 Students*

| Material | Amount Needed With ... | |
|---|---|---|
| | Groups of 3 | Groups of 4 |
| Whiteboards (or chart paper)+ | 10 | 7 |
| Whiteboard markers (or permanent if using chart paper)+ | 20 | 14 |
| Copy of Student Pages (pp. 103–105)* | 28 | 28 |
| Copy of Student Page (p. 106)* | 10 | 7 |
| Copy of Appendix B (p. 366)* | 28 | 28 |

+ Teachers can also have students prepare their arguments in a digital medium (such as PowerPoint or Keynote).
* Teachers can also project these materials onto a screen in order to cut down on paper use.

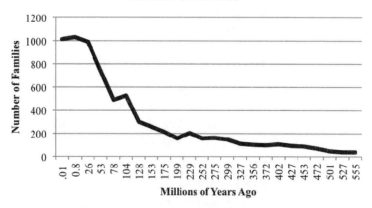

present in the fossil record over time. Fossils can be used to determine the number and type of species found on Earth over time because it provides a record of what was alive in the past.

# SECTION 1: GENERATE AN ARGUMENT

## 8 HISTORY OF LIFE ON EARTH
### TEACHER NOTES

The example argument is good for several reasons. The student's claim (underlined) is sufficient (1/1) and accurate (1/1). The student also uses genuine evidence (in bold) to support the claim because she analyzes and interprets the supplied data set (3/3). The justification of the evidence is complete because she explains why the evidence is important (1/1) but she does not attempt to link the evidence to an important concept or principle (0/1). The author also uses scientific terms correctly (1/1) and avoids using phrases that misrepresent the nature of science (1/1). The organization of the argument is effective, because the arrangement of the sentences does not distract from the development of the main idea (1/1). Finally, there are no grammatical (1/1) or punctuation errors (1/1) in the argument. The overall score for the sample argument, therefore, is 11 out the 12 points possible.

## Standards Addressed in This Activity

This activity can be used to address the following dimensions outlined in *A Framework for K–12 Science Education* (NRC 2012):

### Scientific Practices

- Using mathematics and computational thinking
- Constructing explanations
- Engaging in argument from evidence
- Obtaining, evaluating, and communicating information

### Crosscutting Concepts

- Patterns
- Cause and effect: Mechanism and explanation
- Scale, proportion, and quantity
- Stability and change

### Life Sciences Core Ideas

- From molecules to organisms: Structures and processes
- Biological evolution: Unity and diversity

This activity can be used to address the following standards for literacy in science from the *Common Core State Standards for English Language Arts and Literacy* (NGA and CCSSO 2010):

### Writing

- Text types and purposes
- Production and distribution of writing
- Research to build and present knowledge
- Range of writing

### Speaking and Listening

- Comprehension and collaboration
- Presentation of knowledge and ideas

## References

Benton, M. J. 1993. *The fossil record 2*. London: Chapman & Hall.

## SECTION 1: GENERATE AN ARGUMENT

### HISTORY OF LIFE ON EARTH 8
#### TEACHER NOTES

Benton, M. J. 1995. Diversification and extinction in the history of life. *Science* 268: 52–58.

Campbell, N., and J. Reece. 2002. *Biology*. 6th ed. San Francisco, CA: Benjamin Cummings.

National Governors Association Center (NGA) for Best Practices, and Council of Chief State School Officers (CCSSO). 2010. *Common core state standards for English language arts and literacy*. Washington, DC: National Governors Association for Best Practices, Council of Chief State School.

National Research Council (NRC). 2011. *A framework for K–12 science education: Practices, crosscutting concepts, and core ideas*. Washington, DC: National Academies Press.

# SECTION 1: GENERATE AN ARGUMENT

# SURVIVING WINTER IN THE DUST BOWL (FOOD CHAINS AND TROPHIC LEVELS) 9

In the 1930s, the states of Kansas, Oklahoma, Colorado, and Texas suffered from a severe drought that lasted for almost a decade. Many farmers struggled because of lack of rain, high temperatures, and high winds. These conditions were made even worse by frequent insect infestations and huge dust storms (see Figure 9.1). The dust storms were so bad and happened so often that these states came to be known as the dust bowl. The farmers who stayed on their land were forced to make difficult choices in order to survive in the face of these hardships (see Figure 9.2).

*Figure 9.1. A Dust Storm Approaches Stratford, Texas, in 1935*

*Figure 9.2. A Farmer and His Sons Walking in the Face of a Dust Storm in Cimarron County, Oklahoma*

Imagine that you and the other members of your group are a family of wheat farmers living in Oklahoma, and it is October 15, 1934. It was a very dry year (less than 10 inches of rain fell from January 1, 1934 to October 1, 1934, compared to the average of approximately 42 inches per year), and your crops did not grow well. You and your family planted spring wheat in April and harvested the crop in mid-September. Unfortunately, you were only able to harvest 500 bushels of wheat (1 bushel = 60 pounds), which is much less than 2,800 bushels that you were expecting to harvest (you planted 80 acres of wheat, and you normally are able to harvest 30 to 35 bushels per acre). You only have 500 gallons of potable water left, and you have no way of knowing when it will rain again. You also have a female jersey cow and male bull on your farm, both of which need food and water in order to survive.

## SECTION 1: GENERATE AN ARGUMENT

# 9  SURVIVING WINTER IN THE DUST BOWL

You and the rest of your family decided to use the last of your savings in September to buy the seed and equipment needed to plant a crop of winter wheat. You won't be able to harvest the crop of winter wheat, however, until June (assuming that it grows at all). You therefore need a plan to make sure you and the rest of your family have the food you need to make it through the winter. You have several options:

- Eat the bull. Keep the cow alive but don't feed it. Drink the cow's milk. Eat the cow when the milk production ceases, and then eat the wheat.
- Eat the bull. Keep the cow alive, feed it, and drink the milk. Eat the rest of the wheat.
- Share the wheat with the bull and cow, and keep them alive until the wheat runs out. Then eat the bull and the cow.
- Eat the bull and the cow, and then eat the wheat.

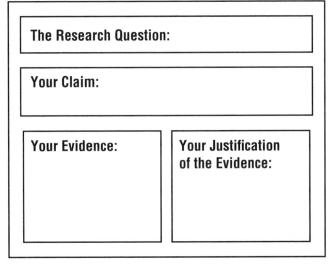

*Figure 9.3. Components of the Whiteboard*

Given all these options (and there are many others), you might be wondering: **What should your group do in order to survive the winter?**

With your group, develop a claim that best answers this research question. Once your group has developed your claim, prepare a whiteboard that you can use to share and justify your ideas. Your whiteboard should include all the information shown in Figure 9.3.

To share your work with others, we will be using a round-robin format. This means that one member of the group will stay at your workstation to share your group's ideas while the other group members go to the other groups one at a time in order to listen to and critique the arguments developed by your classmates.

Remember, as you critique the work of others, you need to decide if their conclusions are valid or acceptable based on the quality of their claim and how well they are able to support their ideas. In other words, you need to determine if their argument is *convincing* or not. One way to determine if their argument is convincing is to ask them some of the following questions:

- How did you analyze or interpret your data? Why did you decide to do it that way?

## SECTION 1: GENERATE AN ARGUMENT

### SURVIVING WINTER IN THE DUST BOWL  9

- How do you know that your analysis of the data is free from errors?
- Why does your evidence support your claim?
- Why did you decide to use that evidence? Why is your evidence important?
- How does your justification of the evidence fit with accepted scientific ideas?
- What are some of the other claims your group discussed before agreeing on your claim, and why did you reject them?

## SECTION 1: GENERATE AN ARGUMENT

### 9 SURVIVING WINTER IN THE DUST BOWL

# Information About Nutritional Values and Dietary Needs

## Table 9.1. Dietary Needs

| Organism | Weight (Pounds) | Calories (Consumed/Day) | Water Intake (Gallons/Day) | Protein (% of Diet) | Carbohydrate (% of Diet) | Fat (% of Diet) |
|---|---|---|---|---|---|---|
| Human Female* | 120–180 | 1200 | 0.4 | 10–35 | 45–65 | 20–35 |
| Human Male* | 150–200 | 1800 | 0.4 | 10–35 | 45–65 | 20–35 |
| Female Cow Lactating** | 800–1000 | 44,000 | 50 | 10–20 | 70–80 | 10–20 |
| Female Cow Dry | 800–1000 | 30,500 | 40 | 10–20 | 70–80 | 10–20 |
| Bull (Male Cow) | 1000–1200 | 46,000 | 45 | 10–20 | 70–80 | 10–20 |

*Humans can survive without food for 4–8 weeks with a minimal activity level (although this is *not* recommended as serious side effects result). However, humans cannot survive more than 3–5 days without potable water.
** A lactating cow produces approximately 6 gallons of milk per day (1 gallon = 128 ounces, 1 ounce of milk = 28.6 grams of milk).

## Table 9.2. Nutritional Information

| Food | Serving Size (in grams) | Calories Per Serving | Total Protein (in grams) | Total Carbohydrates (in grams) | Total Fat (in grams) |
|---|---|---|---|---|---|
| Wheat | 100 | 339.0 | 13.7 | 72.6 | 1.9 |
| Cow** | 453.6 | 662.0 | 95.8 | 0 | 161.0 |
| Milk | 100 | 497 | 26.3 | 38.4 | 26.7 |

* 453.6 grams = 1 pound
** Only 41% of a cow's total weight is consumable.

SECTION 1: GENERATE AN ARGUMENT

Name_____ Date_____

# SURVIVING WINTER IN THE DUST BOWL:
## What Is Your Argument?

In the space below, write an argument in order to persuade another biologist that your claim is valid and acceptable. As you write your argument, remember to do the following:

- State the claim you are trying to support
- Include a sufficient amount of genuine evidence
- Provide a justification of your evidence that explains why the evidence is important and relevant by linking it a specific concept, principle, or an underlying assumption
- Organize your paper in a way that enhances readability
- Use a broad range of words including vocabulary that we have learned
- Make sure your writing has an easy flow and rhythm
- Correct grammar, punctuation, and spelling errors

SECTION 1: GENERATE AN ARGUMENT

# 9 SURVIVING WINTER IN THE DUST BOWL
## TEACHER NOTES

## Purpose

The purpose of this activity is to help students understand food chains, the interactions between trophic levels, the interdependency of organisms, and how energy flows through a system. This activity also helps students learn how to engage in practices such as using mathematics and computational thinking, constructing explanations, arguing from evidence, and communicating information. In addition, this activity is designed to give students an opportunity to learn how to write in science and develop their speaking and listening skills, which are important goals for literacy in science (see Standards Addressed in This Activity for a complete list of the practices, crosscutting concepts, core ideas, and literacy skills that are aligned with this activity).

## The Content and Related Concepts

All living organisms need energy, and all living things get their energy from food. Plants get their energy from the Sun, and the processes of photosynthesis allow them to generate their own food. Plants, therefore, are described as producers. All other living things get their food from consuming other living things. The organisms that consume other living organisms are identified as consumers. There are three types of consumers: herbivores (also known as primary consumers), which eat only plants; carnivores, which eat only other animals; and omnivores, which eat both plants and animals. In addition to these classifications of consumers, the carnivores are further classified into two groups. Secondary consumers are carnivores that eat herbivores, and tertiary consumers are carnivores that eat other carnivores. There are also organisms that consume dead and decaying organisms. These organisms are identified as decomposers (e.g., bacteria and mushrooms). The decomposers speed up the decaying process that releases minerals back into the food chain for absorption by plants as nutrients.

The trophic level of an organism is the position it occupies in a food chain. The first trophic level consists of primary producers such as plants. The second trophic level consists of herbivores or primary consumers. The third trophic level consists of carnivores that eat herbivores, and the fourth level consists of carnivores that eat other carnivores. Each trophic level relates to the one below it by absorbing some of the energy it consumes and as a result, is supported by the trophic level below it. At each stage in the chain, or trophic level, energy is lost due to metabolic activity and other factors (see Figure 9.4). Plants, for example, only convert about 1% of the sunlight they receive into chemical energy, and consumers at each level only covert about 10% of the chemical energy available in their food to actual biomass. As a result of this inefficient

# SECTION 1: GENERATE AN ARGUMENT

## SURVIVING WINTER IN THE DUST BOWL
### TEACHER NOTES 9

*Figure 9.4. A Food Chain That Consists of Four Trophic Levels (Darker arrows represent energy transfer at each level, and the lighter arrows represent energy lost due to metabolic activity.)*

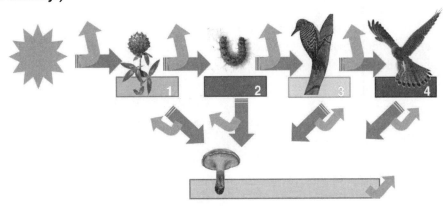

energy transfer, only about 0.001% of the energy available in sunlight is incorporated into the bodies of tertiary consumers.

In this activity, the students must calculate the amount of energy available in the wheat, and then determine how to best allocate that energy given the amount of energy that is lost at each trophic level. They must also consider dietary needs (i.e., amount of protein, fats, and carbohydrates that need to be consumed) and minimal water intake of the livestock and the members of their family. There is no one best answer to the guiding question, especially since the students will need to take into account the needs of each system and the overall goals of the family (for example, if they decide to eat the cows, they will need to purchase more later). What is important, however, is that the students are able to support their claim using genuine evidence and a rationale that explains why the evidence is important.

## Curriculum and Instructional Considerations

### Middle School

Students in the middle grades continue to learn about the comparisons of species and their relationships to each other in an ecosystem through food webs and food chains. There are many resources that provide examples of food webs for various ecosystems; however, these often lack the inclusion of the energy that is transferred and needed to support each of the trophic levels. In addition, students are likely to believe that food is not a scarce resource in an ecosystem and that organisms can change their food source at will (Leach et al. 1992). Students

**SCIENTIFIC ARGUMENTATION IN BIOLOGY: 30 CLASSROOM ACTIVITIES**

## SECTION 1: GENERATE AN ARGUMENT

### 9 SURVIVING WINTER IN THE DUST BOWL
### TEACHER NOTES

are also likely to have the misconception that organisms at the top of a trophic level will have more energy, misunderstanding the storage and use of energy in living organisms. They may believe that some populations of organisms are larger than others in order to meet the demands of food for other populations (Leach et al. 1992).

### High School
Students in the high school grades will focus more of their content on biology and environmental studies than the middle-level students and are therefore more likely to be able to identify connections to the food chains more easily and without much prompting. However, they are not likely to understand the interactions between organisms in a food web that are indirect causal interactions. In addition, high school students may not recognize the concept of matter that is transferred through the chains and are likely to see it as being created and destroyed rather than transferred and conserved, in the same way that energy is transferred and conserved (Smith and Anderson 1986).

This activity would be appropriate as an introduction as well as a summative activity that helps students apply their knowledge to a real-world event. It is an activity that strongly serves as a pre/post activity that will allow students to reflect on the learning. Teachers should allow students to read the activity and discuss their ideas. Then the teacher should provide various lessons related to the food webs, food chains, health and nutrition, and the transfer of energy and matter in the ecosystem. Returning to this activity, students could review their previous responses and develop a more appropriate argument based on the concepts and ideas that they have learned.

## Recommendations for Implementing the Activity

This activity takes approximately 100 minutes of instructional time to complete, but the amount of time devoted to each stage of the activity varies depending on how a teacher decides to spend time in class. For more information about how to implement the activity, see Appendix E on page 369.

Table 9.3 provides information about the type and amount of materials needed to implement this activity in a classroom with 28 students with groups of four and groups of three.

*Table 9.3. Materials Needed to Implement the Activity in a Classroom of 28 Students*

| Material | Amount Needed With ... | |
|---|---|---|
| | Groups of 3 | Groups of 4 |
| Whiteboards (or chart paper)+ | 10 | 7 |
| Whiteboard markers (or permanent if using chart paper)+ | 20 | 14 |
| Copy of Student Pages (pp. 113–116)* | 28 | 28 |
| Copy of Student Page (p. 117)* | 10 | 7 |
| Copy of Appendix B (p. 366)* | 28 | 28 |

+ Teachers can also have students prepare their arguments in a digital medium (such as PowerPoint or Keynote).
* Teachers can also project these materials onto a screen in order to cut down on paper use.

### SECTION 1: GENERATE AN ARGUMENT

## SURVIVING WINTER IN THE DUST BOWL
TEACHER NOTES  **9**

## Assessment

The rubric provided in Appendix B (p. 366) can be used to assess the arguments crafted by each student at the end of the activity. To illustrate how the rubric can be used to score an argument written by a student, consider the following example. This sample argument, which was written by a seventh-grade student, is an example of an argument of moderate quality.

<u>We think that we should keep the bull alive long enough to allow it to breed with the cow.</u> **If we do that, then the cow can have a baby for future food for us. In the time to allow them to breed we would eat very little and we would do little exercise or things that take lots of energy. The wheat we eat and the milk from the cow would provide us with enough energy since the wheat is on the bottom of the food chain. Even though the cow is not the bottom of the food chain it is close and the amount of protein from the milk, combined with the wheat should help us survive.** The problem will be how long we should keep the bull alive to breed with the cow. We will have to think about how long that would be and it depends on how much water we actually have. The bulls and cows need a lot of water and that will be a problem. If the bull is not going to mate within about a week, then we will have to kill him and not try to get her pregnant. We would need to figure out how long the wheat would last for us and the cow before she gives birth. Once she is pregnant, we could kill the bull and let the cow have the wheat while she was pregnant so that her baby would be healthy. By the time she gives birth, the wheat is likely to be nearly gone and we would have to make a decision to either kill the cow or kill the baby. The thing about killing the baby and eating it is that by keeping the cow we would still have milk. But is there still wheat to feed the cow?

The content of the example argument is poor for several reasons. The student's claim (underlined) is sufficient (1/1) but inaccurate (0/1), because it would likely lead to a shortage of food. The student does not use genuine evidence to support the claim because she does not include data (0/1), an analysis of the data (0/1), or an interpretation of the data (0/1). Instead, the author provides a series of reasons as support for the claim (in bold). The justification of the evidence is insufficient, because the author never explains why the "evidence" is important (0/1) nor attempts to link the evidence she uses to an important concept or principle (0/1). However, the author uses scientific terms correctly (1/1) and also uses phrases that are consistent with the nature of science (1/1). The writing mechanics of the sample argument are good. The organization of the argument is appropriate because the arrangement of the sentences does not distract from the development of the main idea (0/1). Finally, there are no grammatical (1/1) or punctuation errors (1/1) in the argu-

## SECTION 1: GENERATE AN ARGUMENT

### 9 SURVIVING WINTER IN THE DUST BOWL
TEACHER NOTES

ment. The overall score for the sample argument, therefore, is 5 out the 12 points possible.

## Standards Addressed in This Activity

This activity can be used to address the following dimensions outlined in *A Framework for K–12 Science Education* (NRC 2012):

## Scientific Practices

- Using mathematics and computational thinking
- Constructing explanations
- Engaging in argument from evidence
- Obtaining, evaluating, and communicating information

## Crosscutting Concepts

- Cause and effect: Mechanism and explanation
- Systems and system models
- Energy and matter: Flows, cycles, and conservation

## Life Sciences Core Ideas

- From molecules to organisms: Structures and processes
- Ecosystems: Interactions, energy, and dynamics

This activity can be used to address the following standards for literacy in science from the *Common Core State Standards for English Language Arts and Literacy* (NGA and CCSSO 2010):

## Writing

- Text types and purposes
- Production and distribution of writing
- Research to build and present knowledge
- Range of writing

## Speaking and Listening

- Comprehension and collaboration
- Presentation of knowledge and ideas

## References

Anderson, C., T. Sheldon, and J. Dubay. 1990. The effects of instruction on college nonmajors' conceptions of respiration and photosynthesis. *Journal of Research in Science Teaching* 27: 761–776.

Bell, B., and A. Brook. 1984. *Aspects of secondary students understanding of plant nutrition.* Leeds, UK: University of Leeds, Centre for Studies in Science and Mathematics Education.

Leach, J., et al. 1992. *Progression in understanding of ecological concepts by pupils aged 5 to 16.* Leeds, UK: The University of Leeds, Centre for Studies in Science and Mathematics Education.

National Governors Association Center (NGA) for Best Practices, and Council of Chief State School Officers (CCSSO). 2010. *Common core state standards for English language arts and literacy*. Washington, DC: National Governors Association for Best Practices, Council of Chief State School.

National Research Council (NRC). 2012. *A framework for K–12 science education: Practices, crosscutting concepts, and core ideas.* Washington, DC: National Academies Press.

Smith, E., and C. Anderson. 1986. *Alternative conceptions of matter cycling in ecosystems.* Paper presented at the annual meeting of the National Association for Research in Science Teaching (NARST), San Francisco, CA.

## SECTION 1: GENERATE AN ARGUMENT

# CHARACTERISTICS OF VIRUSES (CHARACTERISTICS OF LIFE) 10

A virus is a tiny bundle of genetic material—either DNA or RNA—carried in a protein shell called a capsid. Some viruses have an additional layer around this coat called an envelope. The envelope is made of a lipid. Three examples of viruses can be seen in Figure 10.1.

**Figure 10.1. An Adenovirus, a Bacteriophage, and the Influenza Virus (These viruses are between 45 and 200 nanometers.)**

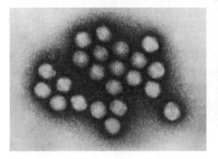
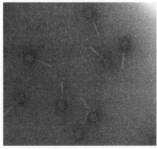
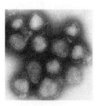

[A nanometer is 1/1,000,000,000 of a meter].

When a virus enters a cell, the information carried in a virus's genetic material enables the virus to force the infected cell to make more copies of the virus. The poliovirus, for example, can make over one million copies of itself inside a single human intestinal cell. A virus is usually very, very small compared to the size of the cell it infects.

Viruses can infect the cells of plants, animals, or even bacteria. Moreover, within an individual species, there may be one hundred or more different types of viruses, which can infect that specific species alone. There are viruses that infect only humans (for example, smallpox), viruses that infect humans and one or two additional kinds of animals (for example, influenza), viruses that infect only a certain kind of plant (for example, the tobacco mosaic virus), and some viruses infect only a particular species of bacteria (for example, the bacteriophage which infects *E. coli*).

These unique traits of viruses have made many scientists wonder: **Should a virus be classified as a living thing?**

With your group, develop a claim that best answers this research question. Once your group has developed your claim, prepare a whiteboard that you can use to share and justify your ideas. Your whiteboard should include all the information shown in Figure 10.2 on page 124.

## SECTION 1: GENERATE AN ARGUMENT

## 10  CHARACTERISTICS OF VIRUSES

To share your work with others, we will be using a round-robin format. This means that one member of the group will stay at your workstation to share your group's ideas while the other group members go to the other groups one at a time in order to listen to and critique the arguments developed by your classmates.

Remember, as you critique the work of others, you need to decide if their conclusions are valid or acceptable based on the quality of their claim and how well they are able to support their ideas. In other words, you need to determine if their argument is *convincing* or not. One way to determine if their argument is convincing is to ask them some of the following questions:

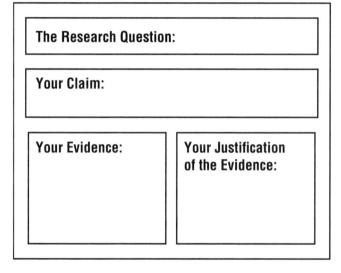

*Figure 10.2. Components of the Whiteboard*

- How did you analyze or interpret your data? Why did you decide to do it that way?
- How do you know that your analysis of the data is free from errors?
- Why does your evidence support your claim?
- Why did you decide to use that evidence? Why is your evidence important?
- How does your justification of the evidence fit with accepted scientific ideas?
- What are some of the other claims your group discussed before agreeing on your claim, and why did you reject them?

Table 10.1. Information About Viruses and Other Objects Found on Earth

| Object and Size[1] | Appearance | Functional "Life" Span | Energy Source | Carbon Source | Waste Production | Responds to External Stimuli | Biomolecules[2] Present in the Object | Form of Reproduction | Genetic Material[3] | Growth? |
|---|---|---|---|---|---|---|---|---|---|---|
| Influenza Virus<br>130 nanometers in diameter | | 10 Years | None | None | None | No | Nucleic Acid<br>Protein | Replication; Requires a host | RNA | No |
| Adenovirus<br>220 nanometers in diameter | | 10 Years | None | None | None | No | Nucleic Acid<br>Protein<br>Lipids | Replication; Requires a host | DNA | No |
| Coriander Seeds<br>3 millimeters in diameter | | 1–10 Years | Organic Compounds[4] | Carbon Dioxide | None | Yes | Nucleic Acid<br>Protein<br>Lipids<br>Carbohydrates | Sexual[5] | DNA and RNA | Yes |
| Amoeba<br>500 micrometers in diameter | | 1–3 Months | Organic Compounds | Carbohydrates | Yes | Yes | Nucleic Acid<br>Protein<br>Lipids<br>Carbohydrates | Asexual[6] | DNA and RNA | Yes |
| Human Red Blood Cell<br>8 micrometers in diameter | | 3–4 Months | Organic Compounds | Carbohydrates | No | No | Nucleic Acid<br>Protein<br>Lipids<br>Carbohydrates | None | None | No |

1. 1 meter = 100 centimeters = 1000 millimeters = 1,000,000 micrometers = 1,000,000,000 nanometers
2. A biomolecule is any molecule that performs an important function in living organisms. Biomolecules are usually composed of hydrogen, carbon, oxygen, nitrogen, prosperous, or sulfur atoms and they are organized into one of four main groups (carbohydrates, proteins, lipids, and nucleic acids).
3. The genetic material of an object is the molecule(s) that play the fundamental role in determining the nature and structure of an organism or cell.
4. Organic compounds are molecules that are composed of carbon such as sugar (which is a type of carbohydrate)
5. *Sexual* refers to a form of reproduction in which two parents give rise to an offspring.
6. *Asexual* refers to a form of reproduction that involves only one parent that produces genetically identical offspring by budding or by the division of a single cell or the entire organism into two parts.

*(continued)*

Table 10.1. Information About Viruses and Other Objects Found on Earth (continued)

| Object and Size | Appearance | Functional "Life" Span | Energy Source | Carbon Source | Waste Production | Responds to External Stimuli | Biomolecules Present in the Object | Form of Reproduction | Genetic Material | Growth? |
|---|---|---|---|---|---|---|---|---|---|---|
| Human White Blood Cell — 10 micrometers in diameter | | 1 Month | Organic Compounds | Carbohydrates | Yes | Yes | Nucleic Acid Protein Lipids Carbohydrates | None | DNA and RNA | No |
| Sponge — 100 centimeters in diameter | | 100–200 Years | Organic Compounds | Carbohydrates | Yes | Yes | Nucleic Acid Protein Lipids Carbohydrates | Sexual and asexual | DNA and RNA | Yes |
| Elodea — 40 centimeters in length | | 2–4 Weeks | Sunlight | Carbon Dioxide | Yes | Yes | Nucleic Acid Protein Lipids Carbohydrates | Sexual and asexual | DNA and RNA | Yes |
| Plasmodium Falciparum — 15 micrometers in length | | 1–2 Months | Organic Compounds | Carbohydrates | Yes | Yes | Nucleic Acid Protein Lipids Carbohydrates | Sexual and asexual but only occurs inside a host | DNA and RNA | Yes |
| E. Coli — 3 micrometers in length | | 1–3 Months | Organic Compounds | Carbohydrates | Yes | Yes | Nucleic Acid Protein Lipids Carbohydrates | Asexual | DNA and RNA | Yes |
| Tube Worms — 1.5 meters in length | | 100–200 Years | Inorganic Compounds | Carbohydrates | Yes | Yes | Nucleic Acid Protein Lipids Carbohydrates | Sexual | DNA and RNA | Yes |
| Dog — 0.75 meters in height | | 15–20 Years | Organic Compounds | Carbohydrates | Yes | Yes | Nucleic Acid Protein Lipids Carbohydrates | Sexual | DNA and RNA | Yes |
| Computer — 45 centimeters in height | | 10–20 years | Electricity | None | Yes | Yes | None | None | None | No |

## SECTION 1: GENERATE AN ARGUMENT

Name_____ Date_____

# CHARACTERISTICS OF VIRUSES:
## What Is Your Argument?

In the space below, write an argument in order to persuade another biologist that your claim is valid and acceptable. As you write your argument, remember to do the following:

- State the claim you are trying to support
- Include genuine evidence (data + analysis + interpretation)
- Provide a justification of your evidence that explains why the evidence is relevant and why it provides adequate support for the claim
- Organize your argument in a way that enhances readability
- Use a broad range of words including vocabulary that we have learned
- Correct grammar, punctuation, and spelling errors

SECTION 1: GENERATE AN ARGUMENT

# 10  CHARACTERISTICS OF VIRUSES
## TEACHER NOTES

## Purpose

The purpose of this activity is to help students understand how biologists distinguish between living and nonliving objects. This activity can also be used to introduce cell theory. This activity will help students learn how to engage in practices such as constructing explanations, arguing from evidence, and communicating information. This activity is also designed to give students an opportunity to learn how to write in science and develop their speaking and listening skills, which are important goals for literacy in science (see Standards Addressed in This Activity for a complete list of the practices, crosscutting concepts, core ideas, and literacy skills that are aligned with this activity).

## The Content and Related Concepts

The state of being alive is hard to define. The following are some of the main criteria that are used by biologists to determine if something is alive or not (Campbell and Reece 2002):

- **Order**: All living things have a highly organized structure and are composed of at least one cell.
- **Use of Energy**: All living things take in energy and transform it to do many kinds of work.
- **Reproduction**: All living things are able to reproduce their own kind through sexual or asexual means. Life comes only from life, which is also known as the principle of biogenesis.
- **Growth and development**: Heritable programs in the form of DNA direct the pattern of growth and development, which results in an organism that has the characteristics of a particular species.
- **Response to stimuli**: All living things are able to respond to an environmental stimulus such as temperature, amount of light, availability of water, or the actions of other living things.
- **Homeostasis**: All living things have regulatory mechanisms that maintain its internal environment within tolerable limits even when the external environment fluctuates. This regulation is called homeostasis.

In this activity, the students are asked to determine if a virus should be classified as a living thing. Based on the criteria outlined above, a virus should not be classified as a living thing. Although viruses are highly organized and respond to stimuli (by being able to highjack a host cell), they do not take in and transform energy to do work, grow or develop, or maintain homeostasis. Viruses are also not composed of at least one cell, and they cannot reproduce sexually or asexually. Instead, they infect a host cell

# SECTION 1: GENERATE AN ARGUMENT

## CHARACTERISTICS OF VIRUSES
### TEACHER NOTES 10

and then take over the cell's organelles in order to replicate itself. All the other objects listed in Table 10.1—with the exception of the computer, the human red blood cell, and the human white blood cell—are living things. These human red blood cell and white blood cells are found within a living thing, but on their own, they cannot be considered alive because these cells are not able to reproduce.

## Curriculum and Instructional Considerations

This activity is best used at the beginning of a school year, because it is a good way to introduce the study of life and address many of the misconceptions students have about living and nonliving things. For example, some students think that movement is a good criterion that can be used to determine if something is alive or not, even though many inanimate objects can move. This activity also fits well as part of a unit on cell theory, because the criterion "composed of a least one cell" is one of the defining characteristics of life on Earth.

The activity, however, is best used as an introductory activity. In this case, students are not told about the various criteria that are used by biologists to determine if something is alive or not prior to starting the activity. The students must determine these criteria for themselves using the information provided in Table 10.1, and then use the criteria they agree on to develop their argument about viruses. This will result in a wide range of claims and rationales during the argumentation session.

The focus of the explicit discussion at the end of the activity should focus on the criteria that biologists use to define life and why it is often difficult to determine if something is alive or not. The teacher should also encourage the students to reflect on the criteria they decided to use and why. The discussion should move to an aspect of the nature of science. For example, a teacher could discuss the difference between data and evidence or how data collection and analysis is guided by current theories using what the students did as an illustrative example. This is also an appropriate time to discuss how the definition of life has changed over time and how science knowledge is not absolute and is able to change over time (i.e., scientific knowledge is durable but tentative).

## Recommendations for Implementing the Activity

This activity takes approximately 100 minutes of instructional time to complete, but the amount of time devoted to each activity varies depending on how a teacher decides to spend time in class. For more information about how to implement the activity, see Appendix E on page 369.

Table 10.2 (p. 130) provides information about the type and amount of materials needed to implement this activity in a classroom with 28 students with groups of four and groups of three.

## Assessment

The rubric provided in Appendix B (p. 366) can be used to assess the arguments crafted by each student at the end of the activity. To illustrate how the rubric can be used to score

## SECTION 1: GENERATE AN ARGUMENT

## 10 CHARACTERISTICS OF VIRUSES
### TEACHER NOTES

*Table 10.2. Materials Needed to Implement the Activity in a Classroom of 28 Students*

| Material | Amount Needed With ... | |
|---|---|---|
| | Groups of 3 | Groups of 4 |
| Whiteboards (or chart paper)+ | 10 | 7 |
| Whiteboard markers (or permanent if using chart paper)+ | 20 | 14 |
| Copy of Student Pages (pp. 123–126)* | 28 | 28 |
| Copy of Student Page (p. 127)* | 10 | 7 |
| Copy of Appendix B (p. 366)* | 28 | 28 |

+ Teachers can also have students prepare their arguments in a digital medium (such as PowerPoint or Keynote).
* Teachers can also project these materials onto a screen in order to cut down on paper use.

an argument written by a student, consider the following example. This sample argument, which was written by a seventh-grade student, is an example of a weak argument.

> <u>Viruses are not living things</u>. **They have no energy source, they require a host to reproduce and they do not grow**. This proves that they are not living things.

Although the student's claim (underlined) is sufficient (1/1) and accurate (1/1), the example argument is poor for several reasons. This student does not include data, analysis of the data, or an interpretation of the analysis in the argument, so there is no evidence (0/3). The justification of the evidence is also inadequate (0/2), because he does not explain why the three characteristics of the virus that he listed are important to consider. The author also uses a phrase (e.g., "This proves") that misrepresents the nature of science (0/1). The writing mechanics of the sample argument are adequate even though the argument is short and lacks detail. The organization of the argument is appropriate because the arrangement of the sentences does not distract from the development of the main idea (1/1). There are also no grammatical (1/1) or punctuation errors (1/1) in the argument. The overall score for the sample argument, therefore, is 5 out the 12 points possible. We decided to include this example to illustrate the inadequacies of the evidence and the rationale. We also included this example to illustrate how some students will provide a short argument in response to the writing prompt if they are not required to do more. Teachers, therefore, need to set high standards for students and hold their students accountable when implementing this type of activity.

## Standards Addressed in This Activity

This activity can be used to address the following dimensions outlined in *A Framework for K–12 Science Education* (NRC 2012):

## Scientific Practices

- Constructing explanations
- Engaging in argument from evidence
- Obtaining, evaluating, and communicating information

## SECTION 1: GENERATE AN ARGUMENT

### CHARACTERISTICS OF VIRUSES
TEACHER NOTES 10

### Crosscutting Concepts
- Patterns
- Structure and function

### Life Sciences Core Ideas
- From molecules to organisms: Structures and processes

This activity can be used to address the following standards for literacy in science from the *Common Core State Standards for English Language Arts and Literacy* (NGA and CCSSO 2010):

### Writing
- Text types and purposes
- Production and distribution of writing
- Research to build and produce knowledge
- Range of writing

### Speaking and Listening
- Comprehension and collaboration
- Presentation of knowledge and ideas

## References

Campbell, N., and J. Reece. 2002. *Biology*. 6th ed. San Francisco, CA: Benjamin Cummings.

National Governors Association Center (NGA) for Best Practices, and Council of Chief State School Officers (CCSSO). 2010. *Common core state standards for English language arts and literacy*. Washington, DC: National Governors Association for Best Practices, Council of Chief State School.

National Research Council (NRC). 2012. *A framework for K–12 science education: Practices, crosscutting concepts, and core ideas*. Washington, DC: National Academies Press.

# EVALUATE ALTERNATIVES

| | |
|---|---|
| ***Framework* Matrix** | 134 |
| **Activity 11:** Spontaneous Generation | 137 |
| *(Cell Theory)* | |
| **Activity 12:** Plant Biomass | 149 |
| *(Photosynthesis)* | |
| **Activity 13:** Movement of Molecules in or out of Cells | 159 |
| *(Osmosis and Diffusion)* | |
| **Activity 14:** Liver and Hydrogen Peroxide | 171 |
| *(Chemical Reactions and Catalysts)* | |
| **Activity 15:** Cell Size and Diffusion | 181 |
| *(Diffusion)* | |
| **Activity 16:** Environmental Influence on Genotypes and Phenotypes | 191 |
| *(Genetics)* | |
| **Activity 17:** Hominid Evolution | 203 |
| *(Macroevolution)* | |
| **Activity 18:** Plants and Energy | 219 |
| *(Respiration and Photosynthesis)* | |
| **Activity 19:** Healthy Diet and Weight | 229 |
| *(Human Health)* | |
| **Activity 20:** Termite Trails | 239 |
| *(Animal Behavior)* | |

# SECTION 2: EVALUATE ALTERNATIVES

# FRAMEWORK MATRIX

| A Framework for K–12 Science Education | Spontaneous Generation | Plant Biomass | Movement of Molecules in or out of Cells | Liver and Hydrogen Peroxide | Cell Size and Diffusion | Environmental Influence on Genotypes and Phenotypes | Hominid Evolution | Plants and Energy | Healthy Diet and Weight | Termite Trails |
|---|---|---|---|---|---|---|---|---|---|---|
| **1. Scientific Practices** | | | | | | | | | | |
| Asking questions | | | | | | | | | | |
| Developing and using models | | | ■ | | ■ | | | ■ | ■ | |
| Planning and carrying out investigations | ■ | ■ | ■ | ■ | ■ | ■ | ■ | ■ | ■ | ■ |
| Using mathematics and computational thinking | | □ | □ | | ■ | ■ | | | ■ | |
| Constructing explanations | | | | | | | | | | |
| Engaging in argument from evidence | ■ | ■ | ■ | ■ | ■ | ■ | ■ | ■ | ■ | ■ |
| Obtaining, evaluating, and communicating information | ■ | ■ | ■ | ■ | ■ | ■ | ■ | ■ | ■ | ■ |
| **2. Crosscutting Concepts** | | | | | | | | | | |
| Patterns | | | | | | □ | ■ | | | |
| Cause and effect: Mechanism and explanation | ■ | ■ | ■ | ■ | ■ | ■ | ■ | ■ | ■ | ■ |
| Scale, proportion, and quantity | | | | | ■ | □ | ■ | | ■ | |
| Systems and system models | | | | | | | ■ | | | |
| Energy and matter: Flows, cycles, and conservation | ■ | ■ | ■ | ■ | ■ | ■ | ■ | ■ | ■ | ■ |
| Structure and function | | ■ | ■ | ■ | ■ | ■ | ■ | ■ | ■ | |
| Stability and change | ■ | | | | | | | | | |

■ = Strong alignment    □ = Weak alignment

# SECTION 2: EVALUATE ALTERNATIVES

| A Framework for K–12 Science Education | Spontaneous Generation | Plant Biomass | Movement of Molecules in or out of Cells | Liver and Hydrogen Peroxide | Cell Size and Diffusion | Environmental Influence on Genotypes and Phenotypes | Hominid Evolution | Plants and Energy | Healthy Diet and Weight | Termite Trails |
|---|---|---|---|---|---|---|---|---|---|---|
| **3. Life Sciences Core Ideas** | | | | | | | | | | |
| From molecules to organisms: Structures and processes | ■ | ■ | ■ | ■ | ■ | | ■ | ■ | ■ | |
| Ecosystems: Interactions, energy, and dynamics | | ■ | | | | | ■ | □ | | |
| Heredity: Inheritance and variation in traits | ■ | | | | | ■ | □ | | | ■ |
| Biological evolution: Unity and diversity | □ | | | | | □ | | | | ■ |
| **Common Core State Standards for English Language Arts and Literacy: Literacy in the Disciplines** | | | | | | | | | | |
| **1. Writing** | | | | | | | | | | |
| Text types and purposes | ■ | ■ | ■ | ■ | ■ | ■ | ■ | ■ | ■ | ■ |
| Production and distribution of writing | ■ | ■ | ■ | ■ | ■ | ■ | ■ | ■ | ■ | ■ |
| Research to build and present knowledge | ■ | ■ | ■ | ■ | ■ | ■ | ■ | ■ | ■ | ■ |
| Range of writing | ■ | ■ | ■ | ■ | ■ | ■ | ■ | ■ | ■ | ■ |
| **2. Speaking and Listening** | | | | | | | | | | |
| Comprehension and collaboration | ■ | ■ | ■ | ■ | ■ | ■ | ■ | ■ | ■ | ■ |
| Presentation of knowledge and ideas | ■ | ■ | ■ | ■ | ■ | ■ | ■ | ■ | ■ | ■ |

■ = Strong alignment   □ = Weak alignment

## SECTION 2: EVALUATE ALTERNATIVES

# SPONTANEOUS GENERATION (CELL THEORY) 11

From the time of the ancient Romans and until the late 19th century, most people believed that some life forms arose from nonliving matter. For example, a 17th-century recipe for the spontaneous generation of mice required placing sweaty underwear and husks of wheat in a large container, and then waiting for about 21 days, during which time it was alleged that husks of wheat would change into mice from exposure to the sweat in the underwear.

Similarly, at that time, it was widely believed that maggots arose spontaneously from rotting meat. However, in 1668, Francesco Redi, an Italian physician, made the first serious challenge to the idea of spontaneous generation. Redi believed that maggots developed from eggs laid by flies. To test his explanation, he set out meat in a variety of flasks—some open to the air, some sealed completely, and others covered with gauze. As he had expected, maggots appeared only in the open flasks in which the flies could reach the meat and lay their eggs.

This was one of the first examples of an experiment in which a researcher identified and controlled variables to test an explanation (i.e., a hypothesis). In spite of his well-executed experiment, however, the belief in spontaneous generation remained strong. For example, many people believed that spontaneous generation was common for microorganisms (living things invisible to the naked eye). In order to create animalcules, as the organisms were called at the time, people only needed to place hay in water and wait a few days before they were able to examine their new creations under the microscope.

The debate over spontaneous generation continued for centuries. In 1745, John Needham proposed what he considered the definitive test. Everyone knew that microorganisms were killed when they were boiled, so he decided to determine whether or not microorganisms would appear spontaneously in a broth after boiling it. He boiled chicken broth, put it into a flask, sealed it, and waited. After a few days, microorganisms grew in the sealed flask (see Figure 11.1). Needham claimed that his experiment demonstrated that microorganisms appear through the process of spontaneous generation.

Lazzaro Spallanzani, however, was not convinced by Needham's results. He suggested that perhaps the microorganisms had entered the broth through the air after the broth was boiled but before it was sealed. To test his alternative explanation, he modified Needham's experiment. He placed the chicken broth in a flask, sealed the flask, drew off the air to create a partial vacuum, and

*Figure 11.1. Needham's Test of Spontaneous Generation*

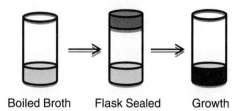

Boiled Broth    Flask Sealed    Growth

## SECTION 2: EVALUATE ALTERNATIVES

## 11 SPONTANEOUS GENERATION

then boiled the broth (see Figure 11.2). No microorganisms grew. Proponents of spontaneous generation, however, argued that Spallanzani had only showed that spontaneous generation cannot happen unless all the elements that are necessary for the formation of new life, such as fresh air, are present.

*Figure 11.2. Spallanzani's Test of Spontaneous Generation*

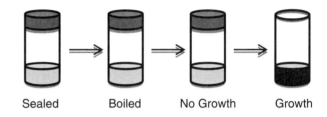

These observations raise an interesting question: **Can life arise from inanimate materials?**

Here are two potential answers to this question:

- **Explanation 1**: Yes, life can arise from inanimate materials, but the elements needed to trigger the formation new life must be available. These elements include fresh air, water, and a source of food.
- **Explanation 2**: No, living things do not come from inanimate materials. All living things come from a preexisting living thing. Living things appear to come from decaying or organic substances, because eggs are laid on or spores land on a food source and then begin to grow.

## Getting Started

You can use the following materials to test these two explanations:

- Glass tubes of different shapes
- Hot plate
- Nutrient broth
- Erlenmeyer flasks
- Rubber stops
- Microscopes
- Slides

With your group, determine which explanation provides the best answer to the research question. You can use as many of the supplies available to you to test your ideas. Make sure that you generate the evidence you will need to support your explanation as you work. You can record your method and any observation you make in the spaces on page 139.

## SECTION 2: EVALUATE ALTERNATIVES
### SPONTANEOUS GENERATION   11

## Our Method

## Our Observations

## Argumentation Session

Once your group has decided which explanation is the most valid or acceptable answer for the research question, prepare a whiteboard that you can use to share and justify your ideas. Your whiteboard should include all the information shown in Figure 11.3.

To share your work with others, we will be using a round-robin format. This means that one member of the group will stay at your workstation to share your group's ideas while the other group members go to the other groups one at a time in order to listen to and critique the arguments developed

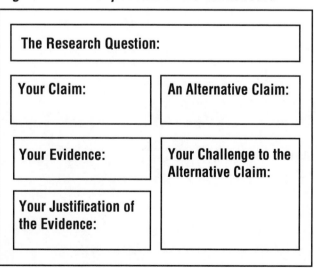

*Figure 11.3. Components of the Whiteboard*

SCIENTIFIC ARGUMENTATION IN BIOLOGY: 30 CLASSROOM ACTIVITIES

## SECTION 2: EVALUATE ALTERNATIVES

## 11  SPONTANEOUS GENERATION

by your classmates. Remember, as you critique the work of others, you need to decide if their conclusions are valid or acceptable based on the quality of their claim and how well they are able to support their ideas. In other words, you need to determine if their argument is *convincing* or not. One way to determine if their argument is convincing is to ask them some of the following questions:

- How did you gather your data? Why did you decide to do it that way?
- How do you know your data is high quality (i.e., free from errors)?
- How did you analyze or interpret your data? Why did you decide to do it that way?
- Why does your evidence support your claim?
- Why did you decide to use that evidence? Why is your evidence important?
- How does your justification of your evidence fit with accepted scientific ideas?

SECTION 2: EVALUATE ALTERNATIVES

Name_____ Date_____

# SPONTANEOUS GENERATION
## What Is Your Argument?

In the space below, write a one- to three-paragraph argument to *support* the explanation that you think is the most valid or acceptable. Your argument must also include a *challenge* to one of the alternative explanations.

As you write your argument, remember to do the following:

- State the explanation you are trying to support
- Include genuine evidence (data + analysis + interpretation)
- Explain why the evidence is important and relevant
- State the explanation you are trying to refute
- Explain why the alternative explanation is invalid or unacceptable
- Organize your argument in a way that enhances readability
- Use a broad range of words including vocabulary that we have learned
- Correct grammar, punctuation, and spelling errors

SECTION 2: EVALUATE ALTERNATIVES

# 11 SPONTANEOUS GENERATION
## TEACHER NOTES

## Purpose
The purpose of this activity is to help students understand that all life on Earth arises only through the reproduction of preexisting life and not from nonliving materials. This "life from life" principle is called biogenesis. This activity also helps students learn how to engage in practices such as using planning and carrying out investigations, arguing from evidence, and communicating information. In addition, this activity is designed to give students an opportunity to learn how to write in science and develop their speaking and listening skills, which are important goals for literacy in science (see Standards Addressed in This Activity for a complete list of the practices, crosscutting concepts, core ideas, and literacy skills that are aligned with this activity).

## The Content and Related Concepts
The debate about spontaneous generation was not settled until 1862 when Louis Pasteur conducted a now-famous experiment. Pasteur began the experiment by heating beef broth to kill any microorganisms that were already present. If a flask of this sterilized broth was then left open, it took just a few days for the broth to become contaminated with a dense growth of microorganisms. The broth, however, remained sterile if it was kept in a sealed flask after heating. In order to test the claim that sealing the flasks made the air unfit for spontaneous generation, Pasteur used a flask with an S-shaped neck. He once again placed broth in the flask, boiled it, and waited. This time, fresh air could reach the broth; however, the bend in the S-shaped neck prevented the particles of dust and microbes in the air from reaching the broth. After months, the broth in the flask with the S-shaped neck remained sterile. The results of this experiment, as a result, overturned the idea of spontaneous generation. It also laid some important groundwork for cell theory.

## Curriculum and Instructional Considerations
There are at least three points within a traditional biology curriculum in which this activity would be appropriate and helpful: characteristics of life, the cell, and evolution. When used as part of a unit on the characteristics of life or cells, the activity can illustrate the tenets of cell theory. If used as a part of a unit on evolution, it can support the idea of common descent and to begin the discussion of the various scientific hypotheses for the origin of life on Earth (such as the four-stage model for the origin of life, which includes the abiotic synthesis of organic monomers, joining of monomers in polymers, the origin of self-replicating molecules that eventually made inheritance possible, and the packaging of these molecules into protobionts. See Campbell and Reece and 2002).

# SECTION 2: EVALUATE ALTERNATIVES

## SPONTANEOUS GENERATION
### TEACHER NOTES 11

This activity is also a good way to teach students about experimental design, the control of variables, the difference between hypotheses (tentative explanations) and predictions (expected results), and other important terminology such as independent and dependent variables. Students will need a basic understanding of these important ideas in order to be able to collect meaningful data during the activity. The focus of the explicit discussion at the end of the activity should focus on the concept of biogenesis and/or cell theory or the current scientific debates about the origin of life on Earth. The explicit discussion should also focus on at least one aspect of the nature of science or the nature of scientific inquiry. For example, a teacher could discuss how scientific explanation must be consistent with observational evidence about nature, or how a hypothesis must be testable in order to be scientific, or even what makes science different from other ways of knowing using what the students did during this activity as an illustrative example.

## Recommendations for Implementing the Activity

This activity takes between 100 and 150 minutes of instructional time to complete, depending on how a teacher decides to spend time in class. Two or three days are also needed for microbes to grow in the nutrient broth.

In Option A, the students are given time to complete all six stages of the lesson during class with a few days between Stages 2 and 3 to allow time for the seeds to germinate and produce leaves (see Figure 11.4, p. 144). Stages 1 and 2 are completed on day 1, Stages 3 and 4 are completed on day 2, and Stages 5 and 6 are completed on day 3. This option for implementing the activity works best in schools where students are not expected to complete much homework or if students need to be encouraged to write more during the school day. It also provides less time for the argumentation sessions, which may or may not be a problem, depending on how confortable the students are with argumentation. After all, some classes are more talkative than others.

In Option B, students complete Stage 1 and begin Stage 2 during class on day 1. The students then complete Stage 3 on day 2 of the lesson. Stages 4 and 5 are completed on day 3, and the final written argument and counterargument (Stage 6) is then assigned as homework and returned the next day.

Due to safety concerns, this activity is best suited for high school students. Be sure to follow all safety requirements and review safe practices with students. For safety tips when handling microorganisms, see NSTA's "Tips for the Safer Handling of Microorganisms in the School Science Laboratory" and have students and parents complete NSTA's "Safety Acknowledgment Form for Working With Microorganisms" listed in Resources on page 147.

Table 11.1 (p. 145) provides information about the type and amount of materials needed to implement this activity in a classroom with 28 students in groups of four and groups of three.

## SECTION 2: EVALUATE ALTERNATIVES

### 11 SPONTANEOUS GENERATION
TEACHER NOTES

*Figure 11.4. Two Options for Implementing the Activity*

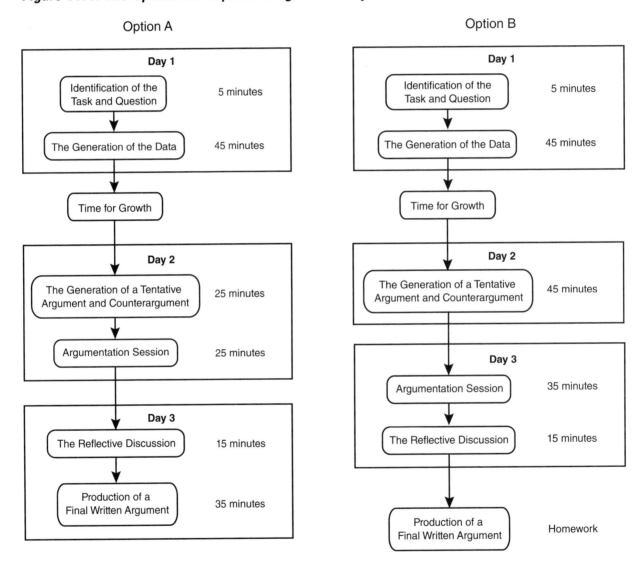

## SECTION 2: EVALUATE ALTERNATIVES

### SPONTANEOUS GENERATION
TEACHER NOTES  **11**

*Table 11.1. Materials Needed to Implement the Activity in a Classroom of 28 Students*

| Material | Amount Needed With ... | |
|---|---|---|
| | Groups of 3 | Groups of 4 |
| Beaker filled with 300 ml of nutrient broth (chicken bouillon mixed with water) | 9 | 6 |
| 250 ml Erlenmeyer flask (four per group) | 36 | 24 |
| Rubber stopper | 9 | 6 |
| One-hole rubber stopper with glass tube | 9 | 6 |
| One-hole rubber stopper with S-shaped glass tube | 9 | 6 |
| Wax pencil (used to label the petri dishes) | 9 | 6 |
| Hot plate | 9 | 6 |
| Microscope | 9 | 6 |
| Slides and coverslips | 36 | 24 |
| Disposable pipettes | 36 | 24 |
| Whiteboards (or chart paper)+ | 10 | 7 |
| Whiteboard markers (or permanent if using chart paper)+ | 20 | 14 |
| Copy of Student Pages (pp. 137–140)* | 28 | 28 |
| Copy of Student Page (p. 141)* | 28 | 28 |
| Copy of Appendix C (p. 367)* | 28 | 28 |

+ Teachers can also have students prepare their arguments in a digital medium (such as PowerPoint or Keynote).
* Teachers can also project these materials onto a screen in order to cut down on paper use.

## Assessment

The rubric provided in Appendix C (p. 367) can be used to assess the argument crafted by each student at the end of the activity. To illustrate how the rubric can be used to score an argument written by a student, consider the following example. This sample, which was written by a high school sophomore, is an example of an argument that is well written but only mediocre in terms of content.

<u>All living things come from a preexisting living thing. Living things appear to come from decaying or organic substances because eggs are</u>

# SECTION 2: EVALUATE ALTERNATIVES

## 11 SPONTANEOUS GENERATION
### TEACHER NOTES

<u>laid on or spores land on a food source and then begin to grow.</u> Our experiment proved that there is no such thing as spontaneous generation. **The only flask that had microorganisms in it at the end of the experiment was the flask with the straight glass tubing.** If spontaneous generation was true, we would have seen growth in all our flasks and we didn't.

The content of the example argument is adequate for several reasons. The student's claim (underlined) is sufficient (1/1) because it provides a complete answer to the research question, and it is accurate (1/1). The student, however, does not use genuine evidence (in bold) to support the claim; there is no analysis of the data (0/1) or an interpretation of the analysis (0/1), although there is some data (1/1). The student also does not include complete justification of the evidence in his argument (0/2). In addition, this student does not provide an adequate challenge to an alternative explanation (0/2). The author uses scientific terms correctly (1/1) but uses phrases that do not reflect the nature of science (0/1). The writing mechanics are acceptable, although the argument is rather short. The organization of the argument is strong, because the arrangement of the sentences aid in the development of the main idea (1/1). There are also no punctuation (1/1) or grammatical errors (1/1) in the argument. The overall score for the sample argument, therefore, is 7 out of the 14 points possible.

## Standards Addressed in This Activity

This activity can be used to address the following dimensions outlined in *A Framework for K–12 Science Education* (NRC 2012):

### Scientific Practices

- Planning and carrying out investigations
- Engaging in argument from evidence
- Obtaining, evaluating, and communicating information

### Crosscutting Concepts

- Cause and effect: Mechanism and explanation
- Energy and matter: Flows, cycles, and conservation
- Stability and change

### Life Sciences Core Ideas

- From molecules to organisms: Structures and processes
- Heredity: Inheritance and variation of traits

This activity can be used to address the following standards for literacy in science from the *Common Core State Standards for English Language Arts and Literacy* (NGA and CCSSO 2010):

### Writing

- Text types and purposes
- Production and distribution of writing

## SECTION 2: EVALUATE ALTERNATIVES

### SPONTANEOUS GENERATION
TEACHER NOTES 11

- Research to build and present knowledge
- Range of writing

## Speaking and Listening

- Comprehension and collaboration
- Presentation of knowledge and ideas

## References

Campbell, N., and J. Reece. 2002. *Biology*. 6th ed. San Francisco, CA: Benjamin Cummings.

National Governors Association Center (NGA) for Best Practices, and Council of Chief State School Officers (CCSSO). 2010. *Common core state standards for English language arts and literacy*. Washington, DC: National Governors Association for Best Practices, Council of Chief State School.

National Research Council (NRC). 2012. *A framework for K–12 science education: Practices, crosscutting concepts, and core ideas*. Washington, DC: National Academies Press.

## Resources

Tips for the Safer Handling of Microorganisms in the School Science Laboratory at *www.nsta.org/pdfs/tipsforsafehandlingofmicroorganisms20120507.pdf*.

NSTA's "Safety Acknowledgment Form for Working With Microorganisms" at *www.nsta.org/pdfs/microorganismsafetyacknowledgment20120507.pdf*

## SECTION 2: EVALUATE ALTERNATIVES

# PLANT BIOMASS (PHOTOSYNTHESIS) 12

Sunflowers are some of the largest and fastest growing plants on Earth. They start as small seeds and eventually grow into 6–15 ft. tall flowering plants.

*Figure 12.1. Sunflower Seeds (Not to Scale) and a Full-Grown Sunflower Plant*

This observation raises an interesting question: **Where does most of the matter that makes up the stem and leaves of a plant come from?**

Here are three possible explanations:
- **Explanation 1**: The matter that makes up the stem and leaves comes from the soil because it contains the minerals and food that a plant needs to survive.
- **Explanation 2**: The matter that makes up the stem and leaves comes from the air because carbon dioxide is the source of carbon for plants.
- **Explanation 3**: The matter that makes up the stem and leaves comes from water because water is used in photosynthesis.

## SECTION 2: EVALUATE ALTERNATIVES

## 12  PLANT BIOMASS

## Getting Started

You can use the following materials to test these three explanations:

- Two to three day-old plants
- Pesticide- and herbicide-free potting soil
- Water
- Graduated cylinders
- 2 L soda bottles
- Electronic balance

**Safety notes**: Wear safety glasses or goggles and aprons. Wash hands with soup and water upon completion.

With your group, determine which explanation provides the best answer to the research question. You can use as many of the supplies available to you to test your ideas. Make sure that you generate the evidence you will need to support your explanation as you work. You can record your method and any observation you make in the space below.

## Our Method

## Our Observations

## SECTION 2: EVALUATE ALTERNATIVES

### PLANT BIOMASS  12

## Argumentation Session

Once your group has decided which explanation is the most valid or acceptable answer to the research question, prepare a whiteboard that you can use to share and justify your ideas. Your whiteboard should include all the information shown in Figure 12.2.

To share your work with others, we will be using a round-robin format. This means that one member of the group will stay at your workstation to share your group's ideas while the other group members go to the other groups one at a time in order to listen to and critique the arguments developed by your classmates. Remember, as you critique the work of others, you need to decide if their conclusions are valid or acceptable based on the quality of their claim and how well they are able to support their ideas. In other words, you need to determine if their argument is *convincing* or not. One way to determine if their argument is convincing is to ask them some of the following questions:

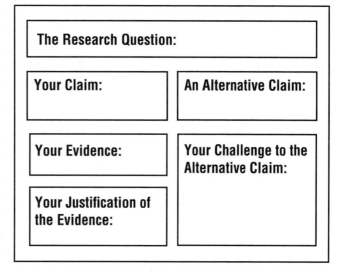

*Figure 12.2. Components of the Whiteboard*

- How did you gather your data? Why did you decide to do it that way?
- How do you know your data is high quality (free from errors)?
- How did you analyze or interpret your data? Why did you decide to do it that way?
- Why does your evidence support your claim?
- Why did you decide to use that evidence? Why is your evidence important?
- How does your justification of your evidence fit with accepted scientific ideas?

**SCIENTIFIC ARGUMENTATION IN BIOLOGY: 30 CLASSROOM ACTIVITIES**

**151**

### SECTION 2: EVALUATE ALTERNATIVES

Name_____   Date_____

# PLANT BIOMASS:
## What Is Your Argument?

In the space below, write a one- to three-paragraph argument to *support* the explanation that you think is the most valid or acceptable. Your argument must also include a *challenge* to one of the alternative explanations. As you write your argument, remember to do the following:

- State the explanation you are trying to support
- Include genuine evidence (data + analysis + interpretation)
- Explain why the evidence is important and relevant
- State the explanation you are trying to refute
- Explain why the alternative explanation is invalid or unacceptable
- Organize your argument in a way that enhances readability
- Use a broad range of words including vocabulary that we have learned
- Correct grammar, punctuation, and spelling errors

SECTION 2: EVALUATE ALTERNATIVES

# PLANT BIOMASS
## TEACHER NOTES 12

## Purpose

The purpose of this activity is to help students understand the synthesis and source of plant biomass as well as the process of photosynthesis. This activity also helps students learn how to engage in practices such using planning and carrying out investigations, arguing from evidence, and communicating information. This activity, in addition, is designed to give students an opportunity to learn how to write in science and develop their speaking and listening skills, which are important goals for literacy in science (see Standards Addressed in This Activity for a complete list of the practices, crosscutting concepts, core ideas, and literacy skills that are aligned with this activity).

## The Content and Related Concepts

Photosynthesis is an extremely complex process. In its simplest form, this important reaction converts $CO_2$ [carbon dioxide] and $H_2O$ [water] into $O_2$ [oxygen] and $C_6H_{12}O_6$ [glucose]. The chemical formula for this process is the following:

$$6CO_2 + 6H_2O \rightarrow C_6H_{12}O_6 + 6O_2$$

It is important to note, however, that water ($H_2O$) does not contribute oxygen to the production of the glucose molecule. Instead, the oxygen that is found in the water molecule ends up as a molecule of oxygen ($O_2$). The oxygen found in the glucose actually comes from carbon dioxide, so the carbon and oxygen atoms found in a molecule of glucose come from the air (and *not* the soil or water as some people think). Once the glucose is produced through the process of photosynthesis, other biochemical pathways are used to convert the glucose into more complex carbohydrates such as starch and cellulose. Starch is a polysaccharide and polysaccharides are just long chains of glucose molecules. Plants convert excess glucose into starch for storage. Cellulose is a type of polysaccharide and is probably the single most abundant organic molecule in the biosphere. It is the major structural material of which plants are made. Wood is largely cellulose, while cotton and paper are almost pure cellulose. Starch and cellulose are often represented as $(C_6H_{10}O_5)n$ with *n* representing the number of glucose molecules found in the polysaccharide (one molecule of $H_2O$ is lost when two glucose molecules are joined together, hence $C_6H_{10}O_5$ instead of $C_6H_{12}O_6$).

Most of the matter that is found in a green plant, therefore, comes from the air. Cellulose, as mentioned earlier, is the primary structural component of green plants and one of the most common organic molecules found on Earth. Plants produce cellulose as they grow and store excess glucose in the form of starch. As a result, plants pull in and use a great deal of carbon from the air as they grow. An acre of 35-year-old loblolly pine trees, for example, can remove about 345 metric tons of $CO_2$ per year from the

## SECTION 2: EVALUATE ALTERNATIVES

### 12  PLANT BIOMASS
### TEACHER NOTES

air, and all this matter ends up as cellulose in these trees (U.S. Department of Energy 1999).

## Curriculum and Instructional Considerations

There are at least three points within a traditional biology curriculum where this activity would be appropriate and helpful: the cell, plant form and function, or ecology. Teachers can use this activity during a unit on cells to introduce the process of photosynthesis or after students have been introduced to photosynthesis to help address the common misconception that most of the biomass of plants comes from the soil or water. When it is used as part of a unit on plant form and function, teachers can connect it to concepts of plant growth and development or plant nutrition (i.e., plants require nine macronutrients and eight micronutrients to grow and develop, but these nutrients do not make a substantial contribution to the actual biomass of the plant). Finally, teachers can use this activity to introduce the concepts of energy flow and the carbon cycle in ecosystems when the activity is integrated into a unit about ecology.

This activity is also a good way to teach students about experimental design, the control of variables, the difference between *hypotheses* (tentative explanations) and *predictions* (expected results), and other important terminology such as *independent* and *dependent variables*. Students will need a basic understanding of these important ideas in order to be able to collect meaningful data during the activity. The focus of the explicit discussion at the end of the activity can also focus on other biochemical pathways and the importance of the short 3-carbon stem molecules that are produced during photosynthesis. For example, the teacher can point out that these 3-carbon molecules are used to create other important biomolecules such as lipids and amino acids and not just carbohydrates. The explicit discussion should also focus on at least one aspect of the nature of science or the nature of scientific inquiry. For example, a teacher could discuss how scientific explanation must be consistent with observational evidence about nature, or how experiments are used to test explanations, the importance of controlling variables during an experiment, or the role of creativity and imagination in science using this activity as an illustrative example.

Teachers can also include a discussion of a classic experiment conducted by Jean Baptista van Helmont (1577–1644) as part of this activity. Many scholars consider this experiment, which was published in *Ortus Medicinae* (in 1648, after Helmont died), to be a milestone in the history of biology, because it marked the start of experimental plant physiology (Hershey 1991). In the following paragraph, van Helmont describes the method he used, the data he collected, and his conclusion:

> I took an earthen pot and in it placed 200 pounds of earth which had been dried out in an oven. This I moistened with rain water, and in it planted a shoot of willow which weighed five pounds. When five years had passed the tree which grew from it weighed 169 pounds and about three

# SECTION 2: EVALUATE ALTERNATIVES

## PLANT BIOMASS
### TEACHER NOTES 12

ounces. The earthen pot was wetted whenever it was necessary with rain or distilled water only. It was very large, and was sunk in the ground, and had a tin plated iron lid with many holes punched in it, which covered the edge of the pot to keep air-borne dust from mixing with the earth. I did not keep track of the weight of the leaves which fell in each of the four autumns. Finally, I dried out the earth in the pot once more, and found the same 200 pounds, less about 2 ounces. Thus, 164 pounds of wood, bark, and roots had arisen from water alone. (quoted in Hershey 1991)

Although Helmont's conclusion was incorrect, this important experiment helped to rule out soil as the source of the biomass of plants. When teachers share this experiment with students as part of the activity, it provides a great opportunity to discuss the nature of scientific inquiry and the nature of scientific knowledge in the context of a major milestone in the history of science.

## Recommendations for Implementing the Activity

This activity takes between 100 and 150 minutes of instructional time to complete, depending on how a teacher decides to spend time in class. This activity also requires two or three weeks for the plants to grow. In Option A, the students are given time to complete all six stages of the lesson during class (see Figure 12.3, p. 156). Stages 1 and 2 are completed on day 1, Stages 3 and 4 are completed on day 2, and Stages 5 and 6 are completed on day 3. This option for implementing the activity works best in schools where students are not expected to complete much homework or if students need to be encouraged to write more during the school day. It also provides less time for the argumentation sessions, which may or may not be a problem, depending on how comfortable the students are with argumentation. After all, some classes are more talkative than others. In Option B, students complete Stage 1 and begin Stage 2 of the lesson during class on day 1. The students then complete Stage 3 on day 2 of the lesson. Stages 4 and 5 are completed on day 3, and the final written argument (Stage 6) is then assigned as homework and returned the next day. Table 12.1 (p. 157) provides information about the type and amount of materials needed to implement this activity in a classroom with 28 students in groups of four and groups of three.

## Assessment

The rubric provided in Appendix C can be used to assess the argument crafted by each student at the end of the activity. To illustrate how the rubric can be used to score an argument written by a student, consider the following example. This sample, which was written by a middle school student, is an example of an argument that is weak in terms of content and mechanics.

> The matter that makes up the stem and leaves in our radishes comes from the air. We put 200 grams of soil in one bottle and paper towels in the other one.

## SECTION 2: EVALUATE ALTERNATIVES

### 12 PLANT BIOMASS
#### TEACHER NOTES

*Figure 12.3. Two Options for Implementing the Activity*

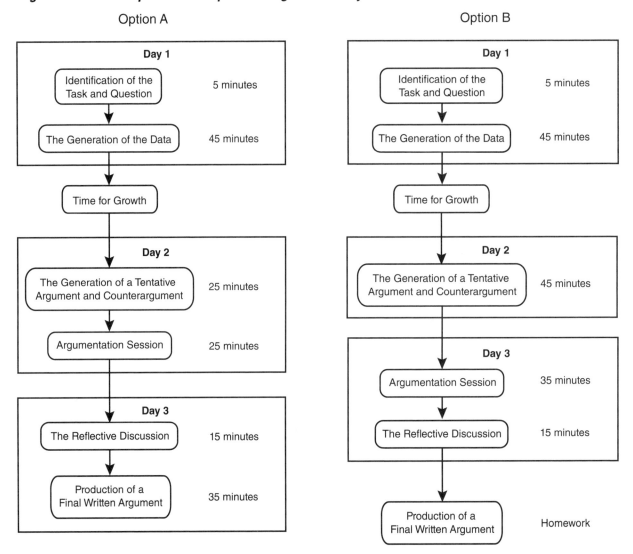

## SECTION 2: EVALUATE ALTERNATIVES

### PLANT BIOMASS
### TEACHER NOTES 12

*Table 12.1. Materials Needed to Implement the Activity in a Classroom of 28 Students*

| Material | Amount Needed With ... | |
|---|---|---|
| | Groups of 3 | Groups of 4 |
| Two- to three-week-old plants (such as Wisconsin Fast Plants™, which germinate in two days and flower in 14; sunflowers, which grow large very quickly; or radishes) | 100 | 100 |
| Potting soil | 1 bag | 1 bag |
| Graduated cylinder | 9 | 6 |
| 2 L soda bottles | 27 | 18 |
| Electronic balance | 9 | 6 |
| Whiteboards (or chart paper)+ | 10 | 7 |
| Whiteboard markers (or permanent markers if using chart paper)+ | 20 | 14 |
| Copy of Student Pages (pp. 149–151)* | 28 | 28 |
| Copy of Student Page (p. 152)* | 28 | 28 |
| Copy of Appendix C (p. 367)* | 28 | 28 |

+ Teachers can also have students prepare their arguments in a digital medium (such as PowerPoint or Keynote).
* Teachers can also project these materials onto a screen in order to cut down on paper use.

We then soaked the 12 radish seeds in water for 20 minutes. We weighed them and dropped 6 of them into each bottle. We added 15 ml of water to each botle and sealed the lids. After to weeks we cut open the botles and weighed the plants, the dirt and the paper towls. **The soil weighed 214 grams and the plants weighed 6 grams.** This proves that the plants come from the air because the soil weighed more because of the water so the water did not go into the plants so it must have come form the air.

The example argument is weak for several reasons. Although the student's claim (underlined) is sufficient (1/1) because it provides a complete answer to the research question and it is accurate (1/1), the student uses incomplete evidence (in bold) to support the claim. There is data (1/1) but no analysis of the data (0/1) and no interpretation of the analysis (0/1). The student does not include a complete justification of the evidence in her argument because she does not explain why the evidence is important (01/1) or link the evidence to a specific principle, concept, or underlying assumption (0/1). She also does not provide an adequate challenge to an alternative expla-

## SECTION 2: EVALUATE ALTERNATIVES

### 12 PLANT BIOMASS
TEACHER NOTES

nation by making the other viewpoint explicit (0/1) and then providing a reason for why it is invalid (0/1). Although the author uses appropriate terms (1/1), she uses phrases that misrepresent the nature of science (0/1). The writing mechanics are also weak. The organization of the argument is acceptable, because the arrangement of the sentences aid in the development of the main idea (1/1), but there are several punctuation (0/1) and grammatical errors (0/1) in the argument. The overall score for the sample argument, therefore, is 5 out of the 14 points possible.

## Standards Addressed in This Activity

This activity can be used to address the following dimensions outlined in *A Framework for K–12 Science Education* (NRC 2012):

### Scientific Practices

- Planning and carrying out investigations
- Engaging in argument from evidence
- Obtaining, evaluating, and communicating information

### Crosscutting Concepts

- Cause and effect: Mechanism and explanation
- Energy and matter: Flows, cycles, and conservation
- Structure and function

### Life Sciences Core Ideas

- From molecules to organisms: Structures and processes
- Ecosystems: Interactions, energy, and dynamics

This activity can be used to address the following standards for literacy in science from the *Common Core State Standards for English Language Arts and Literacy* (NGA and CCSSO 2010):

### Writing

- Text types and purposes
- Production and distribution of writing
- Research to build and present knowledge
- Range of writing

### Speaking and Listening

- Comprehension and collaboration
- Presentation of knowledge and ideas

## References

Hershey, D. 1991. Digging deeper in Helmont's famous willow tree experiment. *The American Biology Teacher* 53 (8): 458–460.

National Governors Association Center (NGA) for Best Practices, and Council of Chief State School Officers (CCSSO). 2010. *Common core state standards for English language arts and literacy*. Washington, DC: National Governors Association for Best Practices, Council of Chief State School.

National Research Council (NRC). 2012. *A framework for K–12 science education: Practices, crosscutting concepts, and core ideas*. Washington, DC: National Academies Press.

### SECTION 2: EVALUATE ALTERNATIVES

# MOVEMENT OF MOLECULES IN OR OUT OF CELLS (OSMOSIS AND DIFFUSION) 13

A student put a drop of blood on a microscope slide and then looked at the cells under a microscope. As you can see in Figure 13.1 below, the magnified red blood cells look like little round balls. After adding a few drops of sugar water to the drop of blood, the student noticed that the cells appeared to become smaller.

*Figure 13.1. Magnified Red Blood Cells*

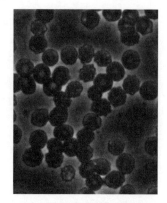

Red Blood Cells

Red Blood Cells After Adding Sugar Water

This observation raises an interesting question: **Why do the red blood cells appear smaller?** Here are three possible explanations:

- **Explanation 1**: Sugar molecules push on the cell membranes and make the cells appear smaller.

- **Explanation 2**: Water molecules move out of the cell because the concentration of water is greater inside the cell than it is outside the cell.

- **Explanation 3**: Sugar molecules enter the cell and take the place of the water. The cells appear smaller because the sugar molecules take up less space.

## SECTION 2: EVALUATE ALTERNATIVES

## 13  MOVEMENT OF MOLECULES IN OR OUT OF CELLS

### Getting Started

You can use the following materials to test these three explanations:

- Sugar (lactose or maltose)
- Water
- A very accurate weighing device
- Dialysis tubing (assume that it behaves just like the membrane of a red blood cell)
- Disposable pipettes
- Benedict's solution (allows you to test for the presence of lactose in water)

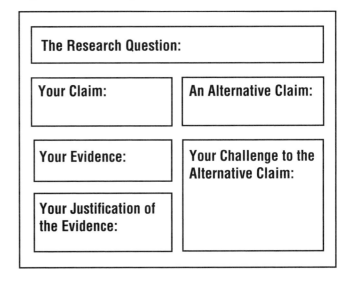

*Figure 13.2. Components of the Whiteboard*

You can construct a model cell by using the dialysis tubing. To do this, place the dialysis tubing in water until it is thoroughly soaked. Remove a section of the soaked tubing from the water and tightly twist one end several times. Then fold the twisted end over and tie it tightly with a string. Now rub the sides of the tubing between your fingers to separate the sides. You can then fill the model cell with any types of solution you wish. Once filled, twist the open end several times and tie it tightly. You can then rinse of the bag with distilled water (to remove anything you might have spilled on it), dry the bag, weigh it, and place it into a beaker filled with a liquid.

If you are interested in testing for the presence of lactose or maltose in a solution, you can conduct a Benedict's test. To conduct a Benedict's test, follow the procedure below:

1. In a test tube, add 40 drops of liquid to be tested.

2. Add 10 drops of Benedict's solution.

3. Carefully heat the test tube by suspending in a hot water bath at approximately 40–50 degrees Celsius for five minutes.

4. Note any color change. If maltose or lactose is present, the solution will turn green, yellow, or brick red, depending on sugar concentration.

**Safety notes**: Wear indirectly vented chemical-splash goggles, aprons, and gloves. Handle the hot water bath with care as a hot water bath can burn skin if splashed. Wash hands with soap and water upon completion.

## SECTION 2: EVALUATE ALTERNATIVES

### MOVEMENT OF MOLECULES IN OR OUT OF CELLS   13

With your group, determine which explanation provides the best answer to the research question. You can use as many of the supplies available to you to test your ideas. Make sure that you generate the evidence you will need to support your explanation as you work. You can record your method and any observations you make in the spaces below.

## Our Method

## Our Observations

## Argumentation Session

Once your group has decided which explanation is the most valid or acceptable answer to the research question, prepare a whiteboard that you can use to share and justify your ideas. Your whiteboard should include all the information shown in Figure 13.2.

To share your work with others, we will be using a round-robin format. This means that one member of the group will stay at your workstation to share your group's ideas while the other group members go to the other groups one at a time in order to listen to and critique the arguments developed by your classmates. Remember, as you critique the work of others, you need to decide if their conclusions are valid or acceptable based on the quality of their claim and how well they are able to support their ideas. In other words, you need to determine if their argument is *convincing* or not. One way to determine if their argument is convincing is to ask them some of the following questions:

## SECTION 2: EVALUATE ALTERNATIVES

### 13  MOVEMENT OF MOLECULES IN OR OUT OF CELLS

- How did you gather your data? Why did you decide to do it that way?
- How do you know your data is high quality (free from errors)?
- How did you analyze or interpret your data? Why did you decide to do it that way?
- Why does your evidence support your claim?
- Why did you decide to use that evidence? Why is your evidence important?
- How does your justification of your evidence fit with accepted scientific ideas?

**SECTION 2: EVALUATE ALTERNATIVES**

Name_____  Date_____

# MOVEMENT OF MOLECULES IN OR OUT OF CELLS:
## What Is Your Argument?

In the space below, write a one- to three-paragraph argument to *support* the explanation that you think is the most valid or acceptable. Your argument must also include a *challenge* to one of the alternative explanations.

As you write your argument, remember to do the following:

- State the explanation you are trying to support
- Include genuine evidence (data + analysis + interpretation)
- Explain why the evidence is important and relevant
- State the explanation you are trying to refute
- Explain why the alternative explanation is invalid or unacceptable
- Organize your argument in a way that enhances readability
- Use a broad range of words including vocabulary that we have learned
- Correct grammar, punctuation, and spelling errors

SECTION 2: EVALUATE ALTERNATIVES

# 13 MOVEMENT OF MOLECULES IN OR OUT OF CELLS
## TEACHER NOTES

## Purpose
The purpose of this activity is to help students understand the process of osmosis and to understand how a cell membrane acts as a selective barrier. This activity also helps students learn how to engage in practices such using planning and carrying out investigations, arguing from evidence, and communicating information. In addition, this activity is designed to give students an opportunity to learn how to write in science and develop their speaking and listening skills, which are important goals for literacy in science (see Standards Addressed in This Activity for a complete list of the practices, crosscutting concepts, core ideas, and literacy skills that are aligned with this activity).

## The Content and Related Concepts
Cell membranes are selectively permeable, which means they can control the flow of substances into and out of a cell. Some substances can diffuse through membranes without work (i.e., passive transport). These substances will spread from areas of high concentration to areas of low concentration. Osmosis is the passive transport of water from a solution of lower solute concentration to one of higher solute concentration. Osmosis causes cells to shrink in hypertonic solutions and swell in hypotonic solutions (see Figure 13.3). The control of water balance, which is called osmoregulation, is essential for the survival of organisms.

In this activity, students use dialysis tubing to make model cells in order to test alternative explanations for why cells appear to shrink when bathed in a solution of sugar water. Students can create cells with a solution of sugar water on the inside of the cell and place the model cell into distilled water to represent a cell in a hypotonic solution. Students can also create a model cell with distilled water on the inside of the bag and then place it in a sugar water solution to represent a hypertonic solution. Students can determine if molecules (either sugar or water) are moving in or out of the cells by massing them before and after the model cell is soaked in distilled water (or a sugar solution). Students can also test for the presence of sugar in a solution by using Benedict's solution (see Recommendations for Implementing the Activity). This indicator turns green or red in the presence of a reducing sugar such as lactose or maltose.

## Curriculum and Instructional Considerations
This activity is best used as part of a unit on cell structure and function. However, it is designed to be an introduction to cell structure and function, so it should be used before students are introduced to concepts such

## SECTION 2: EVALUATE ALTERNATIVES

### MOVEMENT OF MOLECULES IN OR OUT OF CELLS
#### TEACHER NOTES 13

*Figure 13.3. The Net Movement of Water Into and out of Cells in Hypertonic, Isotonic, and Hypotonic Solutions*

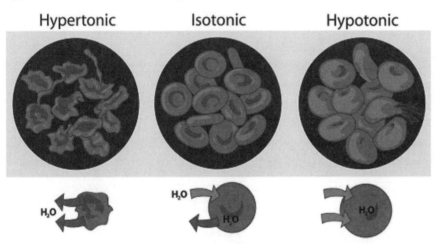

as diffusion, osmosis, and the nature of cell membranes. This activity is also a good way to teach students about experimental design, the control of variables, the difference between *hypotheses* (tentative explanations) and *predictions* (expected results), and other important terminology such as *independent* and *dependent variables*. Students will need a basic understanding of these important ideas in order to be able to collect meaningful data during the activity. The explicit discussion at the end of the activity should focus on the nature of cell membranes, the process of osmosis, and the importance of osmoregulation in organisms. The explicit discussion should also focus on at least one aspect of the nature of science or the nature of scientific inquiry. For example, a teacher could discuss how scientific explanation must be consistent with observational evidence about nature, or how experiments are used to test explanations, the importance of controlling variables during an experiment, or the role of creativity and imagination in science using this activity as an illustrative example.

## Recommendations for Implementing the Activity

This activity takes between 100 and 150 minutes of instructional time to complete, depending on how a teacher decides to spend time in class. In Option A, the students are given time to complete all six stages of the lesson during class (see Figure 13.4, p. 166). Stages 1 and 2 are completed on day 1, Stages 3 and 4 are completed on day 2, and Stages 5 and 6 are completed on day 3. This option for implementing the activity works best in schools where students are not expected to complete much homework or if students need to be encouraged to write more during the

## SECTION 2: EVALUATE ALTERNATIVES

## 13 MOVEMENT OF MOLECULES IN OR OUT OF CELLS
### TEACHER NOTES

*Figure 13.4. Two Options for Implementing the Activity*

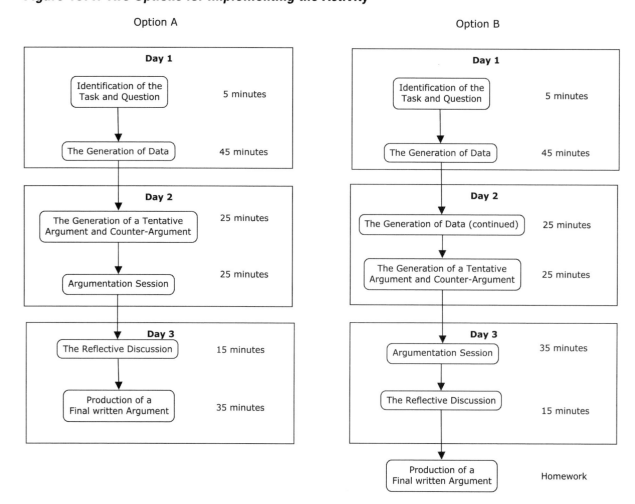

# SECTION 2: EVALUATE ALTERNATIVES

## MOVEMENT OF MOLECULES IN OR OUT OF CELLS
### TEACHER NOTES 13

*Table 13.1. Materials Needed to Implement the Activity in a Classroom of 28 Students*

| Material | Amount Needed With … Groups of 3 | Amount Needed With … Groups of 4 |
|---|---|---|
| 15 cm piece of dialysis tubing | 30 | 21 |
| 200 ml beaker | 30 | 21 |
| 20 g of either maltose or lactose | 10 | 7 |
| Test tube | 30 | 21 |
| Wash bottles filled with distilled water | 10 | 7 |
| Disposable pipettes | 30 | 21 |
| Stopwatch | 10 | 7 |
| Electronic balance | 10 | 7 |
| Hot water bath | 1 | 1 |
| Benedict's solution (in dropper bottles) | 10 | 7 |
| 25 ml graduated cylinder | 10 | 7 |
| Whiteboards (or chart paper)+ | 10 | 7 |
| Whiteboard markers (or permanent markers if using chart paper)+ | 20 | 14 |
| Copy of Student Pages (pp. 159–162)* | 28 | 28 |
| Copy of Student Page (p. 163)* | 28 | 28 |
| Copy of Appendix C (p. 367)* | 28 | 28 |

+ Teachers can also have students prepare their arguments in a digital medium (such as PowerPoint or Keynote).
* Teachers can also project these materials onto a screen in order to cut down on paper use.

school day. It also provide less time for the argumentation sessions, which may or may not be a problem, depending on how comfortable the students are with argumentation. In Option B, students complete Stage 1 and begin Stage 2 of the lesson during class on day 1. The students then complete Stage 3 on day 2 of the lesson. Stages 4 and 5 are completed on day 3 and, the final written argument (Stage 6) is then assigned as homework and returned the next day.

Table 13.1 provides information about the type and amount of materials needed to implement this activity in a classroom with 28 students in groups of four and groups of three.

## SECTION 2: EVALUATE ALTERNATIVES

### 13 MOVEMENT OF MOLECULES IN OR OUT OF CELLS
TEACHER NOTES

Prior to starting the activity, be sure to review with students the Safety Data Sheet (SDS) for Benedict's solution.

## Assessment

The rubric provided in Appendix C (p. 367) can be used to assess the argument crafted by each student at the end of the activity. To illustrate how the rubric can be used to score an argument written by a student, consider the following example. This example, which was written by a high school sophomore, is strong in terms of content and is also well written.

> <u>Cells appear smaller when they are soaked in sugar water because water molecules move out of the cell because the concentration of water is greater inside the cell than it is outside the cell.</u> In our experiment, we created two "cells" made out dialysis tubing by filling them with 15 ml of distilled water each. We then weighed both cells and then placed them in a sugar solution for 20 minutes. We then removed the cells from the solution and weighed them. After we were done weighing them, we cut them open and tested the liquid inside the cell for sugar. **We found that both cells weighed less after they were soaked in sugar water (bag one lost 3.2 grams and bag two lost 3.3 grams). The benedicts test was also negative for both cells.** These results indicate that water moved out of the cells because the bags were lighter (and they would have been the same weight if they did not lose any water). Therefore cells appear smaller when they are soaked in sugar water because they actually lose some of liquid inside them. We also know that the sugar molecules do not enter the cell and take the place of the water (and make the cells appear smaller because they take up less space) because we found no evidence of sugar inside the cells after they were soaked in sugar water for 20 minutes. If the sugar moved into the cell we would have gotten a positive benedicts test.

The example argument is strong for several reasons. The student's claim (underlined) is sufficient (1/1), because it provides a complete answer to the research question, and it is accurate (1/1). The student uses genuine evidence (in bold) to support the claim, because there is data (1/1), an analysis of the data (1/1), and an interpretation of the analysis (1/1). The student includes a complete justification of the evidence in his argument because he does explain why the evidence is important (1/1) by linking it to a specific principle, concept, or underlying assumption (1/1). He also provides an adequate challenge to an alternative explanation by making the other viewpoint explicit (1/1) and providing a reason for why it is invalided (1/1). The author also uses appropriate terms (1/1) and phrases that are consistent with the nature of science (1/1). The writing mechanics are also good. The organization of the argument is strong because the arrangement of the sentences aid in the development of the main idea (1/1). There are also no punctuation (1/1)

## SECTION 2: EVALUATE ALTERNATIVES

### MOVEMENT OF MOLECULES IN OR OUT OF CELLS
#### TEACHER NOTES 13

or grammatical errors (1/1) in the argument. The overall score for the sample argument, therefore, is 14 out of the 14 points possible.

## Standards Addressed in This Activity

This activity can be used to address the following dimensions outlined in *A Framework for K–12 Science Education* (NRC 2012):

## Scientific Practices

- Developing and using models
- Planning and carrying out investigations
- Engaging in argument from evidence
- Obtaining, evaluating, and communicating information

## Crosscutting Concepts

- Cause and effect: Mechanism and explanation
- Energy and matter: Flows, cycles, and conservation
- Structure and function

## Life Sciences Core Ideas

- From molecules to organisms: Structures and processes

This activity can be used to address the following standards for literacy in science from the *Common Core State Standards for English Language Arts and Literacy* (NGA and CCSSO 2010):

## Writing

- Text types and purposes
- Production and distribution of writing
- Research to build and present knowledge
- Range of writing

## Speaking and Listening

- Comprehension and collaboration
- Presentation of knowledge and ideas

## References

National Governors Association Center (NGA) for Best Practices, and Council of Chief State School Officers (CCSSO). 2010. *Common core state standards for English language arts and literacy*. Washington, DC: National Governors Association for Best Practices, Council of Chief State School.

National Research Council (NRC). 2012. *A framework for K–12 science education: Practices, crosscutting concepts, and core ideas*. Washington, DC: National Academies Press.

SECTION 2: EVALUATE ALTERNATIVES

# LIVER AND HYDROGEN PEROXIDE (CHEMICAL REACTIONS AND CATALYSTS) 14

A 1 g piece of fresh liver is placed in a test tube with 1 ml of hydrogen peroxide ($H_2O_2$). Bubbles appear in the test tube immediately after the liver is added, which indicates that a gas is produced when the two substances are mixed (see Figure 14.1). The appearance of the gas indicates that a chemical reaction took place inside the test tube. The procedure was then repeated, but this time, the gas was captured and tested with a glowing splint. The results of the glowing splint test indicate that the gas is oxygen ($O_2$).

A chemical reaction results from a rearrangement of atoms. When two substances are mixed, however, it is not always clear if the molecules in both substances rearrange, or if the reaction is the result of just one set of molecules being altered in some way. Therefore, in order to be able to explain why oxygen gas is produced when liver is mixed with hydrogen peroxide, you will need to determine what happens to the substance that is in the liver and the molecules of hydrogen peroxide when the two substances are mixed.

*Figure 14.1. Oxygen Gas Is Produced When Hydrogen Peroxide and Liver Are Mixed*

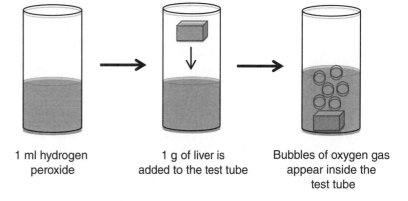

1 ml hydrogen peroxide

1 g of liver is added to the test tube

Bubbles of oxygen gas appear inside the test tube

The guiding question for this investigation is the following: **What happens to the substance in the liver that interacts with the hydrogen peroxide when these two substances are mixed?**

Here are three potential answers to this question:

- **Explanation 1**: The substance in the liver that interacts with hydrogen peroxide is altered when the two substances are mixed; the oxygen gas, as a result, comes from the substance in the liver.

- **Explanation 2**: The substance in the liver that interacts with hydrogen peroxide is not altered when the two substances are mixed; the oxygen gas, as a result, comes from hydrogen peroxide.

- **Explanation 3**: The substance in the liver that interacts with hydrogen peroxide and the hydrogen peroxide are altered when the two substances are mixed; the oxygen gas, as a result, comes from both the hydrogen peroxide and the substance in the liver.

## SECTION 2: EVALUATE ALTERNATIVES

### 14 LIVER AND HYDROGEN PEROXIDE

## Getting Started

You can use the following materials to test these three explanations:

- Test tubes
- 1 g pieces of fresh liver
- Hydrogen peroxide
- Stopwatch
- Other materials as requested

**Safety notes**: Wear indirectly vented chemical-splash goggles, aprons, and gloves. Wash hands with soap and water after activity.

Your task is to determine which explanation provides the best answer to the research question. You can use as many of the supplies available to you to test your ideas. Make sure that you generate the evidence you will need to support your explanation as you work. You can record your method and any observations you make in the spaces below.

## Our Method

## Our Observations

## SECTION 2: EVALUATE ALTERNATIVES

### LIVER AND HYDROGEN PEROXIDE  14

## Argumentation Session

Once your group has decided which explanation is the most valid or acceptable answer to the research question, prepare a whiteboard that you can use to share and justify your ideas. Your whiteboard should include all the information shown in Figure 14.2.

To share your work with others, we will be using a round-robin format. This means that one member of the group will stay at your workstation to share your group's ideas while the other group members go to the other groups one at a time in order to listen to and critique the arguments developed by your classmates. Remember, as you critique the work of others, you need to decide if their conclusions are valid or acceptable based on the quality of their claim and how well they are able to support their ideas. In other words, you need to determine if their argument is *convincing* or not. One way to determine if their argument is convincing is to ask them some of the following questions:

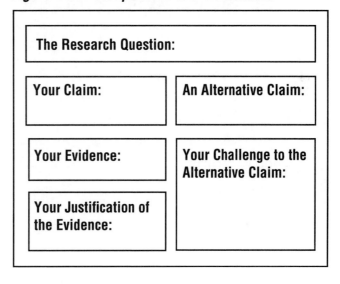

*Figure 14.2. Components of a Whiteboard*

- How did you gather your data? Why did you decide to do it that way?
- How do you know your data is high quality (free from errors)?
- How did you analyze or interpret your data? Why did you decide to do it that way?
- Why does your evidence support your claim?
- Why did you decide to use that evidence? Why is your evidence important?
- How does your justification of your evidence fit with accepted scientific ideas?

## SECTION 2: EVALUATE ALTERNATIVES

Name_____  Date_____

# LIVER AND HYDROGEN PEROXIDE:
## What Is Your Argument?

In the space below, write a one- to three-paragraph argument to *support* the explanation that you think is the most valid or acceptable. Your argument must also include a *challenge* to one of the alternative explanations. As you write your argument, remember to do the following:

- State the explanation you are trying to support
- Include genuine evidence (data + analysis + interpretation)
- Explain why the evidence is important and relevant
- State the explanation you are trying to refute
- Explain why the alternative explanation is invalid or unacceptable
- Organize your argument in a way that enhances readability
- Use a broad range of words including vocabulary that we have learned
- Correct grammar, punctuation, and spelling errors

SECTION 2: EVALUATE ALTERNATIVES

# LIVER AND HYDROGEN PEROXIDE 14
## TEACHER NOTES

## Purpose

The purpose of this activity is to help students understand that most cell functions involve chemical reactions and many reactions (including decomposition and synthesis) are made possible by enzymes. This activity also helps students learn how to engage in practices such using planning and carrying out investigations, arguing from evidence, and communicating information. This activity, in addition, is designed to give students an opportunity to learn how to write in science and develop their speaking and listening skills, which are important goals for literacy in science (see Standards Addressed in This Activity for a complete list of the practices, crosscutting concepts, core ideas, and literacy skills that are aligned with this activity).

## The Content and Related Concepts

Enzymes, which are proteins, are biological catalysts. They speed up reactions by lowering activation energy, allowing bonds of other molecules to break at moderate temperatures without being broken down during the process. Each type of enzyme has a unique active site that combines specifically with its substrate (the reactant molecule on which it acts). Liver contains high levels of an enzyme called catalase. Catalase acts as a catalyst in the conversion of hydrogen peroxide ($H_2O_2$), a powerful and potentially harmful oxidizing agent, into water and molecular oxygen. The reaction of catalase in the decomposition of hydrogen peroxide is the following:

$$2H_2O_2 \xrightarrow{\text{catalase}} 2H_2O + O_2$$

Hydrogen peroxide is a harmful by-product of many normal metabolic processes. It therefore must be converted into other less dangerous substances before it can cause damage to cells, tissues, or organs. All known animals use catalase in every organ, and there are very high concentrations of it found in the liver. When a piece of liver is added to a solution of 3% hydrogen peroxide, the catalase from the liver catalyzes the decomposition of the hydrogen peroxide molecules in the solution, which results in the production of oxygen gas. This process continues until all the hydrogen peroxide is broken down. Catalase is not changed in reaction and can be recycled to break down additional molecules of hydrogen peroxide. The piece of liver, as a result, can be recovered from the water and used again. Catalase works best at a pH of 7, a salt concentration of 0.9%, and at normal body temperature. High temperatures, extremely low and high pH levels, or high salt concentrations will denature the enzyme (change its shape) and cause it to no longer function.

## SECTION 2: EVALUATE ALTERNATIVES

## 14 LIVER AND HYDROGEN PEROXIDE
TEACHER NOTES

## Curriculum and Instructional Considerations

There are at least three points within a traditional biology curriculum in which this activity would be appropriate and helpful: the chemistry of life, the cell, and animal form and function. When used as part of a unit on the chemistry of life, it can be used to emphasize the role of catalysts in chemical reactions. When used as part of a unit on cells, it can be used to introduce the important role of enzymes in cellular function. Finally, if used as part of a unit on animal form and function, it can be used to illustrate how animals regulate the internal environment and maintain homeostasis.

This activity is also an effective way to teach students about experimental design, the control of variables, the difference between hypotheses (tentative explanations) and predictions (expected results), and other important terminology such as *independent* and *dependent variables*. Students will need a basic understanding of these important ideas in order to be able to collect meaningful data during the activity. The explicit discussion at the end of the activity should focus on structure and function of enzymes and their important role in organisms. The explicit discussion should also focus on at least one aspect of the nature of science or the nature of scientific inquiry. For example, a teacher could discuss how scientific explanation must be consistent with observational evidence about nature, or how experiments are used to test explanations, the importance of controlling variables during an experiment, or the role of creativity and imagination in science using this activity as an illustrative example.

## Recommendations for Implementing the Activity

This activity takes between 100 and 150 minutes of instructional time to complete, depending on how a teacher decides to spend time in class. In Option A, the students are given time to complete all six stages of the lesson during class (see Figure 14.3). Stages 1 and 2 are completed on day 1, Stages 3 and 4 are completed on day 2, and Stages 5 and 6 are completed on day 3. This option for implementing the activity works best in schools where students are not expected to complete much homework or if students need to be encouraged to write more during the school day. It also provides less time for the argumentation sessions, which may or may not be a problem, depending on how comfortable the students are with argumentation. In Option B, students complete Stage 1 and begin Stage 2 of the lesson during class on day 1. The students then complete Stage 3 on day 2 of the lesson. Stages 4 and 5 are completed on day 3, and the final written argument (Stage 6) is then assigned as homework and returned the next day.

Table 14.1 (p. 178) provides information about the type and amount of materials needed to implement this activity in a classroom with 28 students in groups of four and groups of three.

Prior to starting the activity, be sure to review with students the SDS for hydrogen peroxide.

## SECTION 2: EVALUATE ALTERNATIVES

### LIVER AND HYDROGEN PEROXIDE
#### TEACHER NOTES 14

*Figure 14.3. Two Options for Implementing the Activity*

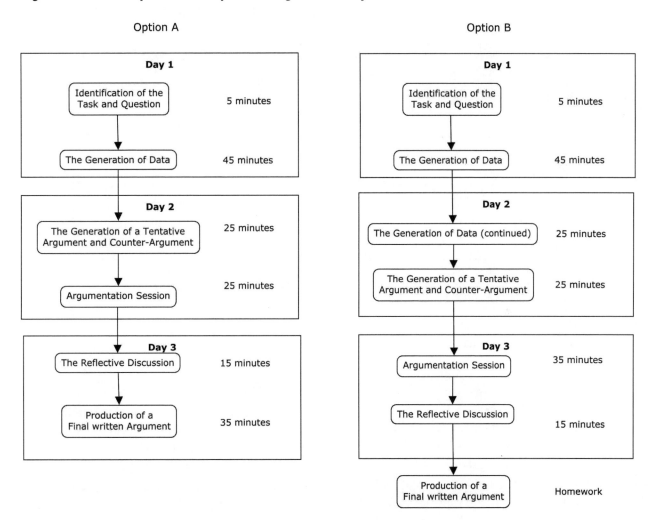

## SECTION 2: EVALUATE ALTERNATIVES

## 14 LIVER AND HYDROGEN PEROXIDE
### TEACHER NOTES

*Table 14.1. Materials Needed to Implement the Activity in a Classroom of 28 Students*

| Material | Amount Needed With ... | |
|---|---|---|
| | Groups of 3 | Groups of 4 |
| 3% Hydrogen peroxide solution (200 ml per group) | 1.8 L | 1.2 L |
| 100 ml Erlenmeyer flask (4 per group) | 36 | 24 |
| Rubber stopper | 9 | 6 |
| Forceps | 9 | 6 |
| Stopwatch | 9 | 6 |
| Pieces of liver (1 g each) | 18 | 12 |
| Whiteboards (or chart paper)+ | 10 | 7 |
| Whiteboard markers (or permanent markers if using chart paper)+ | 20 | 14 |
| Copy of Student Pages (pp. 171–173)* | 28 | 28 |
| Copy of Student Page (p. 174)* | 28 | 28 |
| Copy of Appendix C (p. 367)* | 28 | 28 |

+ Teachers can also have students prepare their arguments in a digital medium (such as PowerPoint or Keynote).
* Teachers can also project these materials onto a screen in order to cut down on paper use.

## Assessment

The rubric provided in Appendix C (p. 367) can be used to assess the argument crafted by each student at the end of the activity. To illustrate how the rubric can be used to score an argument written by a student, consider the following example. This sample, which was written by a high school sophomore, is strong in terms of content but is not very well written.

The molecules found in the hydrogen peroxide are altered but the substance in the liver is not altered when they are mixed. **We filled to flasks with hydrogen peroxide. A piece of liver was then added to the first flask and we saw a lot of bubbles from. After it stopped bubbling we took the piece of liver out of the flask and dropped it into the second flask and it started bubbling to. We then took fresh piece of liver and put into a third flask. Next, we added hydrogen peroxide from the first flask to the third flask and nothing happened.** This shows that the molcules in the liver don't change because we can reuse the liver but we can't resue the hydrogen peroxide. If both the molecules in the liver and the

# SECTION 2: EVALUATE ALTERNATIVES

## LIVER AND HYDROGEN PEROXIDE
### TEACHER NOTES 14

hydrogen peroxide where changed as a result of the reaction then we won't have been able to use either one.

The content of the example argument is rated as strong for several reasons. The student's claim (underlined) is sufficient (1/1), because it provides a complete and accurate (1/1) answer to the research question. The student included evidence (in bold) to support his claim that included data (1/1), an analysis of the data (1/1), and an interpretation of the analysis (1/1). The student, however, did not include a complete justification of the evidence (1/2) in his argument because he did not explain the relevance of the evidence (although he explains why the evidence supports his claim). He also did not include a challenge to an alternative explanation (0/2), The author, however, used phrases and terms that were consistent with the nature of science (2/2). Therefore, the overall content score for the sample argument should be 8 out of the 11 points possible, which makes it a strong argument in terms of content.

The students' writing, on the other hand, was poor. The organization of the argument was sufficient, because the arrangement of the sentences did not distract from the development of the main idea (1/1), but the argument was rather short. There were also several grammatical errors (0/1) and inappropriate use of writing conventions (0/1) in the argument. The overall mechanics score for the sample argument, therefore, is 1 out the 3 points possible.

## Standards Addressed in This Activity

This activity can be used to address the following dimensions outlined in the *A Framework for K–12 Science Education* (NRC 2012):

### Scientific Practices

- Planning and carrying out investigations
- Engaging in argument from evidence
- Obtaining, evaluating, and communicating information

### Crosscutting Concepts

- Cause and effect: Mechanism and explanation
- Energy and matter: Flows, cycles, and conservation
- Structure and function

### Life Sciences Core Ideas

- From molecules to organisms: Structures and processes

This activity can be used to address the following standards for literacy in science from the *Common Core State Standards for English Language Arts and Literacy* (NGA and CCSSO 2010):

## SECTION 2: EVALUATE ALTERNATIVES

### 14 LIVER AND HYDROGEN PEROXIDE
TEACHER NOTES

### Writing

- Text types and purposes
- Production and distribution of writing
- Research to build and present knowledge
- Range of writing

### Speaking and Listening

- Comprehension and collaboration
- Presentation of knowledge and ideas

## References

National Governors Association Center (NGA) for Best Practices, and Council of Chief State School Officers (CCSSO). 2010. *Common core state standards for English language arts and literacy*. Washington, DC: National Governors Association for Best Practices, Council of Chief State School.

National Research Council (NRC). 2012. *A framework for K–12 science education: Practices, crosscutting concepts, and core ideas*. Washington, DC: National Academies Press.

SECTION 2: EVALUATE ALTERNATIVES

# CELL SIZE AND DIFFUSION (DIFFUSION) 15

You have learned that virtually all living cells are dependent on the process of diffusion in order to obtain the essential nutrients they need in order to survive. As cells take in these nutrients, they break them down and use the resulting energy and molecular building blocks to make more cellular components. This causes a cell to grow by increasing in size. However, cells never get too big, even if the organism is rather large. Cells are always small. In other words, the cells of an ant and a horse are, on average, the same size; a horse just has a lot more of them.

These observations raise an interesting question: **Why are cells so small?**

Here are two potential answers to this question:

- **Explanation 1**: Cells that have a larger surface area to volume ratio (surface area divided by volume) are more efficient at diffusing essential nutrients.
- **Explanation 2**: The rate of diffusion (distance traveled divided by time) is related to cell size. Nutrients diffuse through small cells faster than they do in large cells.

## Getting Started

You can test the validity of these different explanations by constructing a model cell by using agar. Agar is a gel-like substance that you can cut into whatever shape or size you want. Agar is a useful material because chemicals can diffuse through it. Your teacher has also added a chemical indicator, which is called bromothymol blue, to this agar. When bromothymol blue comes in contact with an acid (such as vinegar) it turns from blue to yellow. This allows you to see how far an acid diffuses into your model cell over time.

You will have following materials available to use during your investigation:

- Bromothymol blue agar cubes
- Vinegar
- Beakers
- Stopwatch
- Ruler
- A plastic knife (to cut the agar into different-size cubes or cut the cubes open after they have soaked in the vinegar)

## SECTION 2: EVALUATE ALTERNATIVES

### 15 CELL SIZE AND DIFFUSION

**Safety notes**: Wear indirectly vented chemical-splash goggles, aprons, and gloves. Keep bromothymol blue away from flames. Wash hands with soap and water after activity.

With your group, determine which explanation provides the best answer to the research question. You can use as many of the supplies available to you to test your ideas. Make sure that you generate the evidence you will need to support your explanation as you work. You can record your method and any observations you make.

## Our Method

## Our Observations

## Argumentation Session

Once your group has decided which explanation is the most valid, or acceptable answer the research question, prepare a whiteboard that you can use to share and justify your ideas. Your whiteboard should include all the information shown in Figure 15.1.

To share your work with others, we will be using a round-robin format. This means that one member of the group will stay at your workstation to share your group's ideas while the other group members go to the other groups one at a time in order to listen to and critique the arguments developed by your classmates. Remember, as you critique the work of others, you need to decide if their conclusions are valid or acceptable based on the quality of their claim and how well they

## SECTION 2: EVALUATE ALTERNATIVES

### CELL SIZE AND DIFFUSION  15

are able to support their ideas. In other words, you need to determine if their argument is *convincing* or not. One way to determine if their argument is convincing is to ask them some of the following questions:

- How did you gather your data? Why did you decide to do it that way?
- How do you know your data is high quality (free from errors)?
- How did you analyze or interpret your data? Why did you decide to do it that way?
- Why does your evidence support your claim?
- Why did you decide to use that evidence? Why is your evidence important?
- How does your justification of your evidence fit with accepted scientific ideas?

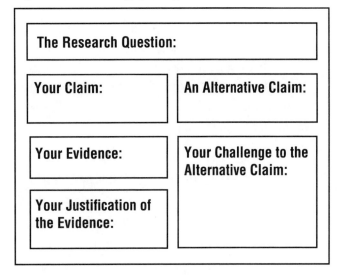

*Figure 15.1. Components of the Whiteboard*

## SECTION 2: EVALUATE ALTERNATIVES

Name_____ Date_____

# CELL SIZE AND DIFFUSION:
## What Is Your Argument?

In the space below, write a one- to three-paragraph argument to *support* the explanation that you think is the most valid or acceptable. Your argument must also include a *challenge* to at least one of the alternative explanations.

As you write your argument, remember to do the following:

- State the explanation you are trying to support
- Include genuine evidence (data + analysis + interpretation)
- Explain why the evidence is important and relevant
- State the explanation you are trying to refute
- Explain why the alternative explanation is invalid or unacceptable
- Organize your argument in a way that enhances readability
- Use a broad range of words including vocabulary that we have learned
- Correct grammar, punctuation, and spelling errors

SECTION 2: EVALUATE ALTERNATIVES

# CELL SIZE AND DIFFUSION
## TEACHER NOTES 15

## Purpose

The purpose of this activity is to help students understand passive transport in cells (i.e., diffusion and osmosis) and how the size of a cell can impact the time it takes for important molecules to diffuse through a cell. This activity also helps students learn how to engage in practices such as using planning and carrying out investigations, arguing from evidence, and communicating information. This activity, in addition, is designed to give students an opportunity to learn how to write in science and develop their speaking and listening skills, which are important goals for literacy in science (see Standards Addressed in This Activity for a complete list of the practices, crosscutting concepts, core ideas, and literacy skills that are aligned with this activity).

## The Content and Related Concepts

Cell membranes are selectively permeable, which means they can control the flow of substances into and out of a cell. Some substances can diffuse through membranes without work (i.e., passive transport). These substances will spread from areas of high concentration to areas of low concentration. Cells must remain small in order for passive transport to work well, because diffusion is a fairly slow process and a cell can grow too large for this process to work efficiently. When volume is too large relative to surface area, substances take longer to travel through the cell. Therefore, small cells with a high surface-area-to-volume ratio are more efficient at moving materials in and out of them than large cells with a low surface-area-to-volume ratio.

In this activity, students use model cells made of agar to explore the relationship between cell size and time it takes for molecules to diffuse through a cell. The agar cells contain a chemical indicator called bromothymol blue. This indicator turns yellow when it comes in contact with an acidic solution (such as vinegar). Therefore, the agar cells can be submerged in vinegar and students will be able to see how far the acid diffuses into the cell. Many students will assume that the "cell" with the largest surface area will be the most efficient at moving materials in and out. When the blocks are submerged in an acidic solution for several minutes, then removed and cut in half, students will be able to measure the penetration depths of the acid. This will allow them to see that the diffusion rate of the vinegar is equal regardless of the size of the cell. What is dramatically different is how much of the interior volume of each cube has been affected when the size of the cell changes. Students can be encouraged to calculate diffuse rate (in mm/sec.) and the percentage of each cell's interior volume that has turned yellow (or the volume of the cell that did not). This will give a fairly direct indi-

## SECTION 2: EVALUATE ALTERNATIVES

## 15 CELL SIZE AND DIFFUSION
TEACHER NOTES

cation of which cell is most likely to "survive." It is important for students to see that smaller cells are better able to move materials in and out (because they have a higher surface-area-to-volume ratio) and a cell could eventually grow to a size in which materials needed for metabolism could not diffuse fast enough for the cell to survive, or waste products could build up to toxic levels.

## Curriculum and Instructional Considerations

This activity is best used as part of a unit on cell structure and function. We recommend that it be used after students learn about diffusion, osmosis, and the nature of cell membranes. This activity is also a good way to teach students about experimental design, the control of variables, the difference between hypotheses (tentative explanations) and predictions (expected results), and other important terminology such as *independent* and *dependent variables*. Students will need a basic understanding of these important ideas in order to be able to collect meaningful data during the activity. The explicit discussion at the end of the activity should focus on relationships between cell size, surface area, volume, surface-area-to-volume ratio, and how passive transport is influenced by the surface-area-to-volume ratio of cells. Teachers can also show pictures of cells with microvilli, which are structures that increase surface area, to make passive transport even more efficient. The explicit discussion should also focus on at least one aspect of the nature of science or the nature of scientific inquiry.

For example, a teacher could discuss how scientific explanation must be consistent with observational evidence about nature, or how experiments are used to test explanations, the importance of controlling variables during an experiment, or the role of creativity and imagination in science using this activity as an illustrative example.

## Recommendations for Implementing the Activity

This activity takes between 100 and 150 minutes of instructional time to complete, depending on how a teacher decides to spend time in class. In Option A, the students are given time to complete all six stages of the lesson during class (see Figure 15.2). Stages 1 and 2 are completed on day 1, Stages 3 and 4 are completed on day 2, and Stages 5 and 6 are completed on day 3. This option for implementing the activity works best in schools where students are not expected to complete much homework or if students need to be encouraged to write more during the school day. It also provides less time for the argumentation sessions, which may or may not be a problem, depending on how comfortable the students are with argumentation. After all, some classes are more talkative than others. In Option B, students complete Stage 1 and begin Stage 2 of the lesson during class on day 1. The students then complete Stage 3 on day 2 of the lesson. Stages 4 and 5 are completed on day 3, and the final written argument (Stage 6) is then assigned as homework and returned the next day.

# SECTION 2: EVALUATE ALTERNATIVES

## CELL SIZE AND DIFFUSION 15
### TEACHER NOTES

*Figure 15.2. Two Options for Implementing the Activity*

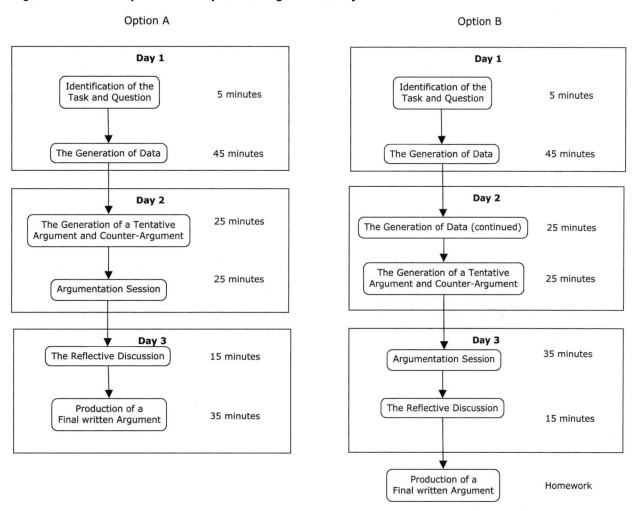

Prior to starting the activity, be sure to review with students the SDS for bromothymol blue and vinegar.

Table 15.1 (p. 188) provides information about the type and amount of materials needed to implement this activity in a class-

## SECTION 2: EVALUATE ALTERNATIVES

## 15 CELL SIZE AND DIFFUSION
TEACHER NOTES

room with 28 students in groups of four and groups of three.

To make the phenolphthalein agar (this recipe makes enough agar for 15 groups), follow the steps. **Safety note**: This recipe is for teachers only.

1. Mix 20 g of agar with 1 L of distilled or deionized water.
2. Heat almost to a boil. Stir frequently until solution is clear.
3. Remove from heat. As the agar mixture cools, add 0.1 g of bromothymol blue and stir. Note: If the mixture is not dark blue, then add more bromothymol blue. If the mixture is green or yellow, you will need to stir in drops of NaOH (or another base) until it turns blue. Wear safety goggles and gloves when handling NaOH.
4. Pour agar into a shallow tray to a depth of 3 cm and allow it to set overnight. A tray measuring 12 cm by 25 cm and at least 3 cm deep will accommodate 1 L of agar mixture. Volume adjustments may be necessary depending on the tray used.

*Table 15.1. Materials Needed to Implement the Activity in a Classroom of 28 Students*

| Material | Amount Needed With … | |
|---|---|---|
| | Groups of 3 | Groups of 4 |
| 3 cm × 3 cm × 5 cm block of bromothymol agar | 10 | 7 |
| 400 ml beaker | 10 | 7 |
| 200 ml of household white vinegar | 10 | 7 |
| Plastic knife | 9 | 6 |
| Stopwatch | 9 | 6 |
| Spoon or tongs | 9 | 6 |
| Ruler | 9 | 6 |
| Whiteboards (or chart paper)+ | 10 | 7 |
| Whiteboard markers (or permanent markers if using chart paper)+ | 20 | 14 |
| Copy of Student Pages (pp. 181–183)* | 28 | 28 |
| Copy of Student Page (p. 184)* | 28 | 28 |
| Copy of Appendix C (p. 367)* | 28 | 28 |

+ Teachers can also have students prepare their arguments in a digital medium (such as PowerPoint or Keynote).
* Teachers can also project these materials onto a screen in order to cut down on paper use.

# SECTION 2: EVALUATE ALTERNATIVES

## CELL SIZE AND DIFFUSION
### TEACHER NOTES 15

5. Cut the agar into 3 cm by 3 cm by 5 cm blocks, one per lab group.

The bromothymol blue agar will turn yellow when it is exposed to any acid. We suggest that students soak the cubes in common household white vinegar. Students can soak all their cubes in the same beaker of vinegar or they can soak each block in an individual beaker.

## Assessment

The rubric provided in Appendix C (p. 367) can be used to assess the argument crafted by each student at the end of the activity. To illustrate how the rubric can be used to score an argument written by a student, consider the following example. This sample, which was written by a high school junior, is weak in terms of content, but it is well written.

> <u>Cells are small because nutrients diffuse through small cells faster than they do in large cells.</u> When we submerged three different sized cubes (one that was 1 cm by 1 cm by 1 cm, one that was 2 cm by 2 cm by 2 cm, and one that was 3 cm by 3 cm by 3 cm) in the sodium hydroxide for 10 minutes each, **the smallest cube was pink all the way through it and the two larger cubes were not.** Therefore, rate of diffusion (distance traveled ÷ time) is related to cell size.

The content of the example argument is weak for several reasons. The student's claim (underlined) is insufficient (0/1) and inaccurate (0/1). The student includes data (1/1), a basic analysis of the data (1/1) and an interpretation of the analysis (1/1); however, the student does not include a justification of the evidence in his argument (0/2). He also does not provide an adequate challenge to an alternative explanation (0/2). The author uses appropriate terms (1/1) and phrases that are consistent with the nature of science (1/1). The writing mechanics are also sufficient. The organization of the argument is strong, because the arrangement of the sentences aid in the development of the main idea (1/1). There are also no punctuation (1/1) or grammatical errors (1/1) in the argument. The overall score for the sample argument, therefore, is 8 out the 14 points possible.

## Standards Addressed in This Activity

This activity can be used to address the following dimensions outlined in *A Framework for K–12 Science Education* (NRC 2012):

### Scientific Practices

- Developing and using models
- Planning and carrying out investigations
- Using mathematics and computational thinking
- Engaging in argument from evidence
- Obtaining, evaluating, and communicating information

### Crosscutting Concepts

- Cause and effect: Mechanism and explanation

## SECTION 2: EVALUATE ALTERNATIVES

### 15 CELL SIZE AND DIFFUSION
**TEACHER NOTES**

- Scale, proportion, and quantity
- Energy and matter: Flows, cycles, and conservation
- Structure and function

## Life Sciences Core Ideas

- From molecules to organisms: Structures and processes

This activity can be used to address the following standards for literacy in science from the *Common Core State Standards for English Language Arts and Literacy* (NGA and CCSSO 2010):

## Writing

- Text types and purposes
- Production and distribution of writing
- Research to build and present knowledge
- Range of writing

## Speaking and Listening

- Comprehension and collaboration
- Presentation of knowledge and ideas

## References

National Governors Association Center (NGA) for Best Practices, and Council of Chief State School Officers (CCSSO). 2010. *Common core state standards for English language arts and literacy*. Washington, DC: National Governors Association for Best Practices, Council of Chief State School.

National Research Council (NRC). 2012. *A framework for K–12 science education: Practices, crosscutting concepts, and core ideas*. Washington, DC: National Academies Press.

SECTION 2: EVALUATE ALTERNATIVES

# ENVIRONMENTAL INFLUENCE ON GENOTYPES AND PHENOTYPES (GENETICS) 16

One of the keys to understanding how traits are passed down from generation to generation is the law of segregation. According to this law, individuals carry two alleles for each trait but only pass down one of these alleles to their offspring. The law of segregation is extremely useful, because it allows scientists to predict the phenotype and genotype of offspring using Punnett squares if the mode of inheritance (such as dominant-recessive, co-dominance, sex-linked) for a particular trait is known. One thing that you might not have thought about as you learned about basic Mendelian genetics, however, is the role that the environment may or may not play in this process.

Often people believe that an individual's traits are determined solely by a person's genes and have nothing to do with environmental influences. But is this really the case? Can the environment influence the genotype or the phenotype of an organism?

This scenario raises an interesting question: **Can the environment influence the genotype or the phenotype of an organism such as the tobacco plant?**

Here are three potential answers to this question:

- **Explanation 1**: Some environmental factors, such as the amount or the intensity of light, have no influence on the genotype or phenotype of tobacco plants.

- **Explanation 2**: Some environmental factors, such as the amount or intensity of light, have no influence on the genotype of tobacco plants, but these factors can alter the phenotype of a tobacco plant as it grows.

- **Explanation 3**: Some environmental factors, such as the amount of light, can alter the genotype and the phenotype of tobacco plants.

## Getting Started

You will be given a vial of approximately 200 tobacco seeds. In tobacco plants, the allele for green leaves is dominant to the allele for white leaves. The seeds that you will be given were produced by pollinating a green (heterozygous) tobacco plant with the pollen from a tobacco plant that was also heterozygous for the green allele. Based on the genotypes of the parent tobacco plants, you should be able to use a Punnett square to predict the ratio of genotypes and phenotypes of the tobacco seeds in your vial. You can then design a controlled experiment to determine which explanation provides the best answer to the research question.

You can use the following materials during your experiment:

## SECTION 2: EVALUATE ALTERNATIVES

## 16  ENVIRONMENTAL INFLUENCE ON GENOTYPES AND PHENOTYPES

- Tobacco seeds (produced by breeding two green plants that were heterozygous for the white allele)
- Petri dishes (will serve as a grow chamber)
- Filter paper (you can keep the tobacco seeds wet using filter paper and the bottom of the petri dish)
- Petri dishes filled with agar, which provide a good medium for germinating seeds (these are optional)

**Safety notes**: Do not touch the light source as it can burn skin. Wash hands with soap and water upon completing the activity.

You can record your method and any observations you make in the spaces below.

## Our Method

## Our Observations

## Argumentation Session

Once your group has decided which explanation is the most valid or acceptable answer to the research question, prepare a whiteboard that you can use to share and justify your ideas. Your white-

## SECTION 2: EVALUATE ALTERNATIVES

# ENVIRONMENTAL INFLUENCE ON GENOTYPES AND PHENOTYPES 16

board should include all the information shown in Figure 16.1.

To share your work with others, we will be using a round-robin format. This means that one member of the group will stay at your workstation to share your group's ideas while the other group members go to the other groups one at a time in order to listen to and critique the arguments developed by your classmates. Remember, as you critique the work of others, you need to decide if their conclusions are valid or acceptable based on the quality of their claim and how well they are able to support their ideas. In other words, you need to determine if their argument is *convincing*

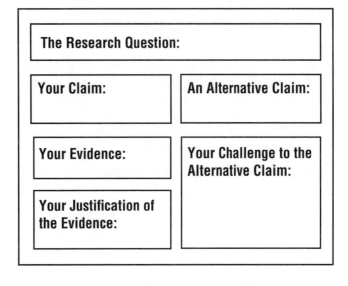

*Figure 16.1. Components of the Whiteboard*

or not. One way to determine if their argument is convincing is to ask them some of the following questions:

- How did you gather your data? Why did you decide to do it that way?
- How do you know your data is high quality (free from errors)?
- How did you analyze or interpret your data? Why did you decide to do it that way?
- Why does your evidence support your claim?
- Why did you decide to use that evidence? Why is your evidence important?
- How does your justification of your evidence fit with accepted scientific ideas?

## SECTION 2: EVALUATE ALTERNATIVES

Name_____ Date_____

# ENVIRONMENTAL INFLUENCE ON GENOTYPES AND PHENOTYPES:
## What Is Your Argument?

In the space below, write a one- to three-paragraph argument to *support* the explanation that you think is the most valid or acceptable. Your argument must also include a *challenge* to one of the alternative explanations. As you write your argument, remember to do the following:

- State the explanation you are trying to support
- Include genuine evidence (data + analysis + interpretation)
- Explain why the evidence is important and relevant
- State the explanation you are trying to refute
- Explain why the alternative explanation is invalid or unacceptable
- Organize your argument in a way that enhances readability
- Use a broad range of words including vocabulary that we have learned
- Correct grammar, punctuation, and spelling errors

SECTION 2: EVALUATE ALTERNATIVES

# ENVIRONMENTAL INFLUENCE ON GENOTYPES AND PHENOTYPES
## 16
### TEACHER NOTES

## Purpose

The purpose of this activity is to help students understand Mendelian genetics and how the environment can influence the phenotypes of organisms. This activity also helps students learn how to engage in practices such as using mathematics and computational thinking, using planning and carrying out investigations, arguing from evidence, and communicating information. This activity, in addition, is designed to give students an opportunity to learn how to write in science and develop their speaking and listening skills, which are important goals for literacy in science (see Standards Addressed in This Activity for a complete list of the practices, crosscutting concepts, core ideas, and literacy skills that are aligned with this activity).

## The Content and Related Concepts

Mendelian genetics is the basis for a great deal of modern research on inheritance. This model consists of three important ideas that are salient to this activity (Campbell and Reece 2002). First and foremost, the fundamental unit of inheritance is the gene, and alternative versions of a gene account for a large portion of the variation in inheritable characters. The gene for a particular inherited character, such as leaf color in tobacco plants, resides at a specific locus (position) on a specific chromosome (see Figure 16.2). Alleles are variants of a particular gene. In Figure 16.2, for example, the leaf color gene exists in two versions: the allele for green leaves and the allele for white leaves. Second, an organism inherits two alleles for each character, one from each

*Figure 16.2. Allele or Alternative Versions of a Gene*

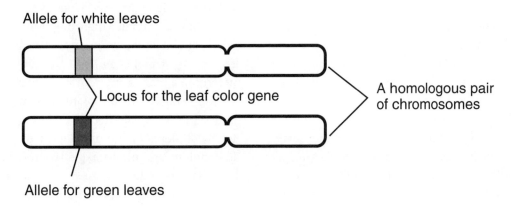

## SECTION 2: EVALUATE ALTERNATIVES

## 16 ENVIRONMENTAL INFLUENCE ON GENOTYPES AND PHENOTYPES
### TEACHER NOTES

parent. This is known as the organism's genotype. In Figure 16.2, the organism is heterozygous, because it has a green allele and a white allele. Third, if the two alleles differ, as is the case in Figure 16.2, then one is fully expressed in the organism's appearance (this version of the gene is called the dominant allele), while the other one has no noticeable effect on the organism's appearance (this version of the gene is called the recessive allele). This organism's phenotype, therefore, refers to the appearance of an organism and the phenotype of an organism is based on its genotype.

An inherited genotype, however, does not guarantee that an organism will have a specific phenotype in all cases. Instead, the appearance of an organism usually falls within a range of phenotypic possibilities because of the impact of the environment. The phenotypic range is called the norm of reaction for a specific genotype (Campbell and Reece 2002). There are cases in which the norm of reaction has no breadth whatsoever; that is, a given genotype mandates a very specific phenotype. An example of a gene that has a small norm of reaction is the MN blood type gene, which determines the type of antigens that found on a person's red blood cells. In contrast, a person's blood count of red and white cells often varies, depending on environmental factors such as altitude of one's home, the person's typical level of physical activity, and the presence or absence of an infection. As a result, the norm of reaction for the genes that control this trait is rather large. Generally, norms of reaction are the largest for polygenic traits (i.e., traits determined by multiple genes) and the smallest for traits that follow a simple dominant-recessive mode of inheritance. Geneticists often describe traits that have a large norm of reaction as multifactorial, which means that many factors (including genetic and environmental) interact to influence the phenotype (Campbell and Reece 2002).

In this activity, the students are given a vial of 200 tobacco seeds that were produced by crossing two tobacco plants that are heterozygous for the white color trait. The white allele is a recessive mutation, and therefore, only plants that inherit two copies of the white allele (homozygous recessive) will have the white genotype (these plants lack the ability to produce chlorophyll). In this case, the 200 tobacco seeds should have a genotypic ratio of 1:2:1 (i.e., a ratio of one homozygous dominant seed and two heterozygous seeds for every seed that is homozygous recessive), and if the environment has no influence on phenotype, the seeds should have a phenotypic ratio of 3:1 (i.e., a ratio of three plants with green leaves to every one plant that has white leaves). The phenotype of plants, however, can be influenced by environmental factors such as the amount or intensity of the available light. If tobacco plants are not exposed to any light after they germinate, then they will not turn green and will therefore appear to be albino regardless of their genotype. Yet, the absence of light in a plant's environment has no impact on the genotype of the plant. The most valid or acceptable explanation of the three provided to the students at the opening of the activity is therefore the second explanation (i.e., some environmental factors, such as the amount or intensity of light, have no influence on the

## SECTION 2: EVALUATE ALTERNATIVES

### ENVIRONMENTAL INFLUENCE ON GENOTYPES AND PHENOTYPES
#### TEACHER NOTES 16

genotype of tobacco plants but these factors can alter the phenotype of a tobacco plant as it grows).

## Curriculum and Instructional Considerations

This activity is best used as part of a unit on genetics. It should be used as a way to introduce students to the environmental influences on traits but only after students have a solid grasp of basic Mendelian genetics. At the very minimum, students should understand the concept of genes, the law of segregation and several different modes of inheritance such as co-dominance, incomplete dominance, and simple dominant-recessive. Students should also understand how to use Punnett squares to predict the phenotypic ratios of offspring from a cross. Students will need a basic understanding of these ideas in order to be able to analyze and interpret the data that will collect during the activity.

Teachers can also use this activity to introduce student to statistical hypothesis testing and the chi-square test (which is employed to test the difference between an actual sample and an expected sample) or encourage students to use a chi-square test as part of their analysis if the students have already learned how to do it earlier in the year. The explicit discussion at the end of the activity should focus on the ways that the environment can influence a phenotype (and in some cases the genotype of the next generation), or on the ongoing debate between the relevant importance of nature or nurture in the traits of individuals. The explicit discussion should also focus on at least one aspect of the nature of science or the nature of scientific inquiry. For example, a teacher could discuss how scientific explanation must be consistent with observational evidence about nature, and must make accurate predictions, when appropriate, about systems being studied or how mathematics plays an important role in scientific inquiry using this activity as an illustrative example.

## Recommendations for Implementing the Activity

This activity takes between 100 and 150 minutes of instructional time to complete, depending on how a teacher decides to spend time in class. This activity also requires approximately one week for the seeds to germinate and produce leaves. In Option A, the students are given time to complete all six stages of the lesson during class with a few days between Stages 2 and 3 to allow time for the seeds to germinate and produce leaves (see Figure 16.3, p 198). Stages 1 and 2 are completed on day 1, Stages 3 and 4 are completed on day 2, and Stages 5 and 6 are completed on day 3. This option for implementing the activity works best in schools where students are not expected to complete much homework or if students need to be encouraged to write more during the school day. It also provides less time for the argumentation sessions, which may or may not be a problem, depending on how comfortable the students are with argumentation. In Option B, students complete Stage 1 and begin Stage 2 of the lesson during class on day 1. The students

# SECTION 2: EVALUATE ALTERNATIVES

## 16 ENVIRONMENTAL INFLUENCE ON GENOTYPES AND PHENOTYPES
### TEACHER NOTES

*Figure 16.3. Two Options for Implementing the Activity*

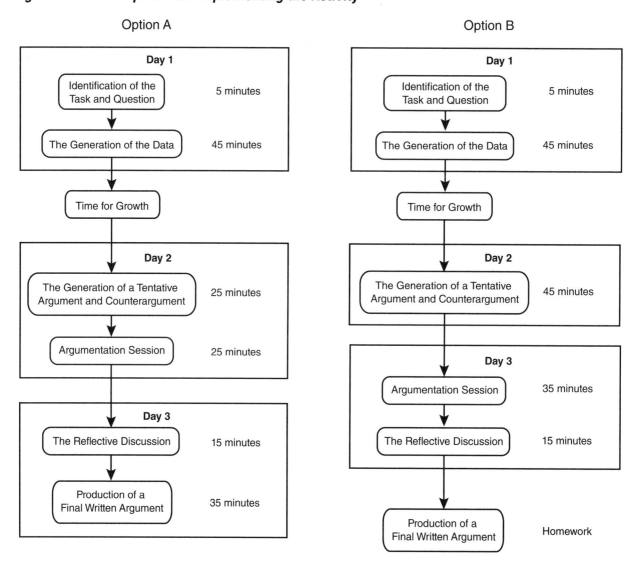

# SECTION 2: EVALUATE ALTERNATIVES

## ENVIRONMENTAL INFLUENCE ON GENOTYPES AND PHENOTYPES
### TEACHER NOTES  16

*Table 16.1. Materials Needed to Implement the Activity in a Classroom of 28 Students*

| Material | Amount Needed With … | |
|---|---|---|
| | Groups of 3 | Groups of 4 |
| Vials of 200 tobacco seeds (3:1 green to white available from biological supply companies such as Wards Biological or Carolina Biological Supply) | 9 | 6 |
| Petri dishes (four per group) | 36 | 24 |
| Filter paper (cut to fit the bottom of the petri dishes) | 36 | 24 |
| Light source | 1 | 1 |
| Permanent markers (used to label the petri dishes) | 9 | 6 |
| Whiteboards (or chart paper)⁺ | 10 | 7 |
| Whiteboard markers (or permanent markers if using chart paper)⁺ | 20 | 14 |
| Copy of Student Pages (pp. 191–193)* | 28 | 28 |
| Copy of Student Page (p. 194)* | 28 | 28 |
| Copy of Appendix C (p. 367)* | 28 | 28 |

⁺ Teachers can also have students prepare their arguments in a digital medium (such as PowerPoint or Keynote).
* Teachers can also project these materials onto a screen in order to cut down on paper use.

then complete Stage 3 on day 2 of the lesson. Stages 4 and 5 are completed on day 3, and the final written argument and counterargument (Stage 6) is then assigned as homework and returned the next day.

Table 16.1 provides information about the type and amount of materials needed to implement this activity in a classroom with 28 students in groups of four and groups of three.

## Assessment

The rubric provided in Appendix C (p. 367) can be used to assess the arguments crafted by each student at the end of the activity. To illustrate how the rubric can be used to score an argument written by a student, consider the following example. This sample, which was written by a high school senior, is a strong argument both in terms of content and mechanics.

<u>The amount of light available clearly has no influence on the genotype of tobacco plants but it can alter the phenotype of a tobacco plant as it grows.</u> **The results of our experiment indicate that 78 of the 100 tobacco plants (78%) that were grown in full light for five days turned green while none of the plants (0%) that were**

**SCIENTIFIC ARGUMENTATION IN BIOLOGY: 30 CLASSROOM ACTIVITIES**  199

## SECTION 2: EVALUATE ALTERNATIVES

### 16 ENVIRONMENTAL INFLUENCE ON GENOTYPES AND PHENOTYPES
TEACHER NOTES

grown in the dark for the same amount of time turned green. **There was no significant difference in the number of observed green plants from what we predicted, $X^2(1) = 0.25$, $p = 0.61$ in the light environment but there was a significant difference in the dark environment, $X^2(1) = 120$, $p < 0.001$. We then took the 100 plants that were grown in the dark and moved them to a well lit spot for three days. 43 of the 70 plants that were still alive began to turn green again (61%).** These results suggest that the tobacco plants we used displayed an expected phenotypic ratio when they were grown in the light but not when grown in the dark. The environment clearly had an influence on the phenotype of these plants. These results also support the idea that the environment does not act on the genotype of the plants because we were able to reverse the effect of growing them in the dark. The plants would not have been able to turn from white to green if their genotype changed.

The example argument is strong for several reasons. The student's claim (underlined) is sufficient (1/1) because it provides a complete and accurate answer to the research question (1/1). The student uses genuine evidence (in bold) to support the claim, because there is data (1/1), an analysis of the data (1/1), and an interpretation of the analysis (1/1). The student includes a complete justification of the evidence in his argument, because he does explain why the evidence is important (1/1) by linking it to a specific principle, concept, or underlying assumption (1/1). He also provides an adequate challenge to an alternative explanation by making the other viewpoint explicit (1/1) and providing a reason why it is invalid (1/1). The author also uses appropriate terms (1/1) and phrases that are consistent with the nature of science (1/1). The writing mechanics are also effective. The organization of the argument is strong because the arrangement of the sentences aid in the development of the main idea (1/1). There are also no punctuation (1/1) or grammatical errors (1/1) in the argument. The overall score for the sample argument, therefore, is 14 out the 14 points possible.

## Standards Addressed in This Activity

This activity can be used to address the following dimensions outlined in *A Framework for K–12 Science Education* (NRC 2012):

### Scientific Practices

- Planning and carrying out investigations
- Using mathematics and computational thinking
- Engaging in argument from evidence
- Obtaining, evaluating, and communicating information

### Crosscutting Concepts

- Patterns

## SECTION 2: EVALUATE ALTERNATIVES

### ENVIRONMENTAL INFLUENCE ON GENOTYPES AND PHENOTYPES
#### TEACHER NOTES 16

- Cause and effect: Mechanism and explanation
- Scale, proportion, and quantity
- Energy and matter: Flows, cycles, and conservation
- Structure and function

## Life Sciences Core Ideas

- Heredity: Inheritance and variation of traits

This activity can be used to address the following standards for literacy in science from the *Common Core State Standards for English Language Arts and Literacy* (NGA and CCSSO 2010):

## Writing

- Text types and purposes
- Production and distribution of writing
- Research to build and present knowledge
- Range of writing

## Speaking and Listening

- Comprehension and collaboration
- Presentation of knowledge and ideas

## References

Campbell, N., and J. Reece. 2002. *Biology*. 6th ed. San Francisco, CA: Benjamin Cummings.

National Governors Association Center (NGA) for Best Practices, and Council of Chief State School Officers (CCSSO). 2010. *Common core state standards for English language arts and literacy*. Washington, DC: National Governors Association for Best Practices, Council of Chief State School.

National Research Coucnil (NRC). 2012. *A framework for K–12 science education: Practices, crosscutting concepts, and core ideas*. Washington, DC: National Academies Press.

## SECTION 2: EVALUATE ALTERNATIVES

# HOMINID EVOLUTION (MACROEVOLUTION) 17

Fossils of hominid skulls found around the world and dated as far back as six million years ago show distinct differences that suggest that there were at least a dozen different species of hominids in our history. The fossils also have many distinct similarities that suggest that there is an evolutionary pathway that would indicate an ancestral relationship between the fossils. We can use fossils to categorize one species of hominid from another, and we can date the age of fossils with a high degree of precision, but our ancestral history is not as easy to determine, especially since, like all creatures, no two hominids were alike.

Figure 17.1 provides information about several different types of hominid fossils that have been found in different locations around the globe. The figure shows the time frame the hominid was present on Earth based on the age of the fossils that have been uncovered at this point in time.

### Figure. 17.1. Hominid Skull Fossils by Age

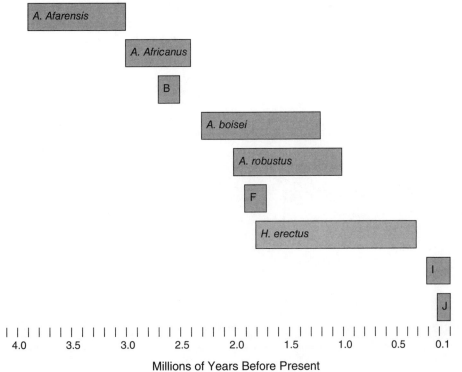

*Note:* Species B is *P. aethiopicus*; species F is *H. habilis*, species I is *H. neanderthalersis*, species J is *H. Sapiens*.

# SECTION 2: EVALUATE ALTERNATIVES

## 17 HOMINID EVOLUTION

These observations raise an interesting question: **What is the phylogenetic relationship of this group of hominids?**

Here are three potential answers to this question:

## Explanation 1

*Figure 17.2. Hominid Phylogenetic Relationship Version A*

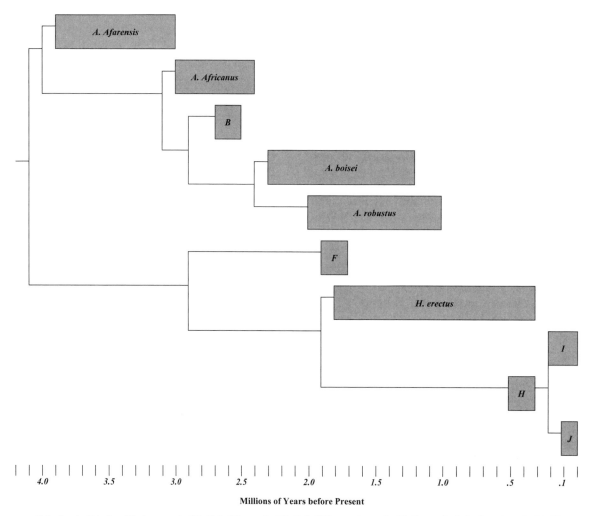

*Note*: Species B is *A. aethiopicus*; species F is *H. habilis*, species H is *H. heidelbergensis*, species I is *H. neanderthalersis*, and species J is *H. sapien*

## SECTION 2: EVALUATE ALTERNATIVES

### HOMINID EVOLUTION 17

## Explanation 2

### Figure 17.3. Hominid Phylogenetic Relationship Version B

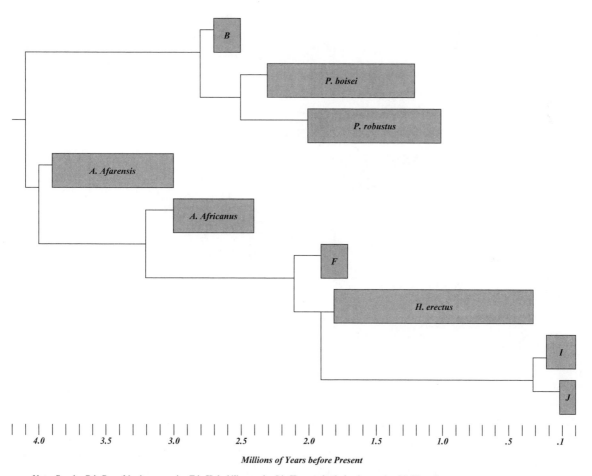

*Note: Species B is P. aethiopicus; species F is H. habilis, species I is H. neanderthalersis, species J is H. sapiens*

## SECTION 2: EVALUATE ALTERNATIVES

## 17 HOMINID EVOLUTION

**Explanation 3**

*Figure 17.4. Hominid Phylogenetic Relationship Version C*

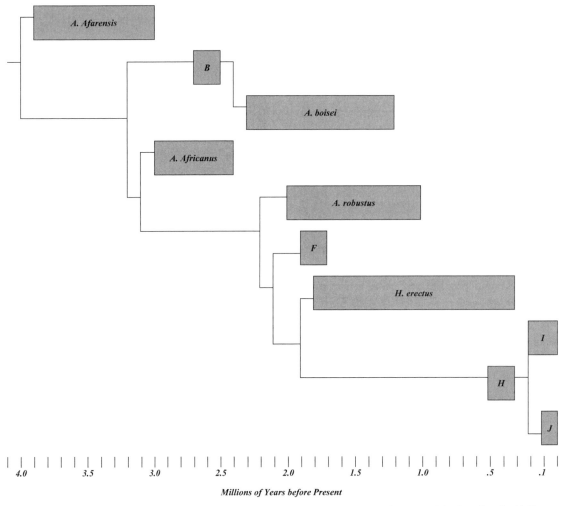

*Note: Species B is A. aethiopicus; species F is H. habilis, species H is H. heidelbergensis, species I is H. neanderthalersis, and species J is H. sapiens*

# SECTION 2: EVALUATE ALTERNATIVES
## HOMINID EVOLUTION 17

## Getting Started

Notice that there are additional species of hominid added to some of the explanations. The presence of additional species in some of the explanations is the result of disagreements among scientists about how certain types of fossils should be classified. A few species of hominid also have different names in some of the explanations for the same reason. There are also question marks in two of the explanations. The question marks represent species that some scientists think existed at that time but have yet to find any fossils.

Your goal in this activity is to determine which explanation provides the best answer to the research question (or develop one of your own if you are dissatisfied with all three explanations).

You can use the following materials to test the merits of these three explanations:

- *Homo habilis* skull
- *Homo erectus* skull
- *Australopithecus africanus* skull
- *Homo neanderthalensis* skull
- *Australopithecus* (or *Paranthropus*) *boisei* skull
- *Australopithecus* (or *Paranthropus*) *robustus* skull
- *Australopithecus afarensis* skull
- *Homo sapiens* skull
- *Pan troglodytes* (Chimpanzee) skull
- *Gorilla gorilla* skull
- Calipers, protractors, and other measurement tools

You can use the following resource:

- Smithsonian National Museum of Natural History online 3-D collection of hominid fossils: *http://humanorigins.si.edu/evidence/3d-collection*

## SECTION 2: EVALUATE ALTERNATIVES

### 17 HOMINID EVOLUTION

Use all the information available to evaluate the explanations. Make sure that you generate the evidence you will need to support the explanation that you feel is most valid and to refute the other explanations as you work. You can record your method and any observations you make in the spaces below.

## Our Method

## Our Observations

## Argumentation Session

Once your group has decided which explanation is the most valid or acceptable answer to the research question, prepare a whiteboard that you can use to share and justify your ideas. Your whiteboard should include all the information shown in Figure 17.5.

To share your work with others, we will be using a round-robin format. This means that one member of the group will stay at your workstation to share your group's ideas while the other group members go to the other groups one at a time in order to listen to and critique the arguments developed by your classmates. Remember, as you critique the work of others, you need to decide if their conclusions are valid or acceptable based on the quality of their claim and how well they are able to support their ideas. In other words, you need to determine if their argument is *convincing* or not. One way to determine if their argument is convincing is to ask them some of the following questions:

## SECTION 2: EVALUATE ALTERNATIVES
### HOMINID EVOLUTION 17

- How did you gather your data? Why did you decide to do it that way?
- How do you know your data is high quality (free from errors)?
- How did you analyze or interpret your data? Why did you decide to do it that way?
- Why does your evidence support your claim?
- Why did you decide to use that evidence? Why is your evidence important?
- How does your justification of your evidence fit with accepted scientific ideas?

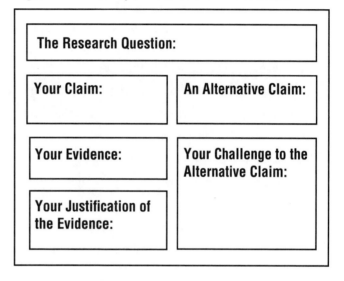

*Figure 17.5. Components of the Whiteboard*

### SECTION 2: EVALUATE ALTERNATIVES

Name_____  Date_____

# HOMINID EVOLUTION:
## What Is Your Argument?

In the space below, write a one- to three-paragraph argument to *support* the explanation that you think is the most valid or acceptable. Your argument must also include a *challenge* to one of the alternative explanations. As you write your argument, remember to do the following:

- State the explanation you are trying to support
- Include genuine evidence (data + analysis + interpretation)
- Explain why the evidence is important and relevant
- State the explanation you are trying to refute
- Explain why the alternative explanation is invalid or unacceptable
- Organize your argument in a way that enhances readability
- Use a broad range of words including vocabulary that we have learned
- Correct grammar, punctuation, and spelling errors

SECTION 2: EVALUATE ALTERNATIVES

# HOMINID EVOLUTION
## TEACHER NOTES 17

## Purpose

The purpose of this activity is to help students understand hominid evolution and to recognize the challenges of identifying evolutionary relationships between hominid species using fossils. This activity also helps students learn how to engage in practices such as planning and carrying out investigations, arguing from evidence, and communicating information. In addition, this activity is designed to give students an opportunity to learn how to write in science and develop their speaking and listening skills, which are important goals for literacy in science (see Standards Addressed in This Activity for a complete list of the practices, crosscutting concepts, core ideas, and literacy skills that are aligned with this activity).

## The Content and Related Concepts

Our family lineage, the hominids, diverged from the chimpanzee lineage about six million years ago. After that initial split, the hominid lineage did not slowly evolve into *Homo sapiens*. Instead, the early hominid lineage gave rise to many different kinds of hominids that are now extinct. By examining fossils of these extinct relatives, scientists have learned a great deal about how this complex hominid tree evolved and how modern humans came to exist.

Fossils are the remains of living organisms that have been preserved from ancient times in rock, amber, or by some other means. It is a common misconception that fossils are formed only from mineralized bones. Fossils can also be imprints created by living organisms. More recently, new techniques have also revealed the preservation of molecular and cellular fossils (PBS 2001). The fossil record is somewhat biased because of the rates of deterioration for different types of matter. For example, since bones and teeth do not deteriorate as easily and quickly as soft tissue and skin, we are more likely to find fossils of life forms that are made of bones than those that aren't (i.e., we find more dinosaur bones than jelly fish fossils). The fossil record is also biased because of the many different types of environments and the requirements for fossilization. Fossilization requires specific environments that prevent deterioration of cells, and some environments have faster rates of erosion than others. For example, water erodes matter quickly, so we are more likely to find fossils in arid deserts than we would in mountainous environments. Furthermore, some areas have been searched for fossils, while others have not, either because of social and governmental barriers or because of ease of accessibility (O'Neil 2010). It is the biases and the missing information that further muddies the understanding of the evolutionary relationships among extinct species.

Scientists use the characteristics and the location of fossil specimens to develop

## SECTION 2: EVALUATE ALTERNATIVES

## 17 HOMINID EVOLUTION
### TEACHER NOTES

explanations about the phylogenetic relationships between extinct species of hominids. These characteristics may include spatial and size relationships between the eyes, nose, and mouth, which can provide clues about how dependent a type of hominid was on a particular sense such as sight or smell. The relationship between size, shape, and location of bones gives some indication of muscle size and nature of locomotion in a type of hominid. The kinds of teeth, worn patterns, and relative jaw size of a hominid also provide information about the kinds of foods that might have been eaten. Scientists can even determine the ages of the hominid fossil samples at the time of death by looking at the extent of tooth erosion and bone growth plates.

In this activity, the students will use replica skulls (or the Smithsonian National Museum of Natural History online 3-D collection of hominid fossils) to evaluate three different explanations for the phylogenetic relationships of several extinct hominid species. The three figures provided on the student handout represent three different explanations for the phylogenetic relationship of hominids that have been proposed by different scientists. These different explanations, therefore, illustrate some of the major disagreements in the field about the phylogenetic relationships of extinct hominids and modern humans.

These disagreements arise because interpretations of the characteristics of extinct hominids are based on incomplete fossil specimens, and these interpretations, as a result, are often contentious. Scientists, for example often disagree about how a specimen should be classified or what a particular characteristic of a specimen suggests about the appearance or behavior of a hominid.

Figure 17.6 represents the current and one of the most widely accepted views of the phylogenetic relationships of hominids and when various traits first evolved. This phylogenetic tree is based on information provided in the book *The Tangled Bank: An Introduction*

*Figure 17.6. Phylogenetic Relationships of Hominids*

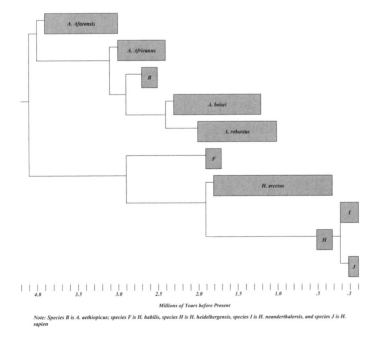

Note: Species B is A. aethiopicus; species F is H. habilis, species H is H. heidelbergensis, species I is H. neanderthalersis, and species J is H. sapien

## SECTION 2: EVALUATE ALTERNATIVES

### HOMINID EVOLUTION
### TEACHER NOTES 17

*to Evolution* (Zimmer 2009). It is important to note, however, that this phylogenetic tree is not definitive, and it will likely change in the future as new hominid fossils are discovered and as scientists reinterpret current specimens in light of theoretical advances.

## Curriculum and Instructional Considerations

This activity is best used during a unit of evolution. Teachers can use this activity to teach students about cladistics, human evolution, and challenges associated with identifying phylogenetic relationships based on the fossil record. This activity is also an effective way to address student misconceptions about evolution and the nature of science or the nature of scientific inquiry. Therefore, the explicit discussion at the end of the activity should focus on common misconceptions related to evolution in general and the evolution of humans in particular. For example, the teacher can point out the common misconception that humans evolved from chimpanzees and then explain why this idea is a misconception based on the information in the activity. The explicit discussion should also focus on at least one aspect of the nature of science or the nature of scientific inquiry. For example, a teacher could discuss how scientific explanation must be consistent with observational evidence about nature, or how inferences and observations are different, and how the interpretation of data is a subjective process using this activity as an illustrative example.

## Recommendations for Implementing the Activity

This activity takes between 150 and 200 minutes of instructional time to complete, depending on how a teacher decides to spend time in class. In Option A, the students are given time to complete all six stages of the lesson during class (see Figure 17.7). Stages 1 and 2 are completed on day 1. Stages 3 and 4 are completed on day 2, and Stages 5 and 6 are completed on day 3. This option for implementing the activity works best in schools where students are not expected to complete much homework or if students need to be encouraged to write more during the school day. It also provides less time for the argumentation sessions. In Option B, students complete Stage 1 and begin Stage 2 of the lesson during class on day 1. The students then complete Stage 2 on day 2 of the lesson. Stages 4 and 5 are completed on day 3, and the final written argument (Stage 6) is then assigned as homework and returned the next day.

Teachers should suggest a list of skull characteristics to examine during the introduction of the activity. These characteristics include the presence or absence of a supraorbital browridge, sagittal crest, and a canine diastema. In addition to looking to determine if a particular feature can be found on the skull, students can also be encouraged to measure the length of the maxilla, height and breadth of the nasal opening, the combined width of the four incisors, and the combined length of the two premolars and three molars. When students focus on these characteristics, it is often easier for them to describe how the skulls are different from one another.

## SECTION 2: EVALUATE ALTERNATIVES

### 17 HOMINID EVOLUTION
### TEACHER NOTES

*Figure 17.7. Two Options for Implementing the Activity*

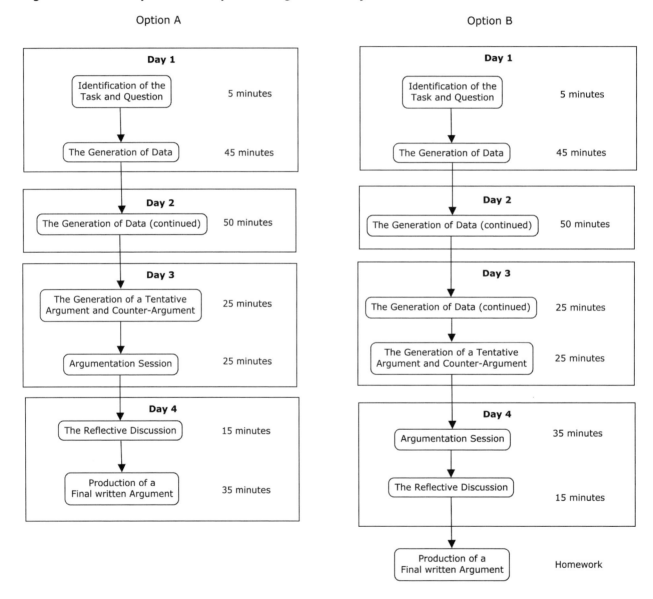

## SECTION 2: EVALUATE ALTERNATIVES

## HOMINID EVOLUTION
### TEACHER NOTES 17

Table 17.1 provides information about the type and amount of materials needed to implement this activity in a classroom with 28 students in groups of four and groups of three. It is important to note that teachers do not need to purchase a set of skulls for each group; one complete set of the 10 skulls is all that is needed. Each group can be given one skull at a time for the purposes of collecting data; the skulls can simply be rotated from group to group. This activity can also be completed without the skulls by having the students access the Smithsonian National Museum of Natural History online 3-D collection of hominid fossils at *http://humanorigins.si.edu/evidence/3d-collection*.

*Table 17.1. Materials Needed to Implement the Activity in a Classroom of 28 Students*

| Material | Amount Needed With ... | |
|---|---|---|
| | Groups of 3 | Groups of 4 |
| *Homo habilis* skull | 1 | 1 |
| *Homo erectus* skull | 1 | 1 |
| Australopithecus africanus skull | 1 | 1 |
| Homo Neanderthalersis skull | 1 | 1 |
| Australopithecus (or *Paranthropus*) boisei skull | 1 | 1 |
| Australopithecus (or *Paranthropus*) robustus skull | 1 | 1 |
| Australopithecus afarensis skull | 1 | 1 |
| *Homo sapiens* skull | 1 | 1 |
| Pan troglodytes (Chimpanzee) skull | 1 | 1 |
| *Gorilla gorilla* (skull) | 1 | 1 |
| Calipers, protractors, and other measurement tools | 9 | 7 |
| A computer with internet access | 9 | 7 |
| Whiteboards (or chart paper)⁺ | 10 | 7 |
| Whiteboard markers (or permanent if using chart paper)⁺ | 20 | 14 |
| Copy of Student Pages (pp. 203–209)* | 28 | 28 |
| Copy of Student Page (p. 210)* | 28 | 28 |
| Copy of Appendix C (p. 367)* | 28 | 28 |

⁺ Teachers can also have students prepare their arguments in a digital medium (such as PowerPoint or Keynote).
* Teachers can also project these materials onto a screen in order to cut down on paper use.

## SECTION 2: EVALUATE ALTERNATIVES

### 17 HOMINID EVOLUTION
### TEACHER NOTES

## Assessment

The rubric in Appendix C (p. 367) can be used to assess the arguments crafted by each student at the end of the activity. The rubric includes categories for the adequacy and conceptual quality of the claim, the appropriate use of evidence, the sufficiency of the rationale, and the overall quality of the writing. We strongly recommend that teachers use the Comments or Suggestions sections to give students detailed feedback so they will understand what they did wrong, why it is wrong, and ways they can improve their performance next time. To illustrate how to score the arguments, consider the following example:

*Our claim:* We think that with the information that we were provided and the 3 possible choices for hominid evolution, that the most likely explanation is number 1.

*Our method:* We looked at the 3 choices and tried to find one place that was different in all three pathways to focus on. We thought that if we could figure out that one difference then we could make a choice. In the three explanations the Australopithecus robustus (AR) is different for all three. In explanation 1 AR is directly related to Australopithecus aethiopicus (AAE). In explanation 2 AR is related to both AR and Australopithecus africanus (AAF). In explanation 3 AR is related to only AAF and does not have a relation to Australopithecus boisei (ABO). We looked at the skull features of all four species to decide who AR was most like.

*Our evidence:* AR is least like AAF for the skull features because AR has a pretty big ridge on top but AAF is really smooth and round and the head is bigger. Also, AAF has a bigger forehead. AAE and ABO both have ridges and they are smaller skulls and they have small foreheads. Also, the eyes of AR are more like AAE, then like ABO and least like AAF. The skull features are important because they can tell how smart the species were and they can tell how they used their senses. AAF was probably the smartest and AR was probably about as smart as AAE and ABO with a little difference between ABO. (Smartness: AR≤AAE<ABO<AAF). This seems like a good way to see evolutionary changes. This would show that AR is most closely related to AAE and that AR is least closely related to AAF.

*Our challenge:* We said earlier that AR was furthest in relationship to AAF because of their size of brain (skull) so this would mean that it is less likely that explanation 3 would be right than explanation 2. We also said that based on their food that AR was similar to ABO so that would mean that explanation 2 is more likely than 3 even though we still think explanation 1 is right. Because sometimes species will crossbreed if

## SECTION 2: EVALUATE ALTERNATIVES

### HOMINID EVOLUTION
### TEACHER NOTES 17

they are in similar places and using similar resources, explanation 2 is not impossible but then we thought that AR would be more similar to AAF. Since AAF is so different from AR compared to ABO, we don't think that explanation 2 is likely. The reason we think it is the least likely is because AAF is really more like us than any of the other three as far as skull features go and smartness. We think that supports that explanation 3 is least likely since AAF is not really so directly along the path for our ancestors.

The example argument is strong for several reasons. The group's claim (underlined) is sufficient (1/1) but not accurate (0/1). The group uses genuine evidence (in bold) to support the claim, because there is data (1/1), an analysis of the data (1/1), and an interpretation of the analysis (1/1). The group includes a vague justification of the evidence in the argument; they explain why the evidence is important (1/1) but do not attempt to link it to a specific principle, concept, or underlying assumption (0/1). However, the group does provide an adequate challenge to an alternative explanation by making another explanation explicit (1/1) and then providing a reason for why it is invalid (1/1). This group also uses appropriate terms (1/1) and phrases that are consistent with the nature of science (1/1). The writing mechanics are also effective. The organization of the argument is strong, because the arrangement of the sentences aid in the development of the main idea (1/1). There are also no major punctuation (1/1) or grammatical errors (1/1). The overall score for the sample argument, therefore, is 12 out the 14 points possible.

## Standards Addressed in This Activity

This activity can be used to address the following dimensions outlined in *A Framework for K–12 Science Education* (NRC 2012):

### Scientific Practices

- Planning and carrying out investigations
- Engaging in argument from evidence
- Obtaining, evaluating, and communicating information

### Crosscutting Concepts

- Cause and effect: Mechanism and explanation
- Energy and matter: Flows, cycles, and conservation
- Structure and function

### Life Sciences Core Ideas

- From molecules to organisms: Structures and processes
- Ecosystems: Interactions, energy, and dynamics

This activity can be used to address the following standards for literacy in science from the *Common Core State Standards for English Language Arts and Literacy* (NGA and CCSSO 2010):

## SECTION 2: EVALUATE ALTERNATIVES

## 17 HOMINID EVOLUTION
TEACHER NOTES

### Writing

- Text types and purposes
- Production and distribution of writing
- Research to build and present knowledge
- Range of writing

### Speaking and Listening

- Comprehension and collaboration
- Presentation of knowledge and ideas

## References

Institute of Human origins [IHO]. 2008. Becoming human. IHO. *www.becominghuman.org/node/interactive-documentary*.

National Governors Association Center (NGA) for Best Practices, and Council of Chief State School Officers (CCSSO). 2010. *Common core state standards for English language arts and literacy*. Washington, DC: National Governors Association for Best Practices, Council of Chief State School.

National Research Council (NRC). 2012. *A framework for K–12 science education: Practices, crosscutting concepts, and core ideas*. Washington, DC: National Academies Press.

O'Neil, D. 2010. Interpreting the fossil records. *http://anthro.palomar.edu/time/time_1.htm*.

Public Broadcast System (PBS). 2001. Online lessons for students: Learning evolution. WGBH Educational Foundation and Clear Blue Sky Productions, Inc. *www.pbs.org/wgbh/evolution/educators/lessons/lesson5/act2.html*.

Ridley, M. 2003. Evolution. 3rd ed. Wiley-Blackwell Publishing. *www.blackwellpublishing.com/ridley*.

Zimmer, C. 2009. *The tangled bank: An introduction to evolution*. Greenwood Village, CO: Roberts and Company.

## SECTION 2: EVALUATE ALTERNATIVES

# PLANTS AND ENERGY (RESPIRATION AND PHOTOSYNTHESIS) 18

We all know that we need oxygen in order to survive. We know this because the air we inhale contains approximately 21% oxygen ($O_2$), but the air we exhale only contains approximately 15% oxygen. We use the oxygen that we get from the air to convert sugar into energy. In fact, we can only survive for a few minutes without oxygen—it is *that* important.

We also know that other animals, such as dogs, birds, and even whales, use $O_2$ to produce energy. We can see them inhale and exhale air out of their lungs just as we do. Even animals that don't have lungs, such as snails and crickets, use oxygen to produce energy and give off $CO_2$ as a waste product. It is a unifying characteristic of all animals.

These observations and underlying explanations raise an interesting question: Do plants use $O_2$ to convert the sugar (which they produce using photosynthesis) into energy and release $CO_2$ as a waste product as animals do?

Here are three potential answers to this question:

- **Explanation 1**: Plants do not use oxygen as we do. Plants only take in carbon dioxide and give off oxygen as a waste product because of photosynthesis. This process produces all of the energy a plant needs, so they do not need oxygen at all.

- **Explanation 2**: Plants take in carbon dioxide during photosynthesis in order to make sugar, but they also use oxygen to convert the sugar into energy. As a result, plants release carbon dioxide as a waste product all the time just as animals do.

- **Explanation 3**: Plants release carbon dioxide all the time because they are always using oxygen to convert sugar to energy just as animals do. Plants, however, also take in carbon dioxide and release oxygen when exposed to light.

## Getting Started
You can use the following materials to test these three explanations:

- Elodea (an aquatic plant with leaves)
- Aquatic snails
- Water
- Test tubes
- Rubber stoppers (that fit tightly in the test tubes) or parafilm

## SECTION 2: EVALUATE ALTERNATIVES

### 18  PLANTS AND ENERGY

- Two large beakers (to hold the test tubes)
- Light source (and a dark place to store some of your test tubes)
- Bromthymol blue (BTB) (BTB is a pH indicator, which means it changes color as the pH of a liquid varies. It is yellow in acidic conditions, green in neutral conditions, and blue in basic conditions. When $CO_2$ reacts with water, it produces a weak acid—the more $CO_2$, the more acidic the water gets.)
- Phenol red (Phenol red is an indicator solution. It is yellow in acidic conditions, pink in basic conditions, and orange in neutral conditions. The pH of water is affected by the presence of $CO_2$. If $CO_2$ concentrations increase, phenol red will change from pink to yellow.)

**Safety notes**: Wear indirectly vented chemical-splash goggles, aprons, and gloves. Do not touch the light source as it can burn skin. Wash hands with soap and water after completing the activity.

With your group, determine which explanation provides the best answer to the research question. You can use as many of the supplies available to you to test your ideas. Make sure that you generate the evidence you will need to support an explanation and refute the others as you work. You can record your method and any observations you make in the spaces below.

## Our Method

## Our Observations

# SECTION 2: EVALUATE ALTERNATIVES

## PLANTS AND ENERGY    18

## Argumentation Session

Once your group has decided which explanation is the most valid or acceptable answer to the research question, prepare a whiteboard that you can use to share and justify your ideas. Your whiteboard should include all the information shown in Figure 18.1

To share your work with others, we will be using a round-robin format. This means that one member of the group will stay at your workstation to share your group's ideas while the other group members go to the other groups one at a time in order to listen to and critique the arguments developed by your classmates. Remember, as you critique the work of others, you need to decide if their conclusions are valid or acceptable based on the quality of their claim and how well they are able to support their ideas. In other words, you need to determine if their argument is *convincing* or not. One way to determine if their argument is convincing is to ask them some of the following questions:

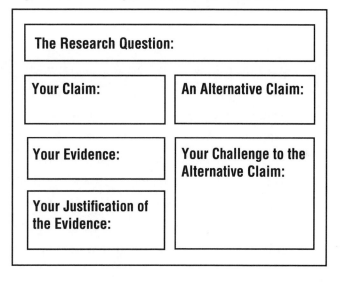

*Figure 18.1. Components of the Whiteboard*

- How did you gather your data? Why did you decide to do it that way?
- How do you know your data is high quality (free from errors)?
- How did you analyze or interpret your data? Why did you decide to do it that way?
- Why does your evidence support your claim?
- Why did you decide to use that evidence? Why is your evidence important?
- How does your justification of your evidence fit with accepted scientific ideas?

## SECTION 2: EVALUATE ALTERNATIVES

Name_____   Date_____

# PLANTS AND ENERGY:
## What Is Your Argument?

In the space below, write a one- to three-paragraph argument to *support* the explanation that you think is the most valid or acceptable. Your argument must also include a *challenge* to one of the alternative explanations. As you write your argument, remember to do the following:

- State the explanation you are trying to support
- Include genuine evidence (data + analysis + interpretation)
- Explain why the evidence is important and relevant
- State the explanation you are trying to refute
- Explain why the alternative explanation is invalid or unacceptable
- Organize your argument in a way that enhances readability
- Use a broad range of words including vocabulary that we have learned
- Correct grammar, punctuation, and spelling errors

SECTION 2: EVALUATE ALTERNATIVES

# PLANTS AND ENERGY
## TEACHER NOTES 18

## Purpose
The purpose of this activity is to help students understand how plants use photosynthesis to convert carbon dioxide into sugar and then use cellular respiration to convert the sugar into a useable form of energy. This activity also helps students learn how to engage in practices such as planning and carrying out investigations, arguing from evidence, and communicating information. In addition, this activity is designed to give students an opportunity to learn how to write in science and develop their speaking and listening skills, which are important goals for literacy in science (see Standards Addressed in This Activity for a complete list of the practices, crosscutting concepts, core ideas, and literacy skills that are aligned with this activity).

## The Content and Related Concepts
An important process in plants and animals is cellular respiration. Cellular respiration involves several metabolic reactions that are used to transfer biochemical energy from nutrients into adenosine triphosphate (ATP) (an energy storage molecule). The energy stored in ATP can then be used to drive processes requiring energy, including biosynthesis, locomotion, or the transport of molecules across cell membranes. Cellular respiration can be either aerobic (with oxygen) or anaerobic (without oxygen). The overall process of aerobic cellular respiration is often represented with the following reaction:

$$C_6H_{12}O_6 \text{ (aq)} + 6\ O_2 \text{ (g)} \rightarrow 6\ CO_2 \text{ (g)} + 6\ H_2O \text{ (l)}$$

During the multiple-step process of aerobic respiration (which includes glycolysis and the Krebs cycle), 38 ATP molecules are produced as a by-product of the intermediate reactions. The source of the sugar for respiration in plants is the process of photosynthesis (which requires light). As a result, during the day, plants both respire and photosynthesize; during the night, they only respire.

## Curriculum and Instructional Considerations
There are several places in a biology curriculum where this activity would be appropriate. It could be used to begin a unit on cellular processes or used as part of a unit about the flow, cycles, and conversation of energy and matter in a biological system. If implemented in the middle of a biology course, this activity provides an opportunity to identify differences between plants and animals beyond a simple definition, emphasizing the different ways that living organisms obtain and use energy. The activity could also be used to emphasize the cycle of carbon in an ecosystem during a unit

## SECTION 2: EVALUATE ALTERNATIVES

### 18 PLANTS AND ENERGY
#### TEACHER NOTES

on the environment. In this way, the activity allows the teachers to emphasize the interactions of living organisms and the relationships among living organisms in an environment. Teachers can also use this activity to help students develop inquiry skills and the understanding of the nature of scientific inquiry.

The focus of the explicit discussion at the end of the activity should focus on the way energy and matter flows, cycles, and is conserved in plants. The explicit discussion should also focus on at least one aspect of the nature of science or the nature of scientific inquiry. For example, a teacher could discuss how scientific explanation must be consistent with observational evidence about nature, and must make accurate predictions, when appropriate, about systems being studied or how mathematics plays an important role in scientific inquiry using what the students did as an illustrative example.

## Recommendations for Implementing the Activity

This activity takes between 100 and 150 minutes of instructional time to complete, depending on how a teacher decides to spend time in class. The activity also requires about 24 hours for the experiments to run. In Figure 18.2, Option A gives students time to complete all six stages of the lesson during class with one day between Stages 2 and 3 to allow time for carbon dioxide to build up in the test tubes. Stages 1 and 2 are completed on day 1, Stages 3 and 4 are completed on day 2, and Stages 5 and 6 are completed on day 3. This option for implementing the activity works best in schools where students are not expected to complete much homework or if students need to be encouraged to write more during the school day. It also provide less time for the argumentation sessions, which may or may not be a problem, depending on how comfortable the students are with argumentation. In Option B, students complete Stage 1 and begin Stage 2 of the lesson during class on day 1. The students then complete Stage 3 on day 2 of the lesson. Stages 4 and 5 are completed on day 3 and the final written argument and counter argument (Stage 6) is then assigned as homework and returned the next day.

# SECTION 2: EVALUATE ALTERNATIVES

## PLANTS AND ENERGY
### TEACHER NOTES 18

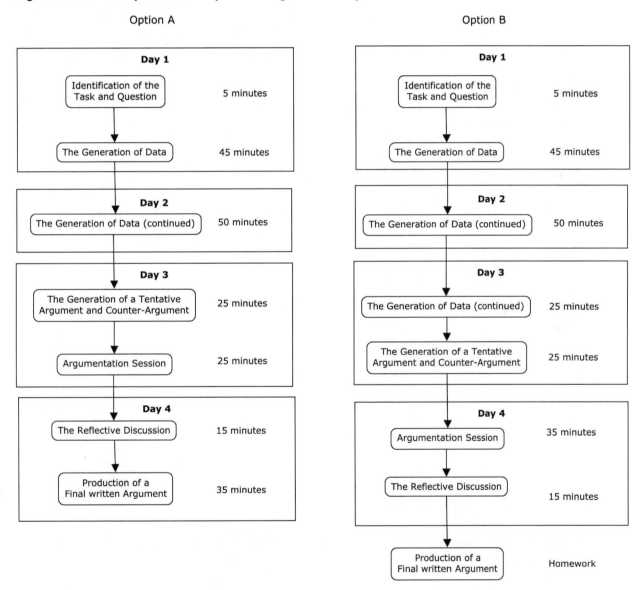

*Figure 18.2. Two Options for Implementing the Activity*

## SECTION 2: EVALUATE ALTERNATIVES

### 18 PLANTS AND ENERGY
### TEACHER NOTES

*Table 18.1. Materials Needed to Implement the Activity in a Classroom of 28 Students*

| Material | Amount Needed With ... | |
| --- | --- | --- |
| | Groups of 3 | Groups of 4 |
| Water (aged tap or spring) | 1 L | 1 L |
| Aquatic snails (two per group) | 20 | 14 |
| Elodea (anacharis) sprigs (four per group) | 40 | 32 |
| Light source | 1 | 1 |
| Bromthymol blue (BTB) indicator solution | 9 | 6 |
| Phenol red indicator solution | 9 | 6 |
| Large beakers (two per group) | 20 | 14 |
| Test tubes (eight per group) | 80 | 54 |
| Rubber stoppers that fit snuggly in mouth of the test tubes (eight per group) or parafilm | 80 | 54 |
| Whiteboards (or chart paper)+ | 10 | 7 |
| Whiteboard markers (or permanent markers if using chart paper)+ | 20 | 14 |
| Copy of Student Pages (pp. 219–221)* | 28 | 28 |
| Copy of Student Page (p. 222)* | 28 | 28 |
| Copy of Appendix C (p. 367)* | 28 | 28 |

+ Teachers can also have students prepare their arguments in a digital medium (such as PowerPoint or Keynote).
* Teachers can also project these materials onto a screen in order to cut down on paper use.

Prior to starting the activity, be sure to review with students the SDS for bromthymol blue and phenol red.

Table 18.1 provides information about the type and amount of materials needed to implement this activity in a classroom with 28 students in groups of four and groups of three.

## Assessment

The included rubric can be used to assess the arguments crafted by each student at the end of the activity. The rubric includes categories for the adequacy and conceptual quality of the claim, the appropriate use of evidence, the sufficiency of the rationale, and the overall

## SECTION 2: EVALUATE ALTERNATIVES

### PLANTS AND ENERGY
### TEACHER NOTES 18

quality of the writing. We strongly recommend that teachers use the Comments or Suggestions sections to give students detailed feedback so they will understand what they did wrong, why it is wrong, and ways they can improve their performance next time. To illustrate how to score the arguments consider the following example written by a ninth-grade student:

> Our group agreed that plants consume oxygen and carbon dioxide but the amounts that they consume changes because of the environment. Plants also produce or release oxygen and the amounts are also because of the environment. Because plants release and consume both oxygen and carbon dioxide, we think that explanation #2 is the best supported. **We know that plants consume carbon dioxide because that is something that is always explained in classes. We need plants on earth to take away our carbon dioxide so that we don't all die because we actually need only oxygen to breath. And too much carbon dioxide is making our planet too hot so we need plants.** We didn't think that oxygen would be consumed by the plants so we experimented on that by comparing the change of BTB with a different amount of plants in tubes in the dark and different amounts of plants in tubes in the light. The plants in the dark should have been all the same since the plants need sunlight to do photosynthesis (they consume carbon dioxide). **But we found out that the plants in the dark were different shades off yellow, which meant that the plants consume oxygen and give off carbon dioxide.** Because our plants consume oxygen that means that the explanation 1 is not correct. **And also even though we didn't test it other kinds of plants we know that some other plants consume other kinds of things like bugs. A Venus Fly Trap east bugs.** So explanation 1 can't be right. The 3rd explanation wasn't right because plants need light to use carbon dioxide and that means it can't be always using carbon dioxide. In the dark it would stop. We didn't experiment this, but Jessica's group used snails and plants in the dark and they said that the plants weren't feeding the snails correctly.

The student's claim (underlined) is insufficient (0/1), because it provides an incomplete and vague answer to the research question. The claim also includes some inaccurate elements (0/1). The student uses evidence (in bold) to support the claim by providing data (1/1), an analysis of the data (1/1), and an interpretation of the analysis (1/1). The student includes an incomplete justification of the evidence in his argument; he provides a vague rationale for why the evidence is included (1/1) but never really explains why measuring changes in pH levels would be important (0/1). He provides a good challenge to an alternative explanation by making it explicit (1/1) and providing a reason for why it is not valid (1/1). However, the author

# SECTION 2: EVALUATE ALTERNATIVES

## 18 PLANTS AND ENERGY
### TEACHER NOTES

also uses inappropriate phrases (e.g., "the 3rd explanation was right") that misrepresent the nature of science (1/2). The writing mechanics of the sample argument are also adequate. The organization of the argument is acceptable, because the arrangement of the sentences aid in the development of the main idea (0/1). There are, however, several punctuation (0/1) and grammatical errors (0/1) in the argument. The overall score for the sample argument, therefore, is 7 out of the 14 points possible.

## Standards Addressed in This Activity

This activity can be used to address the following dimensions outlined in *A Framework for K–12 Science Education* (NRC 2012):

### Scientific Practices
- Developing and using models
- Planning and carrying out investigations
- Engaging in argument from evidence
- Obtaining, evaluating, and communicating information

### Crosscutting Concepts
- Cause and effect: Mechanism and explanation
- System and system models
- Energy and matter: Flows, cycles, and conservation
- Structure and function

### Life Sciences Core Ideas
- From molecules to organisms: Structures and processes
- Ecosystems: Interactions, energy, and dynamics

This activity can be used to address the following standards for literacy in science from the *Common Core State Standards for English Language Arts and Literacy* (NGA and CCSSO 2010):

### Writing
- Text types and purposes
- Production and distribution of writing
- Research to build and present knowledge
- Range of writing

### Speaking and Listening
- Comprehension and collaboration
- Presentation of knowledge and ideas

## References

National Governors Association Center (NGA) for Best Practices, and Council of Chief State School Officers (CCSSO). 2010. *Common core state standards for English language arts and literacy*. Washington, DC: National Governors Association for Best Practices, Council of Chief State School.

National Research Coucnil (NRC). 2012. *A framework for K–12 science education: Practices, crosscutting concepts, and core ideas*. Washington, DC: National Academies Press.

SECTION 2: EVALUATE ALTERNATIVES

# HEALTHY DIET AND WEIGHT (HUMAN HEALTH) 19

One of the biggest health problems in the United States is obesity (see Figure 19.1). Many people are overweight, and being overweight increases their risk of cancer, diabetes, and early death. The risks are particularly severe for children, because being overweight as a child makes a person more prone to obesity as an adult.

*Figure 19.1. Percentage of the Population in Different Countries That Is Considered Obese (Body Mass Index > 30)*

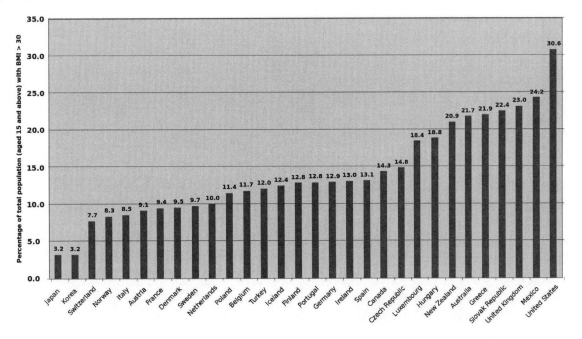

There are many possible treatments for obesity. Dieting is the most popular approach. However, there are many different types of diets. Some dieticians, for example, recommend that you avoid carbohydrates like starch and sugar and eat foods that are high in protein. Other dieticians recommend a diet that is low in fat. Some dieticians even suggest that a low protein, high carb diet is the best way to lose weight. In addition to these types of approaches, you can also diet by simply limiting the amount of calories you consume each day, which is a more traditional approach to dieting. Dieticians

**SCIENTIFIC ARGUMENTATION IN BIOLOGY: 30 CLASSROOM ACTIVITIES**

## SECTION 2: EVALUATE ALTERNATIVES

### 19   HEALTHY DIET AND WEIGHT

and personal trainers also suggest that you can lose weight by increasing the amount of calories you burn each day by exercising more.

This discussion raises an interesting question: **Which method is the most effective way to lose weight or decrease body mass index (BMI) over a 12-month period?**

Here are four potential answers to this question:

- **Explanation 1**: Keep your same calorie intake, but eat foods that are high in protein and fat but low in carbohydrates. Maintain your same activity level.
- **Explanation 2**: Keep your same calorie intake, but eat foods that are low in protein and fat but high in carbohydrates. Maintain your same activity level.
- **Explanation 3**: Eat any type of food you want, but reduce your total calorie intake so it is less than the number of calories you burn each day. Maintain the same activity level.
- **Explanation 4**: Eat any type of food you want, and keep your total calorie intake the same but increase your activity level so you burn more calories that you consume.

## Getting Started

In order to determine which explanation is the best answer to the research question, you can use an online simulation called Eating and Exercise. This simulation allows you to set the height, weight, and activity level (e.g., sedentary lifestyle, active lifestyle) of an individual. You can then set the amount of calories that the individual will eat per day and the type of foods that are eaten by that individual. You can also have the individual engage in certain types of exercise each day (beyond the activities associated with their typical lifestyle). Once these parameters are set, you can track how the individual's weight and BMI changes over the course of a year. To access this simulation, use the following link:

*http://phet.colorado.edu/en/simulation/eating-and-exercise*

Make sure that you use the simulation to generate the data that you will need to evaluate the various explanations. You will then be able to transform the data you collect into evidence that you can use to support one of the explanations and to challenge the others. You can record your method and your observations, or the data you collect in the spaces below and on page 231.

## Our Method

## SECTION 2: EVALUATE ALTERNATIVES
### HEALTHY DIET AND WEIGHT 19

## Our Observations

## Argumentation Session

Once your group has decided which explanation is the most valid or acceptable answer to the research question, prepare a whiteboard that you can use to share and justify your ideas. Your whiteboard should include all the information shown in Figure 19.2.

To share your work with others, we will be using a round-robin format. This means that one member of the group will stay at your workstation to share your group's ideas, while the other group members go to the other groups one at a time in order to listen to and critique the arguments developed by your classmates. Remember, as you critique the work of others, you need to decide if their conclusions are valid or acceptable based on the quality of their claim and how well they are able to support their ideas. In other words, you need to determine if their argument is *convincing* or not. One way to determine if their argument is convincing is to ask them some of the following questions:

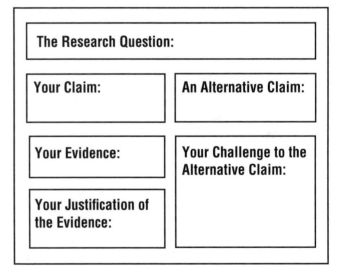

Figure 19.2. Components of the Whiteboard

- How did you gather your data? Why did you decide to do it that way?
- How do you know your data is high quality (free from errors)?
- How did you analyze or interpret your data? Why did you decide to do it that way?
- Why does your evidence support your claim?
- Why did you decide to use that evidence? Why is your evidence important?
- How does your justification of your evidence fit with accepted scientific ideas?

## SECTION 2: EVALUATE ALTERNATIVES

Name_____  Date_____

# HEALTHY DIET AND WEIGHT:
## What Is Your Argument?

In the space below, write a one- to three-paragraph argument to *support* the explanation that you think is the most valid or acceptable. Your argument must also include a *challenge* to one of the alternative explanations. As you write your argument, remember to do the following:

- State the explanation you are trying to support
- Include genuine evidence (data + analysis + interpretation)
- Explain why the evidence is important and relevant
- State the explanation you are trying to refute
- Explain why the alternative explanation is invalid or unacceptable
- Organize your argument in a way that enhances readability
- Use a broad range of words including vocabulary that we have learned
- Correct grammar, punctuation, and spelling errors

SECTION 2: EVALUATE ALTERNATIVES

# HEALTHY DIET AND WEIGHT
## TEACHER NOTES 19

## Purpose

The purpose of this activity is to help students understand the relationship between diet, exercise, weight, and body mass index. This activity also helps students learn how to engage in scientific practices such using planning and carrying out investigations, arguing from evidence, and communicating information. In addition, this activity is designed to give students an opportunity to learn how to write in science and develop their speaking and listening skills, which are important goals for literacy in science (see Standards Addressed in This Activity for a complete list of the practices, crosscutting concepts, core ideas, and literacy skills that are aligned with this activity).

## The Content and Related Concepts

The relationships among diet, exercise, weight, and body mass index are complex. Students, however, should understand the importance of eating a balanced diet in order ensure proper nutrition and how consuming more calories than are used will lead to weight gain. Calories are a measure of energy available in food. When people consume more calories than what is needed to maintain metabolic activity, the body converts the excess into body fat. This fat serves as an energy store for later times when food is not as plentiful. The excess body fat, however, leads to weight gain and a higher body mass index. Once a person's body mass index reaches 30, he or she is considered obese, which in turn puts him or her at higher risk of heart disease, diabetes, and cancer. When people consume fewer calories than is what needed for metabolic activity, the body breaks down body fat in order to make up the difference. As a result, people tend to lose weight.

Students also need to understand that some foods contain more calories than others. For example, one banana has approximately 172 calories, 1 oz. of peanuts has 160 calories, and one slice of pepperoni pizza has 300 calories. Students, therefore, need to understand that both the amount and type of food they eat is directly related to their daily caloric intake.

## Curriculum and Instructional Considerations

There are at least three points within a traditional biology curriculum in which this activity would be appropriate and helpful. The first is during a unit on the chemistry of life. The second is during a unit on cellular processes. The third place is during a unit on human systems. When it is used as part of a unit on the chemistry of life, teachers can focus on the different types of biomolecules found in food and the number of calories found in different foods. Teachers can also use this activity during a unit on cellular processes to highlight

SCIENTIFIC ARGUMENTATION IN BIOLOGY: 30 CLASSROOM ACTIVITIES

## SECTION 2: EVALUATE ALTERNATIVES

## 19 HEALTHY DIET AND WEIGHT
### TEACHER NOTES

the different biochemical pathways that are used to break down the biomolecules found in food into basic building blocks and how these building blocks can then be converted into other biomolecules and tissues. Teachers can also focus on the biochemical pathways that are used inside the body to transform chemical energy in food into a form that is useable in the human body. When the activity is used as part of a unit about human systems, teachers can focus on nutrition and why obesity is related to poor cardiovascular health and an increased risk of diabetes.

This activity is also an effective way to teach students about experimental design, the control of variables, the difference between hypotheses (tentative explanations) and predictions (expected results), and other important terminology such as *independent* and *dependent variables*. Students will need a basic understanding of these important ideas in order to be able to collect meaningful data during the activity. Teachers can also discuss how scientists often use models, such as the online simulation, to test ideas. The explicit discussion at the end of the activity should focus on the relationship between eating, exercise, and body mass index. The explicit discussion should also focus on at least one aspect of the nature of science or the nature of scientific inquiry. For example, a teacher could discuss the ways in which scientific explanations must be consistent with observational evidence about nature, the importance of controlling variables during an experiment, or the role of models in science using this activity as an illustrative example.

## Recommendations for Implementing the Activity

This activity takes between 100 and 150 minutes of instructional time to complete, depending on how a teacher decides to spend time in class. In Option A, the students are given time to complete all six stages of the lesson during class (see Figure 19.3). Stages 1 and 2 are completed on day 1, Stages 3 and 4 are completed on day 2, and Stages 5 and 6 are completed on day 3. This option for implementing the activity works best in schools where students are not expected to complete much homework or if students need to be encouraged to write more during the school day. It also provide less time for the argumentation sessions. In Option B, students complete Stage 1 and begin Stage 2 of the lesson during class on day 1. The students then complete Stage 3 on day 2 of the lesson. Stages 4 and 5 are completed on day 3, and the final written argument (Stage 6) is then assigned as homework and returned the next day.

Table 19.1 (p. 236) provides information about the type and amount of materials needed to implement this activity in a classroom of 28 students in groups of four and groups of three.

## Assessment

The rubric provided in Appendix C (p. 367) can be used to assess the argument crafted by each student at the end of the activity. To illustrate how the rubric can be used to score an argument written by a student, consider the following example. This sample, which was written by an eighth-grade student, is weak in terms of content and mechanics.

# SECTION 2: EVALUATE ALTERNATIVES

## HEALTHY DIET AND WEIGHT
### TEACHER NOTES 19

*Figure 19.3. Two Options for Implementing the Activity*

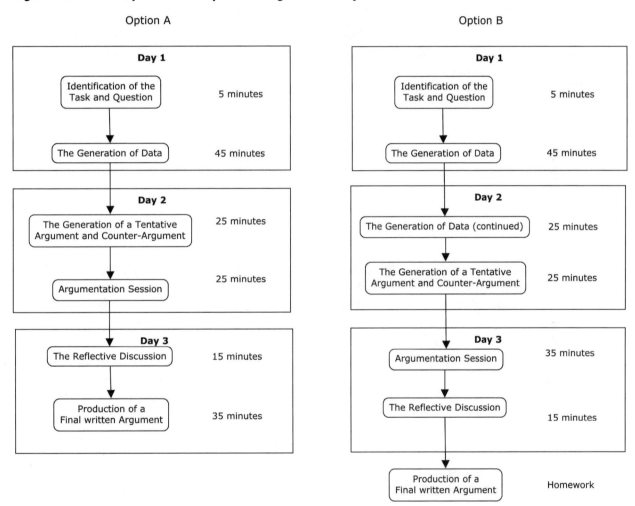

## SECTION 2: EVALUATE ALTERNATIVES

### 19 HEALTHY DIET AND WEIGHT
### TEACHER NOTES

*Table 19.1. Materials Needed to Implement the Activity in a Classroom of 28 Students*

| Material | Amount Needed With ... | |
|---|---|---|
| | Groups of 3 | Groups of 4 |
| Computer with internet access | 9 | 7 |
| Whiteboards (or chart paper)+ | 9 | 7 |
| Whiteboard markers (or permanent markers if using chart paper)+ | 20 | 14 |
| Copy of Student Pages (pp. 229–231)* | 28 | 28 |
| Copy of Student Page (p. 232)* | 28 | 28 |
| Copy of Appendix C (p. 367)* | 28 | 28 |

+ Teachers can also have students prepare their arguments in a digital medium (such as PowerPoint or Keynote).
* Teachers can also project these materials onto a screen in order to cut down on paper use.

The best way to lose weight is explanation number 1. **We used the online simulation to test two different diets. The first diet was foods high in protein (all meat) and the second diet was all carbohydrates (pasta and bread) and the individual lost more weight by eating all meat. This proves that a diet high in protein is the best way to lose weight.** We felt that its important to examine how much a person loses because it allows us to prove which diet is best.

I don't think the explanation number 2 is right. As I said before, the individual lose more weight when eating all protein so a diet high in carbohydrates can't be good for you. People should avoid carbs.

The content of the example argument is rated as weak for several reasons. The student's claim (underlined) is insufficient (0/1), because it only refers to the number of the explanation on the handout and does not make the actual explanation explicit. The claim is also inaccurate (0/1). The student uses information (in bold) to support his claim that is not really evidence; it is little more than an unsubstantiated inference (0/3). This student also does not discuss other factors that could have resulted in weight loss (such as decreased caloric intake or increased exercise). The student provides a vague justification of the evidence (1/1) but does not attempt to link the evidence to a specific scientific principle or underlying assumption (0/1). The student includes a challenge to an alternative explanation, but this challenge does not make the explanation that was being changed explicit (0/1). In addition, the student does not provide additional evidence to challenge the alternative explanation (0/1). The student also uses phrases that misrepresent the nature of science (0/1), although scientific terms are used correctly (1/1). Although the student's writing overall is

# SECTION 2: EVALUATE ALTERNATIVES

## HEALTHY DIET AND WEIGHT
### TEACHER NOTES 19

poor, the organization of the argument is sufficient, because the arrangement of the sentences does not distract from the development of the main idea (0/1). The argument is rather short, however, and there are also many spelling (0/1) and grammatical errors (0/1). The overall score for the sample argument, therefore, is 2 out of the 14 points possible.

## Standards Addressed in This Activity

This activity can be used to address the following dimensions outlined in *A Framework for K–12 Science Education* (NRC 2012):

## Scientific Practices

- Developing and using models
- Planning and carrying out investigations
- Using mathematics and computational thinking
- Engaging in argument from evidence
- Obtaining, evaluating, and communicating information

## Crosscutting Concepts

- Cause and effect: Mechanism and explanation
- Scale, proportion, and quantity
- Energy and matter: Flows, cycles, and conservation
- Structure and function

## Life Sciences Core Ideas

- From molecules to organisms: Structures and processes

This activity can be used to address the following standards for literacy in science from the *Common Core State Standards for English Language Arts and Litearcy* (NGA and CCSSO 2010):

## Writing

- Text types and purposes
- Production and distribution of writing
- Research to build and present knowledge
- Range of writing

## Speaking and Listening

- Comprehension and collaboration
- Presentation of knowledge and ideas

## References

National Governors Association Center (NGA) for Best Practices, and Council of Chief State School Officers (CCSSO). 2010. *Common core state standards for English language arts and literacy.* Washington, DC: National Governors Association for Best Practices, Council of Chief State School.

National Research Council (NRC). 2012. *A framework for K–12 science education: Practices, crosscutting concepts, and core ideas.* Washington, DC: National Academies Press.

## SECTION 2: EVALUATE ALTERNATIVES

# TERMITE TRAILS (ANIMAL BEHAVIOR) 20

Termites are small, soft-bodied, usually pale-colored insects (see Figure 20.1). They are social creatures and live in a caste system. Termites create colonies in the ground or in wood. Their food consists primarily of wood or other plant-based material. The workers are sterile and are responsible for the main work of the colony: collecting food; expanding the size of the colony; and feeding the queen, soldiers, and young.

Figure 20.1. Termites

Termites also exhibit some interesting behaviors. To illustrate, consider the following example. Take a blue felt pen and draw a figure eight (see Figure 20.2). Then place several termites on the piece of paper in the middle of the figure eight and watch what happens. After a few minutes, you should see the termites following the lines.

This observation raises an interesting question: **Why do the termites follow the lines?**

Here are three potential answers to this question:

- **Explanation 1**: Termites navigate by sight. They are attracted to certain colors such as blue. As a result, termites follow the line because it is blue.

- **Explanation 2**: Termites are cooperative and navigate by sight. Termites leave a trail when they go looking for food so they can find their way back to the colony. So termites will follow any line they come across.

- **Explanation 3**: Termites are cooperative and navigate by smell. The ink in the pen contains a chemical that smells the same as the pheromones that are secreted by the termites. As a result, the termites follow the line, because it smells like a trail left by another termite.

## Getting Started

You can use the following materials to test these explanations.

- Petri dish
- Paper
- Paper Mate or BIC ballpoint pen
- Other types of pens (felt tip or rollerball) and pencils
- Termites
- Forceps (tweezers) and paintbrush
- Scissors

Figure 20.2. Figure Eight

**Safety note**: Wash hands with soap and water after completing the activity.

## SECTION 2: EVALUATE ALTERNATIVES

## 20  TERMITE TRAILS

With your group, determine which explanation provides the best answer to the research question. You can use as many of the supplies available to you to test your ideas. Make sure that you generate the evidence that you will need to support your explanation (and refute the alternatives) as you work. You can record your method and any observations you make in the spaces below.

## Our Method

## Our Observations

## Argumentation Session

Once your group has decided which explanation is the most valid or acceptable answer to the research question, prepare a whiteboard that you can use to share and justify your ideas. Your whiteboard should include all the information shown in Figure 20.3.

To share your work with others, we will be using a round-robin format. This means that one member of the group will stay at your workstation to share your group's ideas while the other group members go to the other groups one at a time in order to listen to and critique the arguments developed by your classmates. Remember, as you critique the work of others, you need to decide if their conclusions are valid or acceptable based on the quality of their claim and how well they are able to support their ideas. In other words, you need to determine if their argument is *convincing* or not. One way to determine if their argument is convincing is to ask them some of the following questions:

## SECTION 2: EVALUATE ALTERNATIVES

### TERMITE TRAILS  20

- How did you gather your data? Why did you decide to do it that way?
- How do you know your data is high quality (free from errors)?
- How did you analyze or interpret your data? Why did you decide to do it that way?
- Why does your evidence support your claim?
- Why did you decide to use that evidence? Why is your evidence important?
- How does your justification of your evidence fit with accepted scientific ideas?

*Figure 20.3. Components of the Whiteboard*

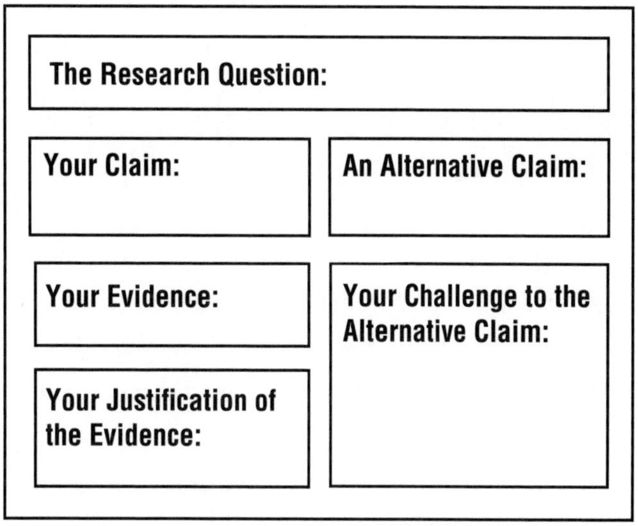

| The Research Question: | |
|---|---|
| Your Claim: | An Alternative Claim: |
| Your Evidence: | Your Challenge to the Alternative Claim: |
| Your Justification of the Evidence: | |

## SECTION 2: EVALUATE ALTERNATIVES

Name_____  Date_____

# TERMITE TRAILS:
## What Is Your Argument?

In the space below, write a one- to three-paragraph argument to *support* the explanation that you think is the most valid or acceptable. Your argument must also include a *challenge* to one of the alternative explanations. As you write your argument, remember to do the following:

- State the explanation you are trying to support
- Include genuine evidence (data + analysis + interpretation)
- Explain why the evidence is important and relevant
- State the explanation you are trying to refute
- Explain why the alternative explanation is invalid or unacceptable
- Organize your argument in a way that enhances readability
- Use a broad range of words including vocabulary that we have learned
- Correct grammar, punctuation, and spelling errors

SECTION 2: EVALUATE ALTERNATIVES

# TERMITE TRAILS
## TEACHER NOTES 20

## Purpose

The purpose of this activity is to help students understand that some organisms, such as termites, have instinctive inherited behaviors. These inherited instinctive behaviors, which evolved over time through the process of natural selection, had a positive impact on the reproductive success of the organism in the past (and may continue to have a positive impact on the reproductive success of the organism if environmental conditions do not change). This activity also helps students learn how to engage in practices such as planning and carrying out investigations, arguing from evidence, and communicating information. In addition, this activity is designed to give students an opportunity to learn how to write in science and develop their speaking and listening skills, which are important goals for literacy in science (see Standards Addressed in This Activity for a complete list of the practices, crosscutting concepts, core ideas, and literacy skills that are aligned with this activity).

## The Content and Related Concepts

Behavior can be interpreted in terms of proximate causes, which are immediate interactions with the environment, or in terms of ultimate causes, which are evolutionary (because natural selection preserves behaviors that enhance fitness). Animal behavior often stems from a combination of genetic programming (i.e., an innate behavior) and environmental experiences (i.e., learning). In this activity, the termite response (i.e., trail following) to an external stimulus (i.e., a line made with ink from a ballpoint pen) is an example of a fixed action pattern. A fixed action pattern is a type of innate behavior that is triggered by a specific environmental trigger. Termites will follow lines made on paper by ballpoint pens (we recommend using Paper Mate or BIC), because the ink contains a chemical that is similar to the trail pheromones used by termites to lead colony members to food sources. Termites exhibit the trail following behavior because it increases their chances of survival. Therefore, the behavior was preserved over time by the process of natural selection.

## Curriculum and Instructional Considerations

This activity is best used during a unit on behavior to introduce the ways in which organisms, populations, and communities respond to external factors and the relationships among behavior, evolution, and ecology. This activity is also an effective way to teach students about experimental design, the control of variables, the difference between hypotheses (tentative explanations) and predictions (expected results), and other important terminology such as independent and dependent variables.

## SECTION 2: EVALUATE ALTERNATIVES

## 20  TERMITE TRAILS
## TEACHER NOTES

Students will need a basic understanding of these important ideas in order to be able to collect meaningful data during the activity. The explicit discussion at the end of the activity should focus on classic concepts in behavior (e.g., proximate and ultimate causes, how natural selection can influence behavior when it has a genetic basis). The explicit discussion should also focus on at least one aspect of the nature of science or the nature of scientific inquiry. For example, a teacher could discuss how scientific explanations must be consistent with observational evidence about nature, or how experiments are used to test explanations, the importance of controlling variables during an experiment, or the role of creativity and imagination in science using this activity as an illustrative example.

## Recommendations for Implementing the Activity

This activity takes between 100 and 150 minutes of instructional time to complete, depending on how a teacher decides to spend time in class. In Option A, the students are given time to complete all six stages of the lesson during class (see Figure 20.4). Stages 1 and 2 are completed on day 1, Stages 3 and 4 are completed on day 2, and Stages 5 and 6 are completed on day 3. This option for implementing the activity works best in schools where students are not expected to complete much homework or if students need to be encouraged to write more during the school day. It also provides less time for the argumentation sessions, which may or may not be a problem, depending on how comfortable the students are with argumentation. In Option B, students complete Stage 1 and begin Stage 2 of the lesson during class on day 1. The students then complete Stage 3 on day 2 of the lesson. Stages 4 and 5 are completed on day 3, and the final written argument (Stage 6) is then assigned as homework and returned the next day.

Table 20.1 (p. 246) provides information about the type and amount of materials needed to implement this activity in a classroom of 28 students in groups of four and groups of three. Termites can be ordered from biological supply companies such as Ward's Natural Science (*http://wardsci.com*) or Carolina Biological Supply (*www.carolina.com*). Termites should be returned to a moist dark vial after about 20 minutes of observation, because they tend to dehydrate fairly quickly. Termites can be refrigerated in order to prolong their viability (but allow them to warm up to room temperature before students use them). Do not use sluggish termites (they are probably unhealthy or near death).

## SECTION 2: EVALUATE ALTERNATIVES

### TERMITE TRAILS
### TEACHER NOTES 20

*Figure 20.4. Two Options for Implementing the Activity*

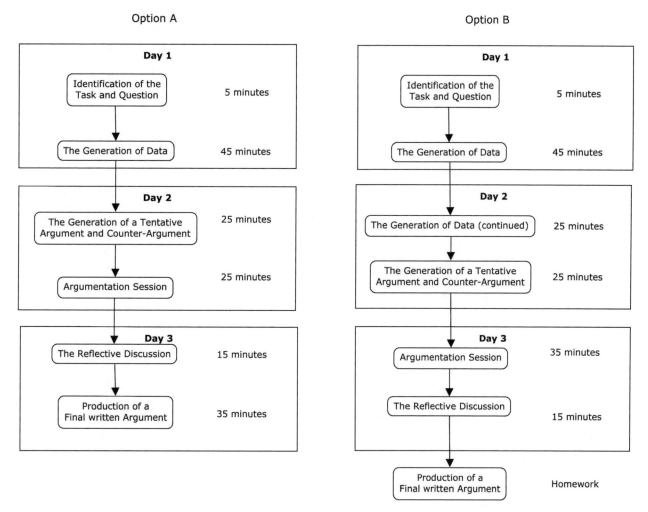

## SECTION 2: EVALUATE ALTERNATIVES

## 20 TERMITE TRAILS
### TEACHER NOTES

*Table 20.1. Materials Needed to Implement the Activity in a Classroom of 28 Students*

| Material | Amount Needed With ... | |
| --- | --- | --- |
| | Groups of 3 | Groups of 4 |
| Petri dish | 10 | 7 |
| Pieces of unlined printer paper | 10 | 7 |
| Paper Mate or BIC ballpoint pen | 10 | 7 |
| Other types of pens (felt tip or rollerball) in various colors | 30 | 21 |
| Crayons or pencils in various colors | 30 | 21 |
| Termites | 50 | 35 |
| Tweezers | 10 | 7 |
| Beaker | 10 | 7 |
| Paintbrush (used to move termites without killing them) | 10 | 7 |
| Whiteboards (or chart paper)+ | 10 | 7 |
| Whiteboard markers (or permanent markers if using chart paper)+ | 20 | 14 |
| Copy of Student Pages (pp. 239–241)* | 28 | 28 |
| Copy of Student Page (p. 242)* | 28 | 28 |
| Copy of Appendix C (p. 367)* | 28 | 28 |

+ Teachers can also have students prepare their arguments in a digital medium (such as PowerPoint or Keynote).
* Teachers can also project these materials onto a screen in order to cut down on paper use.

## Assessment

The rubric provided in Appendix C (p. 367) can be used to assess the argument crafted by each student at the end of the activity. To illustrate how the rubric can be used to score an argument written by a student, consider the following example. This sample, which was written by a high school sophomore, is strong in terms of content and is also well written.

We decided to determine if the termites navigate by sight or by smell. <u>We think the termites navigate by smell.</u> We created five treatment conditions. In all five conditions, we used a pen or a pencil to draw a circle for the termites to follow. In the first condition we used a black papermate ballpoint pen. In the second condition we used a blue

## SECTION 2: EVALUATE ALTERNATIVES

### TERMITE TRAILS
### TEACHER NOTES 20

papermate ballpoint pen. In the third condition we used a black sharpie and in the forth we used a blue sharpie. Finally, in the last condition we used a pencil to draw the circle. We then put termites on each piece of paper and watched them to see if they would start the line following behavior for four minutes each. **The termites followed the line when we used the blue and black papermate pens. The termites did not follow the lines when we used a sharpie or a pencil.** The differences in termite behavior in the five conditions indicate that only a certain type of stimulus will trigger the line following behavior. These observations also indicate that the termites use smell to navigate and that there are attracted to the ink in the papermate pens. If the termites navigated by sight and follow any line they come across, then they should have followed all five lines we made on the pieces of paper. Likewise, if termites navigate by sight and are attracted to certain colors, then we should have seen them follow the black lines only or the blues lines only regardless of the type of pen. Since our observations don't match these predictions, we can rule out the two 'navigate by sight' explanations for the termite's behavior.

The example argument is strong for several reasons. The student's claim (underlined) is sufficient (1/1), because it provides a complete answer to the research question and it is accurate (1/1). The student uses genuine evidence (in bold) to support the claim; there is data (1/1), an analysis of the data (1/1), and an interpretation of the analysis (1/1). The student includes a complete justification of the evidence in his argument; he does explain why the evidence was important (1/1) by linking it to a specific principle, concept, or underlying assumption (1/1). He also provides an adequate challenge to an alternative explanation by making the other viewpoint explicit (1/1) and providing a reason for why it is invalidated (1/1). The author also uses appropriate terms (1/1) and phrases that are consistent with the nature of science (1/1). The writing mechanics are also effective. The organization of the argument is strong because the arrangement of the sentences aid in the development of the main idea (1/1). There are also no punctuation (1/1) or grammatical errors (1/1) in the argument. The overall score for the sample argument, therefore, is 14 out the 14 points possible.

## Standards Addressed in This Activity

This activity can be used to address the following dimensions outlined in *A Framework for K–12 Science Education* (NRC 2012):

### Scientific Practices

- Planning and carrying out investigations
- Engaging in argument from evidence

# SECTION 2: EVALUATE ALTERNATIVES

## 20 TERMITE TRAILS
### TEACHER NOTES

- Obtaining, evaluating, and communicating information

## Crosscutting Concepts

- Cause and effect: Mechanism and explanation

## Life Sciences Core Ideas

- Heredity: Inheritance and variation of traits
- Biological evolution: Unity and diversity

This activity can be used to address the following standards for literacy in science from the *Common Core State Standards for English Language Arts and Literacy* (NGA and CCSSO 2010):

## Writing

- Text types and purposes
- Production and distribution of writing
- Research to build and present knowledge
- Range of writing

## Speaking and Listening

- Comprehension and collaboration
- Presentation of knowledge and ideas

## References

National Governors Association Center (NGA) for Best Practices, and Council of Chief State School Officers (CCSSO). 2010. *Common core state standards for English language arts and literacy*. Washington, DC: National Governors Association for Best Practices, Council of Chief State School.

National Research Council (NRC). 2012. *A framework for K–12 science education: Practices, crosscutting concepts, and core ideas.* Washington, DC: National Academies Press.

# REFUTATIONAL WRITING

**Framework Matrix** ................................................................. 250

**Activity 21:** Misconception About Theories and Laws ............. 253
*(Nature of Science)*

**Activity 22:** Misconception About the Nature of
Scientific Knowledge ................................................................. 261
*(Nature of Science)*

**Activity 23:** Misconception About the Work of Scientists ....... 269
*(Nature of Science)*

**Activity 24:** Misconception About the Methods of Scientific
Investigations ............................................................................. 277
*(Nature of Science)*

**Activity 25:** Misconception About Life on Earth ...................... 285
*(Evolution)*

**Activity 26:** Misconception About Bacteria ............................. 293
*(Microbiology)*

**Activity 27:** Misconception About Interactions That
Take Place Between Organisms ................................................ 301
*(Ecology)*

**Activity 28:** Misconception About Plant Reproduction ........... 309
*(Botany)*

**Activity 29:** Misconception About Inheritance of Traits .......... 315
*(Genetics)*

**Activity 30:** Misconception About Insects ............................... 321
*(Ecology)*

# SECTION 3: REFUTATIONAL WRITING

# FRAMEWORK MATRIX

| A Framework for K–12 Science Education | Misconception About Theories and Laws | Misconception About the Nature of Scientific Knowledge | Misconception About the Work of Scientists | Misconception About the Methods of Scientific Investigations | Misconception About Life on Earth | Misconception About Bacteria | Misconception About Interactions That Take Place Between Organisms | Misconception About Plant Reproduction | Misconception About Inheritance of Traits | Misconception About Insects |
|---|---|---|---|---|---|---|---|---|---|---|
| **1. Scientific Practices** | | | | | | | | | | |
| Asking questions | | | | | | | | | | |
| Developing and using models | | | | | | | | | | |
| Planning and carrying out investigations | | | | | | | | | | |
| Using mathematics and computational thinking | | | | | | | | | | |
| Constructing explanations | | | | | | | | | | |
| Engaging in argument from evidence | ■ | ■ | ■ | ■ | ■ | ■ | ■ | ■ | ■ | ■ |
| Obtaining, evaluating, and communicating information | ■ | ■ | ■ | ■ | ■ | ■ | ■ | ■ | ■ | ■ |
| **2. Crosscutting Concepts** | | | | | | | | | | |
| Patterns | | | | | | | | ■ | ■ | |
| Cause and effect: Mechanism and explanation | | | | | ■ | ■ | ■ | ■ | ■ | ■ |
| Scale, proportion, and quantity | | | | | | | | | | |
| Systems and system models | | | | | | | | | | |
| Energy and matter: Flows, cycles, and conservation | | | | | | ■ | ■ | | | ■ |
| Structure and function | | | | | | | | | ■ | ■ |
| Stability and change | | | | | ■ | | | | | |

■ = Strong alignment     ☐ = Weak alignment

## SECTION 3: REFUTATIONAL WRITING

| A Framework for K–12 Science Education | Misconception About Theories and Laws | Misconception About the Nature of Scientific Knowledge | Misconception About the Work of Scientists | Misconception About the Methods of Scientific Investigations | Misconception About Life on Earth | Misconception About Bacteria | Misconception About Interactions That Take Place Between Organisms | Misconception About Plant Reproduction | Misconception About Inheritance of Traits | Misconception About Insects |
|---|---|---|---|---|---|---|---|---|---|---|
| **3. Life Sciences Core Ideas** | | | | | | | | | | |
| From molecules to organisms: Structures and processes | | | | | | | | ■ | □ | |
| Ecosystems: Interactions, energy, and dynamics | | | | | | ■ | ■ | | | ■ |
| Heredity: Inheritance and variation of traits | | | | | | | | ■ | ■ | |
| Biological evolution: Unity and diversity | | | | | ■ | | | | | |
| **Common Core State Standards in English and Language Arts: Literacy in the Disciplines** | | | | | | | | | | |
| **1. Writing** | | | | | | | | | | |
| Text types and purposes | ■ | ■ | ■ | ■ | ■ | ■ | ■ | ■ | ■ | ■ |
| Production and distribution of writing | ■ | ■ | ■ | ■ | ■ | ■ | ■ | ■ | ■ | ■ |
| Research to build and present knowledge | ■ | ■ | ■ | ■ | ■ | ■ | ■ | ■ | ■ | ■ |
| Range of writing | ■ | ■ | ■ | ■ | ■ | ■ | ■ | ■ | ■ | ■ |
| **2. Speaking and Listening** | | | | | | | | | | |
| Comprehension and collaboration | | | | | | | | | | |
| Presentation of knowledge and ideas | | | | | | | | | | |

■ = Strong alignment    □ = Weak alignment

SECTION 3: REFUTATIONAL WRITING

# MISCONCEPTION ABOUT THEORIES AND LAWS (NATURE OF SCIENCE) 21

Many people believe theories turn into laws. In other words, some people believe that a theory is a well-supported hypothesis and a law is a theory that has been proven true. Write a one- to two-page refutational essay to convince someone who thinks that theories turn into laws that this idea is a misconception. As you write your paper, remember to

- clearly state the misconception that you are trying to refute;
- include several specific facts, details, reasons, and/or examples that show or demonstrate why the misconception is inaccurate;
- explain how theories and laws are different types of knowledge in science;
- include several specific facts, details, reasons, and/or examples that show or demonstrate how theories and laws are different types of knowledge in science;
- present your ideas in a clear and logical order, including an introduction, body, and conclusion;
- use a variety of words and well-constructed sentences to create tone and voice;
- include at least five references in your essay, and be sure to cite your references correctly; and
- correct errors in capitalization, punctuation, sentence structure, spelling, and grammar.

You will have _____ to research this topic, and then plan, write, edit, and produce a final product. You must use at least five different reference materials (e.g., your textbook, online resources, and so on) during this process. Be sure to cite all your references in the text of your essay and include a reference section.

You must complete your **research** by _____.

Your **prewrite** (an outline, a concept map, and so on) is due on _____.

Your **initial draft** of your essay is due on _____.

Your **final draft** of your essay is due on _____.

SECTION 3: REFUTATIONAL WRITING

# 21 MISCONCEPTION ABOUT THEORIES AND LAWS
## TEACHER NOTES

### Purpose

The purpose of this activity is to help students understand that theories do not become laws in science and that there is no hierarchal relationship between these two types of scientific knowledge. This activity is also designed to address many of the Common Core State Standards for English Language Arts and Literacy, which have a strong emphasis on literacy in science. These standards include writing arguments focused on discipline-specific content; writing in a clear and coherent manner; and developing and strengthening students' papers through a process of planning, revising, editing, and rewriting. *The Common Core State Standards for English Language Arts and Literacy* (NGA and CCSSO 2010) also calls for students to be able to conduct a short research project, gather relevant information from multiple print and digital sources, assess the credibility and accuracy of each source, and quote or paraphrase the data and conclusions of others while avoiding plagiarism. This writing assignment provides an opportunity for students to develop these skills in the context of science.

### The Content and Related Concepts

The terms *hypothesis*, *theory*, and *scientific* law are not only misused as interchangeable terms but also misunderstood as hierarchical terms that somehow develop into one another. While there is some development in terms of scientific processes, these terms do not become or grow into one another. In other words, if you were to conduct an experiment, and that experiment supports your hypothesis, you would not have a theory. Instead, you would have a supported hypothesis. Similarly, if you had a theory, and you tested it again, ending with the same results, you would not have a law.

The terms *theory* and *law* have specific meaning in science. Scientific laws describe generalized relationships, observed or perceived, of natural phenomena under certain conditions. An example of a scientific law is Charles's law, which describes how gases expand when heated ($V_1/T_1 = V_2/T_2$). Scientific theories, in contrast, are well-substantiated explanations of some aspect of the natural world. An example of a scientific theory is the molecular kinetic theory of matter, which suggests that all matter consists of molecules that are in constant and random motion. Theories do not become laws even with additional evidence; they explain laws. Scientists, for example, often use the molecular kinetic theory of matter to explain why gases expand when they are heated (as described by Charles's law). However, it is important to note that not all scientific laws have accompanying explanatory theories. Theories are often built on multiple ideas with complex variable relationships. They are applied or used as a means to

# SECTION 3: REFUTATIONAL WRITING

## MISCONCEPTION ABOUT THEORIES AND LAWS
### TEACHER NOTES 21

describe a mechanism or as an explanation of a phenomenon. Theories are not proven true. Instead, often there are competing theories for the same phenomena with evidence that can support more than one theory.

## Curriculum and Instructional Considerations

This activity can be used at any point in a traditional biology curriculum. We recommend, however, that it is used after teachers introduce students to the theories and laws in science. Teachers can then have students conduct research about this topic by gathering relevant information from multiple print and digital sources. The students, however, should be taught how to assess the credibility and accuracy of each source they use, how to quote or paraphrase the information they find, and how to use a standard format for citation in their paper. Students also need to be taught to avoid plagiarism. Students must have an opportunity to conduct research and write during this activity in order for it to address the Common Core State Standards for English Language Arts and Literacy.

Students are likely to think that if something is a theory, it is merely a guess, it is unproved, and that it could even lack credibility (Wilson 2007). The confusion of these terms is usually from classroom instruction but can also be developed through media sources and science reports. Science teachers spend much time discussing hypotheses and theories. These terms are the most widely used. Laws are stated as facts and are often shown hierarchically in reliability. Students are not so readily exposed to the idea that there is a difference in the relationship that exists in reliability (i.e., it happens all the time) and in development (i.e., it requires a great deal of evidentiary support from many sources).

Knowing that scientific knowledge changes may be a common rhetoric for students, but their understanding is usually tied to an idea that science changes mainly in facts and mostly through the invention of improved technology for observation and measurement. They are not likely to understand that changes are due to new observations or reinterpretations of previous observations (Aikenhead 1987; Lederman and O'Malley 1990). Students have difficulty distinguishing between theories and the evidence for a theory (Kuhn 1991, 1992) but can develop this understanding in the middle grades (Rosebery, Warren, and Conant 1992).

## Recommendations for Implementing the Activity

This activity takes 100–400 minutes of instructional time to complete, depending on how a teacher decides to spend time in class. See Appendix F (pp. 370–371) for more information about implementing this refutational writing activity.

# SECTION 3: REFUTATIONAL WRITING

## 21 MISCONCEPTION ABOUT THEORIES AND LAWS
### TEACHER NOTES

## Assessment

This activity can be used to identify prior knowledge and knowledge development if it is implemented as a preactivity to the science unit (or course) and then as a postactivity of the same unit (or course). The rubric in Appendix D (p. 368) can be used to not only assess the students' essays but also to compare with previous work to determine changes in their ideas, their writing skills, and their skills in developing a scientific argument. The rubric includes focuses on the content, the structure of the argument, and Mechanics of the essay. The mechanics section of the rubric is well aligned with the Common Core State Standards for English Language Arts and Literacy (NGA and CCSSO 2010). We strongly recommend that teachers use the Comments or Suggestions sections to give students detailed feedback so they will understand what they did wrong, why it is wrong, and ways they can improve their performance next time.

The following sample of a ninth-grade response is provided with scores using the rubric (Figure 21.1) to help understand how the rubric can be applied:

(a) I am going to explain how it is not true that a hypothesis is not a theory that is tested true, and a theory is not tested true to make a law. (b) Sometimes people think that a hypothesis can turn into a theory if there is a test and the test is positive. And they think that a theory that is tested lots of times for a long time by a lot of people is how a science law is made. It can be very confusing, but that is not really how science works.

I am going to help explain this by giving the definitions first and then I am going to give an example. A hypothesis is when you think you have an idea about what will happen before you test. But it really isn't uneducated like some people say. What that means is that you didn't test it yet but you will and that is why it is a hypothesis. (c) The hypothesis works for a single thing in a test. An example is when you want to find out if a certain gas will make your car drive farther than another gas. You <u>might could</u> guess that one gas is better or that both gas types are the same. (d) Even if you test it and you get it right that only means that you had a hypothesis that was supported by the test results. It does not mean that you have made a theory or that you have proven a theory.

A lot of times people will say things like "<u>thats</u> my theory" but what they really should say is "<u>thats</u> my hypothesis" because first they didn't even test it and second they didn't really have a theory to start with.

# SECTION 3: REFUTATIONAL WRITING

## MISCONCEPTION ABOUT THEORIES AND LAWS
### TEACHER NOTES 21

*Figure 21.1. Activity 21 Student Sample Scored Rubric*

| Aspect of the Essay | Point Value 0 | Point Value 1 | Comments/Suggestions |
|---|---|---|---|
| **The claim** | | | |
| The claim that is being advanced is clear | | 1 | a, k |
| The claim that is being refuted (the misconception) is clear | | 1 | b, l |
| **The evidence** | | | |
| Describes the evidence that supports the claim that is being advanced | | 1 | g |
| Describes the evidence as examples, applications, observations, etc. rather than provided as sets of facts | | 1 | g, h, i |
| Multiple sources used to support the argument | 0 | | p and q? Not reliable? |
| Interpretation of the literature is correct | 0 | | None really given. |
| **The justification** | | | |
| Explains why the evidence supporting the claim that is being advanced is important or relevant | | 1 | j |
| Links the evidence to important concepts or principles | 0 | | d attempts but not clear |
| **The challenge** | | | |
| Explains why the claim being refuted is inaccurate | 0 | | vague |
| Explains how or why the misconception may have been developed | | 1 | f, o |
| **Language of science** | | | |
| Use of scientific terms is correct | | 1 | |
| Does not use rhetorical references that misrepresent the nature of science or scientific inquiry | | 1 | |
| **Mechanics** | | | |
| The order and arrangement of the sentences enhances the development of the main idea (organization) | | 1 | |
| The author used complete sentences, proper subject-verb agreement, and kept the tense constant (grammar) | 0 | | Lots of errors in the underlined text |
| The author used appropriate spelling, punctuation, and capitalization (conventions) | 0 | | Lots of errors in the underlined text |
| **Word choice and voice** | | | |
| The author employs a broad range of words and uses the right word (e.g., *affect* vs. *effect*, *their* vs. *there*, etc.). | 0 | | i.e.: "Maybe a lot of hypothesizes make a theory." |
| The sentences are written in an active (rather than passive) voice. | 0 | | i.e.: "... how the theory worked kind of." |
| **Total score** | | 9 / 17 | |

# SECTION 3: REFUTATIONAL WRITING

## 21 MISCONCEPTION ABOUT THEORIES AND LAWS
### TEACHER NOTES

To have a theory does mean that a test had to be done and that the test had to be right. **(e)** But it isn't really that simple because a theory is lots of ideas that were tested to make a big idea right. **(f)** Everybody likes to use gravity for <u>a example</u> but that is more confusing because gravity is like <u>a every day</u> thing that people think is true. It is really a theory since we can't see how it works but we know it does. **(g)** If I had to find a good example I would say that the theory of global warming makes theories easier to understand. Just think about all the ways that people try to prove that there is global warming. They try to test how much seawater rises. And they try to measure how much the ice caps and glaciers are melting. And they look at how many storms there are and how strong they are. And there are tests to measure how much <u>carbondioxide</u> is in the air and how hot the temperatures are all over the place. If all of these were put together to show one idea about the earth, then that would be a theory. **(h)** So global warming is a theory. Theories are complicated because they have lots of different parts to it. **(i)** Like a car. It works because the wheels move and the gas goes through the engine and the motor fires and the car pushes out the gas. <u>Lots of</u> parts make a car and each one of those parts can be a test for something pacific.

**(j)** That is why a hypothesis doesn't really become a theory. It can help add to a theory when it gets tested but not by itself. Maybe a lot of hypothesizes make a theory.

**(k)** Just like there is confusion on the hypothesis and the theory there is confusion on the law and the theory. **(l)** Some people say that a theory is tested and then it becomes a <u>law but that</u> is not true because the theories are really for something different than the laws. **(m)** <u>The theories are for</u> how we explain how something worked. Like when you think of the car you <u>might could</u> think about why the car moved better and find out it was the gas that was in the <u>pacific</u> engine and that <u>pacific</u> car. **(n)** So the theory is helping you to know why or helping you to explain why the car <u>worked good.</u>

But the scientific law is different because it is how the theory worked kind of. **(o)** I think that knowing what the difference is between the scientific law and the theory is even harder because nobody talks about them the right way. Lots of people say that the

# SECTION 3: REFUTATIONAL WRITING

## MISCONCEPTION ABOUT THEORIES AND LAWS
### TEACHER NOTES 21

theory turns into the law when it is tested true but that isn't true. **(p)** I even checked on yahoo answers and that is what people were saying but they were wrong. The scientific law is made from experiments too but the answers for the experiment can be turned into an equation. **(q)** On "How Stuff Works" an example of a Scientific law is $E=MC^2$ because the equation tells you what will happen but not why it happens. That is the biggest reason that the theories and the laws aren't the same. It is like they kind of tell different things about what you are learning when you experiment.

So the next time somebody tells you that they have a theory, you can ask tell them that they really have an hypothesis even if they tested it and it was right.

## Standards Addressed in This Activity

This activity can be used to address the following dimensions outlined in *A Framework for K–12 Science Education* (NRC 2012):

### Scientific Practices

- Engaging in argument from evidence
- Obtaining, evaluating, and communicating information

This activity can be used to address the following standards for literacy in science from the *Common Core State Standards for English Language Arts and Literacy* (NGA and CCSSO 2010):

### Writing

- Text types and purposes
- Production and distribution of writing
- Research to build and present knowledge
- Range of writing

## References

Aikenhead, G. S. 1987. High school graduates' beliefs about science-technology-society III. Characteristics and limitations of scientific knowledge. *Science Education* 71: 459–487.

Kuhn, D. 1991. *The skills of argument*. Cambridge: Cambridge University Press.

Kuhn, D. 1992. *Thinking as argument*. Harvard Educational Review 62: 155–178.

Lederman, N., and M. O'Malley. 1990. Students' perceptions of the tentativeness in science: Development, use, and sources of change. *Science Education* 74: 225–239.

National Governors Association Center (NGA) for Best Practices, and Council of Chief State School Officers (CCSSO). 2010. *Common core state standards for English language arts and literacy*. Washington, DC: National Governors Association for Best Practices, Council of Chief State School.

National Research Council (NRC). 2012. *A framework for K–12 science education:*

## SECTION 3: REFUTATIONAL WRITING

## 21 MISCONCEPTION ABOUT THEORIES AND LAWS
TEACHER NOTES

*Practices, crosscutting concepts, and core ideas*. Washington, DC: National Academies Press.

Rosebery, A., B. Warren, and F. Conant. 1992. Appropriating scientific discourse: Findings from language minority classrooms. *The Journal of Learning Sciences* 2 (1): 61–64.

Wilson, J. 2007. Scientific laws, hypothesis, and theories. Wilstar.com. *http://wilstar.com/theories.htm*.

SECTION 3: REFUTATIONAL WRITING

# MISCONCEPTION ABOUT THE NATURE OF SCIENTIFIC KNOWLEDGE (NATURE OF SCIENCE) 22

Many people believe that scientific knowledge is absolute. In other words, some people think of scientific knowledge as something that has been proven true and as a result, it cannot or will not change. For these people, scientific knowledge is not tentative; scientific knowledge is absolute.

Write a one- to two-page refutational essay to convince someone who thinks that scientific knowledge is absolute and never changes that their idea is a misconception. As you write your paper, remember to

- clearly state the misconception that you are trying to refute;
- include several specific facts, details, reasons, and/or examples that show or demonstrate why the misconception is inaccurate;
- explain how scientific knowledge is durable but why it can change;
- include several specific facts, details, reasons, and/or examples that show or demonstrate how scientific knowledge is tentative and can change over time;
- present your ideas in a clear and logical order, including an introduction, body, and conclusion;
- use a variety of words and well-constructed sentences to create tone and voice;
- include at least five references in your essay, and be sure to cite your references correctly; and
- correct errors in capitalization, punctuation, sentence structure, spelling, and grammar.

You will have _____ to research this topic, and then plan, write, edit, and produce a final product. You must use at least five different reference materials (e.g., your textbook, online resources, and so on) during this process. Be sure to cite all your references in the text of your essay and include a reference section.

You must complete your **research** by _____.

Your **prewrite** (an outline, a concept map, and so on) is due on _____.

Your **initial draft** of your essay is due on _____.

Your **final draft** of your essay is due on _____.

SECTION 3: REFUTATIONAL WRITING

# 22 MISCONCEPTION ABOUT THE NATURE OF SCIENTIFIC KNOWLEDGE
## TEACHER NOTES

## Purpose

The purpose of this activity is to help students understand that scientific knowledge is tentative, although it is well-supported by evidence. This activity is also designed to address many of the Common Core State Standards for English Language Arts and Literacy, which have a strong emphasis on literacy in science. These standards include writing arguments focused on discipline-specific content, writing in a clear and coherent manner, and developing and strengthening students' papers through a process of planning, revising, editing, and rewriting. The *Common Core State Standards for English Language Arts and Literacy* (NGA and CCSSO 2010) also calls for students to be able to conduct a short research project, gather relevant information from multiple print and digital sources, assess the credibility and accuracy of each source, and quote or paraphrase the data and conclusions of others while avoiding plagiarism. This writing assignment provides an opportunity for students to develop these skills in the context of science.

## The Content and Related Concepts

The history of science is filled with examples of how scientific knowledge has experienced both evolutionary and revolutionary changes. It is important to understand that scientific knowledge can be well-supported by empirical evidence, but it can never be proven true in an absolute sense due to the many limitations associated with scientific research. Scientific knowledge, as a result, should not be described as absolute. Instead, scientific knowledge is best described as durable and robust but tentative, because it is abandoned or modified in light of new evidence or ideas or the reconceptualization of existing evidence or ideas. People can have confidence in the body of scientific knowledge because it reflects the scientific community's most current and valid descriptions of and explanations for natural phenomena. But people also need to keep in mind that these descriptions and explanations might one day be modified or abandoned.

Changes in scientific knowledge (i.e., current theories) are influenced by new discoveries as well as by social and religious factors. For example, it was once accepted that ulcers were a factor of weakness or stress. If someone was diagnosed with an ulcer, it was perceived as a sign of weakness, and the person was expected to find some way to relieve the stress. Research into preventing or curing ulcers did not progress very quickly. Instead, medications were prescribed to relax those with ulcers, and it was suggested that people change jobs or occupations. When it was found that people who had ulcers also had a particular kind of bacteria in their stomach, the medical research switched their direction of curing ulcers by addressing

# SECTION 3: REFUTATIONAL WRITING

## MISCONCEPTION ABOUT THE NATURE OF SCIENTIFIC KNOWLEDGE
### TEACHER NOTES 22

the bacteria rather than a person's mental or emotional state. Many people still believe that ulcers are stress related and therefore an illness that is a fault of the patient.

New discoveries are made all the time, and old ideas are modified with new technology and new social perspectives. For example, the idea of spontaneous generation—in which life could be created by rotten meat—was once an accepted idea. Aristotle described this ability for living things to spontaneously generate because they were all "full of soul." Spontaneous generation could support the religious belief that the origin of life could come from nonliving matter. This idea changed when Louis Pasteur placed meat in a jar covered with a screen to prevent flies from entering the jar. His observations, which involved simple methods, changed ideas that had long been held and supported by religious beliefs. People began to recognize the life cycles of living things.

Another example of an old idea that was supported by social and religious ideas is bloodletting. It was believed that by making a person bleed, diseases could be prevented and cured. This was a practice that lasted for over 2,000 years and was based on women's menstrual cycles: Since women had menstrual cycles that were thought to be a cleaning of evil and bad humors, releasing blood from other parts of the body would help balance the humors. The humors were representations of the four basic elements: earth, wind (air), fire, and water. It was further believed that the blood went to the extremities of the body and became stagnant, which was in part why it needed to be let out. It wasn't until there were more discoveries about the circulation of blood and menstrual cycles did some of these ideas change.

Another example is the common sneeze. Prior to understanding what caused a sneeze, many cultures had different ideas about what a sneeze was. For example, the Romans and Egyptians regarded the head as the main location of intelligence, emotions, and spirit. They thought a sneeze was a changing of personality (Beal 2007). Epidemics in the middle ages led Christians to believe that sneezing was the blowing out of the devil and that by making an agreement with God, you could prevent an illness that would lead to death, thus the origin of the phrase "God bless you" after someone sneezed (Beal 2007). With development of science and scientific tools (e.g., the microscope), understanding of how our body works (i.e., dust and mucus in the nose, which cause it to itch), and the understanding of pollen and allergies, the old ideas about sneezes changed.

These kinds of examples are easily found when we look up myths and superstitions. But students may not consider these examples as relevant to what they are learning now in science or to current events. They may have the idea that everything there is to know about science is already known. This comes from the classroom activities, called cookbook labs, in which students are given experiments as a predetermined formula with a right or wrong outcome. Students need to understand that the experiments and theories that they read about and practice in a classroom were often initially derived from trial and error, and the progress of understanding current ideas has developed out of errors rather than out of right answers.

# SECTION 3: REFUTATIONAL WRITING

## 22 MISCONCEPTION ABOUT THE NATURE OF SCIENTIFIC KNOWLEDGE
### TEACHER NOTES

With the design of most classroom instruction in which students earn grades by being right, it is difficult for students to accept that science is more productive, more often about looking for a null hypothesis, and learning from being wrong. Many students are not exposed to science as original inquiry. Labs that meet objectives and rely on formulas for correct outcomes send the message that science is absolute and that it is about being right. Understanding that the nature of science is based on original inquiry, which includes multiple possible results to provide multiple solutions, will be essential to dispelling the misconception that scientific ideas are absolute.

In addition, students are often given laws, theories, formulas, and definitions of terms as static ideas and concepts that are to be memorized and are then asked to duplicate and apply these in the labs with predetermined results. This perpetuates the idea that science is about learning definitions and truths, and as such, science is absolute. To suggest that the ideas are not facts—and to suggest that ideas in science could even possibly be accepted as being wrong in the future—is a difficult concept to accept when textbooks are given as books of facts to memorize, when teachers provide lectures and labs as experts who should know all answers, and when students are expected to write and memorize what teachers say to pass a test (Schleigh 2011).

It is still a common misperception that scientists are "crazy," confusing, don't agree with one another, and are always changing their minds (Schleigh and Keeton 2011; Leblebicioglu et al. 2011). These negative perceptions prevent students from learning more about science, add to their confusion about the topics, and interfere in their motivation and interest in considering a career in science (Schleigh 2011; She 1998; Boylan et al. 1992; Leblebicioglu et al. 2011).

To be able to recognize that science is not absolute but always changing is an important tenet in understanding the nature of science. To apply healthy skepticism about information students hear in everyday news requires an ability to evaluate evidence to determine the acceptability of the new discoveries about old ideas, and a willingness to evaluate the evidence to determine the acceptability of the new discoveries in changing old ideas and understandings. This is what is meant by being scientifically literate.

## Curriculum and Instructional Considerations

This activity can be used at any point in a traditional biology curriculum. However, we recommend that it is used after teachers introduce students to the methods of science or the various practices of science. Teachers can then have students conduct research about this topic by gathering relevant information from multiple print and digital sources. The students, however, should be taught how to assess the credibility and accuracy of each source they use, how to quote or paraphrase the information they find, and how to use a standard format for citation in their paper. Students also need to be taught to avoid plagiarism. Students must have an opportunity to conduct research and write during this activity in order for it to

# SECTION 3: REFUTATIONAL WRITING

## MISCONCEPTION ABOUT THE NATURE OF SCIENTIFIC KNOWLEDGE
### TEACHER NOTES 22

address the Common Core State Standards for English Language Arts and Literacy.

## Middle School
Students are introduced to science as activities that often emphasize reading of textbooks to do research and then reporting that information. Experiments may follow with instruction usually focusing on an understanding of a scientific method for an experimental design. Students learn about variables, developing questions, and writing procedures. The lessons and labs are often provided using a well-known outcome so that students can develop an understanding of relationships between variables and learn about foundational theories and concepts to build from later. Lessons and labs may also focus on getting a correct answer for a good grade. Students in the middle grades should instead learn about theories and their development and about the data that is relevant in supporting those theories. They should be learning that scientific knowledge is subject to modification as new information challenges prevailing theories and as a new theory leads to looking at old observations in a new way (AAAS 2009).

## High School
Students have already learned how to design an experiment and isolate variables. They have some basic understanding of theories that they have memorized, and they are often engaged in lab activities that allow them to confirm those theories. This substantiates the previous ideas about objectivity and absolute truths that they learned in the middle grades. Students have most likely found success in the science classroom if they have been able to complete an experiment, get the expected results from a cookbook lab, and if they have been able to memorize the terms and concepts to pass a test. Inquiry is limited to them testing themselves to see if they can replicate a concept or theory in an experiment.

High school students should also be learning about the features of science that involve continuity and persistence of change. They should be examining how different theories can fit for the same situation and that the purpose of testing and retesting is to reevaluate ideas, not just to confirm them. This ongoing process of developing ideas and scientific knowledge helps develop an understanding for how the world works without perpetuating the misconception for an absolute truth (AAAS 2009).

## Recommendations for Implementing the Activity
This activity takes between 100 and 400 minutes of instructional time to complete, depending on how a teacher decides to spend time in class. See Appendix F (pp. 370–371) for more information on how to implement this activity.

## Assessment
This activity can be used to identify prior knowledge and knowledge development if it is implemented as a preactivity prior to the start of a course (or unit) then following the course (or unit). The rubric in Appendix D (p. 368) can be used to assess the students' essays and to compare previous work to determine changes in their ideas, writing skills, and skills in devel-

# SECTION 3: REFUTATIONAL WRITING

## 22 MISCONCEPTION ABOUT THE NATURE OF SCIENTIFIC KNOWLEDGE
### TEACHER NOTES

oping a scientific argument. The rubric focuses on the content, the structure of the argument, and mechanics of the essay. The Mechanics section of the rubric is well aligned with the Common Core State Standards for English Language Arts and Literacy (NGA and CCSSO 2010). We strongly recommend that teachers use the Comments or Suggestions sections to give students detailed feedback so they will understand what they did wrong, why it is wrong, and ways they can improve their performance next time.

The following response written by a sixth-grade student is an example of how a student's response might be scored (see Figure 22.1) in terms of the structure and content:

> (a) Many people think that science is about a lot of facts and stuff that is always true. (b) But it isn't. (c) Science is actually lots of theories and even though those might seem like they are true they aren't really true. (d) It is only what people think right now. (e) When scientists get ideas or think they know something, it is because they did an experiment. (f) But experiments don't always find out everything there is to know about something so later on when the experiment is done again, it might find another answer and that answer could still be right. (g) An example we talked about is when people thought that life started from meat that was rotten. (h) They observed that when they had rotten meat that flies would suddenly show up from nowhere. Some people did tests to show that the flies really were from the rotting meat. (i) Other people had advertisements that they could make animals out of different kinds of materials. (j) Mr. Pasteur showed that the fact that life from rotten meat wasn't true when he tested it in a different way than what other scientists did. (k) When they tested it they didn't cover the meat but he did. So by changing the way he did the tests he could see that the things that the people thought were true really weren't. (l) Sometimes people think that the science is changing because the scientists were just wrong to start with. (m) But that isn't really always what happens. (n) Sometimes science changes because of how scientists decide to think about it. (o) In class we did a paper towel experiment. (p) Our teacher asked us to decide which paper towel was the best and we had to have an experiment to prove it. (q) But it was really hard for everyone to talk about it because even though we all did good experiments, we did them different because we thought the word best meant something different. (r) It can be very confusing if science is always changing and it is hard to learn because you can't just memorize the answers. (s) But that is what science is about. It is about ideas that change and trying to find out the answer in lots of different ways. (t) And everyone has to agree or it isn't going to be right. (u) But it is ok not to agree until all of the information is in and everyone shares.

# SECTION 3: REFUTATIONAL WRITING

## MISCONCEPTION ABOUT THE NATURE OF SCIENTIFIC KNOWLEDGE
### TEACHER NOTES 22

*Figure 22.1. Activity 22 Student Sample Scored Rubric*

| Aspect of the Essay | Point Value 0 | Point Value 1 | Comments or Suggestions |
|---|---|---|---|
| **The claim** | | | |
| The claim that is being advanced is clear | | 1 | c, d |
| The claim that is being refuted (the misconception) is clear | | 1 | a |
| **The evidence** | | | |
| Describes the evidence that supports the claim that is being advanced | | 1 | g, h, i, j, k, |
| Describes the evidence as examples, applications, observations, etc. rather than provided as sets of facts | | 1 | g, h, j, k, o |
| Multiple sources used to support the argument | 0 | | None specified |
| Interpretation of the literature is correct | | 1 | e, f, n, s but wrong at t |
| **The justification** | | | |
| Explains why the evidence supporting the claim that is being advanced is important or relevant | | 1 | n |
| Links the evidence to important concepts or principles | 0 | | |
| **The challenge** | | | |
| Explains why the claim being refuted is inaccurate | 0 | | |
| Explains how or why the misconception may have been developed | | 1 | l |
| **Language of science** | | | |
| Use of scientific terms is correct | 0 | | Not clear how terms are understood |
| Does not use rhetorical references that misrepresent the nature of science or scientific inquiry | 0 | | t, u |
| **Mechanics** | | | |
| The order and arrangement of the sentences enhances the development of the main idea (organization) | 0 | | Organization of paragraphs is not well structured |
| The author used complete sentences, proper subject-verb agreement, and kept the tense constant (grammar) | 0 | | Sentence structures are not always correct |
| The author used appropriate spelling, punctuation, and capitalization (conventions) | | 1 | |
| **Word choice and voice** | | | |
| The author employs a broad range of words and uses the right word (e.g., *affect* vs. *effect*, *their* vs. *there*, etc.). | 0 | | |
| The sentences are written in an active (rather than passive) voice. | 0 | | |
| **Total score** | 8 / 17 | | |

## SECTION 3: REFUTATIONAL WRITING

## 22  MISCONCEPTION ABOUT THE NATURE OF SCIENTIFIC KNOWLEDGE
### TEACHER NOTES

**(v)** So science is not absolute. **(w)** It always changes and it depends how you look at it on whether it is going to change and sometimes when we get new tools and technology. **(x)** So don't believe this myth anymore. **(y)** And have fun with science.

## Standards Addressed in This Activity

This activity can be used to address the following dimensions outlined in *A Framework for K–12 Science Education* (NRC 2012):

## Scientific Practices

- Engaging in argument from evidence
- Obtaining, evaluating, and communicating information

This activity can be used to address the following standards for literacy in science from the *Common Core State Standards for English Language Arts and Literacy* (NGA and CCSSO 2010):

## Writing

- Text types and purposes
- Production and distribution of writing
- Research to build and present knowledge
- Range of writing

## References

American Association for the Advancement of Science (AAAS). 2009. *Benchmarks for science literacy*. Washington, DC: AAAS.

Beal, J. 2007. *Sneezing myths you'll never forget*. Catalogs.com Info Library. www.catalogs.com/info/history/sneezing-myths.html.

Boylan, C. R., D. M. Hill, A. R. Wallace, and A. E. Wheeler. 1992. Beyond stereotypes. *Science Education* 76: 465–476.

Leblebicioglu, G., D. Metin, E. Yardimci, and P. S. Cetin. 2011. The effect of informal and formal interaction between scientists and children at a science camp on their image of scientists. *Science Education International* 22 (3): 158–174.

National Governors Association Center (NGA) for Best Practices, and Council of Chief State School Officers (CCSSO). 2010. *Common core state standards for English language arts and literacy*. Washington, DC: National Governors Association for Best Practices, Council of Chief State School.

National Research Council (NRC). 2012. *A framework for K–12 science education: Practices, crosscutting concepts, and core ideas*. Washington, DC: National Academies Press.

She, H. C. 1998. Gender and grade level differences in Taiwan students' stereotypes of science and scientists. *Research in Science and Technological Education* 16 (2): 125–135.

Schleigh, S. P. 2011. *A review of middle grade students' perceptions of science and scientists*. Forthcoming.

Schleigh, S. P., and D. Keeton. 2011. *The role of science competitions in motivating students to think and do science*. Forthcoming.

SECTION 3: REFUTATIONAL WRITING

# MISCONCEPTION ABOUT THE WORK OF SCIENTISTS (NATURE OF SCIENCE) 23

Many people think that the work of scientists is procedural in nature. In other words, some people think that scientists use the same techniques to collect or analyze data or follow the same routine for all their investigations. These people, as a result, think that the work of a scientist requires little creativity or imagination.

Write a one- to two-page refutational essay to convince someone who thinks that science is procedural in nature and does not require creativity and imagination that this idea is a misconception. As you write your paper, remember to

- clearly state the misconception that you are trying to refute;
- include several specific facts, details, reasons, and/or examples that show or demonstrate why the misconception is inaccurate;
- explain why scientific investigations require so much creativity and imagination;
- include several specific facts, details, reasons, and/or examples that show how creativity and imagination are important aspects of scientific investigation;
- present your ideas in a clear and logical order, including an introduction, body, and conclusion;
- use a variety of words and well-constructed sentences to create tone and voice;
- include at least five references in your essay and be sure to cite your references correctly; and
- correct errors in capitalization, punctuation, sentence structure, spelling, and grammar.

You will have _____ to research this topic, and then plan, write, edit, and produce a final product. You must use at least five different reference materials (e.g., your textbook, online resources, and so on) during this process. Be sure to cite all your references in the text of your essay and include a reference section.

You must complete your **research** by _____.

Your **prewrite** (an outline, a concept map, and so on) is due on _____.

Your **initial draft** of your essay is due on _____.

Your **final draft** of your essay is due on _____.

SECTION 3: REFUTATIONAL WRITING

# 23 MISCONCEPTION ABOUT THE WORK OF SCIENTISTS
## TEACHER NOTES

## Purpose

The purpose of this activity is to help students understand the important roles that creativity and imagination play in the work of scientists. This activity is also designed to address many of the *Common Core State Standards for English Language Arts and Literacy*, which have a strong emphasis on literacy in science. These standards include writing arguments focused on discipline-specific content, writing in a clear and coherent manner, and developing and strengthening students' papers through a process of planning, revising, editing, and rewriting. The Common Core State Standards for English Language Arts and Literacy (NGA and CCSSO 2010) also calls for students to be able to conduct a short research project, gather relevant information from multiple print and digital sources, assess the credibility and accuracy of each source, and quote or paraphrase the data and conclusions of others while avoiding plagiarism. This writing assignment provides an opportunity for students to develop these skills in the context of science.

## The Content and Related Concepts

Albert Einstein has been quoted as saying, "Imagination is more important than knowledge." New scientific knowledge is often developed out of the actions of creative thinking. For example, much of the technology we have today was first imagined as a fictional concept or event that has later followed as a current-day tool. The creative thought from science fiction, "what if?" is also an underlying question that drives much of science processes. There are many resources and documentaries that provide examples of the connection between the creative thought of science fiction and investigations and discoveries in science (see Resource).

In addition, there are many ideas in science that are supported by inference and then demonstrated and tested with models. The development of these models requires creativity in imagining what the model should be representing and how. For example, the double helix model commonly referred to as DNA was not directly observed; rather, it was a model that was imagined from inferences and investigations. The atom has not been directly observed; however, with the observations of results from investigations, there have been several suggestions of models that would support the observations that have been made.

Although scientists must rely on logic and reason, scientific ideas do not emerge automatically from data or by crunching numbers. Scientists must rely on their imagination to invent new ways to explain how the world works and then to figure out how the new ideas can be put to the test. Scientists must use their imagination and creativity to develop

## SECTION 3: REFUTATIONAL WRITING

### MISCONCEPTION ABOUT THE WORK OF SCIENTISTS
TEACHER NOTES 23

new ways to collect data, to analyze data, and to interpret results. As a result, scientists use their imagination all the time, and their work is every bit as creative as writing poetry, composing music, or designing buildings. Science, therefore, is a blend of logic, imagination, and creativity.

## Curriculum and Instructional Considerations

This activity can be used at any point in a traditional biology curriculum. However, we recommend that it is used after teachers introduce students to the methods of science or the various practices of science, or during a unit in which the development of a model to represent an event or concept is important. This is an appropriate time to explain to students how scientists must be creative and use their imagination in order to be successful in science. Teachers can then have students conduct research about this topic by gathering relevant information from multiple print and digital sources. The students, however, should be taught how to assess the credibility and accuracy of each source they use, how to quote or paraphrase the information they find, and how to use a standard format for citation in their paper. Students also need to be taught to avoid plagiarism. Students must have an opportunity to conduct research and write during this activity in order for it to address the Common Core State Standards for English Language Arts and Literacy.

Traditional science courses have been taught with a strong focus on the memorization of definitions, formulas and theories, and the repetition of procedural labs. There is some value in practicing lab procedures as those found in a cookbook lab, and there are important points in the curriculum in which providing direct information can build student knowledge. However, students should have an opportunity to discuss the methods, tools, and results of the lab, and they should have an opportunity to design their own investigations and models. This will facilitate an understanding for the creativity that can drive the scientific endeavor.

## Middle School

Students are being introduced to the processes of science as a scientific method for an experimental design. Traditional labs focus on following directions to confirm a known outcome. Students are not likely to understand the value and purpose of revising an experiment, and they are also highly focused on getting a correct answer for a grade. Although following procedures can be important for some experiments, it can also lead to misconceptions about how science is done and what the nature of science is. Students are likely to think that there is only one way of doing science, which includes following a specific procedure in order to get a right answer. This removes the sense of creativity and opportunities for discovery and invention

# SECTION 3: REFUTATIONAL WRITING

## 23 MISCONCEPTION ABOUT THE WORK OF SCIENTISTS
### TEACHER NOTES

(AAAS 2009). With the elimination of creativity in science, students are likely to lose interest in science (Schleigh, Messer, and Miles 2011).

### High School

In the high school classroom, many of the labs in which students engage are designed by the instructor or the textbook. Often this is to eliminate safety concerns since many high school labs involve chemicals and sophisticated equipment. In addition, it may be assumed that students have already learned how to design an experiment. Therefore, the focus of the labs in the classroom is to observe a known outcome to confirm a science concept. Students are graded on their ability to complete the labs with the expected outcome, receiving a correct answer. Students are not permitted to use their creativity, or to deviate from the procedures that have been given to them. Just as with middle school, the misconception in high school that science is procedural is substantiated in the cookbook labs and students continue to lose interest in doing science and in pursuing science careers (Schleigh, Messer, and Miles 2011).

## Recommendations for Implementing the Activity

This activity takes between 100 and 400 minutes of instructional time to complete, depending on how a teacher decides to spend time in class. See Appendix F (pp. 370–371) for more information on how to implement this activity.

## Assessment

This activity can be used to identify prior knowledge and knowledge development if it is implemented as a preactivity to a science unit (or course) and then as a postactivity of the same unit (or course). The rubric in Appendix D (p. 368) can be used to assess the students' essays and to compare previous work to determine changes in their ideas, writing skills, and skills in developing a scientific argument. The rubric focuses on the content, the structure of the argument, and mechanics of the essay. The Mechanics section of the rubric is well aligned with the Common Core State Standards for English Language Arts and Literacy. We strongly recommend that teachers use the Comments or Suggestions sections to give students detailed feedback so they will understand what they did wrong, why it is wrong, and ways they can improve their performance next time.

The response written by a seventh-grade student is a weak example in terms of the structure and content. The scoring is provided in the rubric in Figure 23.1:

> **(a)** I think that science sometimes should have procedures but I think that it is also good when there aren't procedures. **(b)** Sometimes procedures get in the way and <u>scientist don't</u> get

# SECTION 3: REFUTATIONAL WRITING

## MISCONCEPTION ABOUT THE WORK OF SCIENTISTS
### TEACHER NOTES 23

*Figure 23.1. Activity 23 Student Sample Scored Rubric*

| Aspect of the Essay | Point Value 0 | Point Value 1 | Comments or Suggestions |
|---|---|---|---|
| **The claim** | | | |
| The claim that is being advanced is clear | | 1 | a |
| The claim that is being refuted (the misconception) is clear | 0 | | Not stated |
| **The evidence** | | | |
| Describes the evidence that supports the claim that is being advanced | 0 | | No evidence; all opinions |
| Describes the evidence as examples, applications, observations, etc. rather than provided as sets of facts | 0 | | d?, h? |
| Multiple sources used to support the argument | 0 | | None provided |
| Interpretation of the literature is correct | 0 | | None included, i? |
| **The justification** | | | |
| Explains why the evidence supporting the claim that is being advanced is important or relevant | 0 | | d (but very weak) |
| Links the evidence to important concepts or principles | 0 | | |
| **The challenge** | | | |
| Explains why the claim being refuted is inaccurate | 0 | | |
| Explains how or why the misconception may have been developed | | 1 | e |
| **Language of science** | | | |
| Use of scientific terms is correct | 0 | | h, |
| Does not use rhetorical references that misrepresent the nature of science or scientific inquiry | 0 | | e, f |
| **Mechanics** | | | |
| The order and arrangement of the sentences enhances the development of the main idea (organization) | 0 | | No real structure. Not enough elaboration |
| The author used complete sentences, proper subject-verb agreement, and kept the tense constant (grammar) | 0 | | Underlined sections |
| The author used appropriate spelling, punctuation, and capitalization (conventions) | | 1 | |
| **Word choice and voice** | | | |
| The author employs a broad range of words and uses the right word (e.g., *affect* vs. *effect*, *their* vs. *there*, etc.). | 0 | | Simplified sentence structure and word choice |
| The sentences are written in an active (rather than passive) voice. | 0 | | |
| **Total score** | 3 / 17 | | |

## SECTION 3: REFUTATIONAL WRITING

## 23 MISCONCEPTION ABOUT THE WORK OF SCIENTISTS
### TEACHER NOTES

to discover things. **(c)** <u>They just do the same things over and over again.</u> **(d)** But science has to be making discoveries not just doing the same thing. **(e)** We do the same things in school because we are still learning how to be scientists. **(f)** <u>But when we grow up, we might be</u> great scientists and we won't have to repeat what other people do. **(g)** We can <u>just discover</u> things. **(h)** Inventions are more like trial and error so they aren't procedures. **(i)** <u>And </u>that is an example of how science is not always just following the steps that someone gives you. **(j)** Science is <u>funner</u> when we get to be creative and do it the way we want instead of worrying about the steps to follow. **(k)** But it is important to write down everything anyways so that other people can check your work and see if you are right with your discoveries.

## Standards Addressed in This Activity

This activity can be used to address the following dimensions outlined in *A Framework for K–12 Science Education* (NRC 2012):

## Scientific Practices

- Engaging in argument from evidence
- Obtaining, evaluating, and communicating information

This activity can be used to address the following standards for literacy in science from the *Common Core State Standards for English Language Arts and Literacy* (NGA and CCSSO 2010):

## Writing

- Text types and purposes
- Production and distribution of writing
- Research to build and present knowledge
- Range of writing

## References

American Association for the Advancement of Science (AAAS). 2009. *Benchmarks for science literacy*. Washington, DC: AAAS.

National Governors Association Center (NGA) for Best Practices, and Council of Chief State School Officers (CCSSO). 2010. *Common core state standards for English language arts and literacy*. Washington, DC: National Governors Association for Best Practices, Council of Chief State School.

National Research Council (NRC). 2012. *A framework for K–12 science education: Practices, crosscutting concepts, and core ideas*. Washington, DC: National Academies Press.

## SECTION 3: REFUTATIONAL WRITING

### MISCONCEPTION ABOUT THE WORK OF SCIENTISTS
#### TEACHER NOTES 23

Schleigh, S. P., and A. Manda. Forthcoming. *What science teachers and scientists know about science*.

Schleigh, S. P., T. Messer, and R. Miles. Forthcoming. *The relationship between creativity and science: A progression of interest*.

## Resource

History Channel. 2009. The universe: Science fiction, science fact. YouTube. *www.youtube.com/watch?v=h3HSthDIe2U&feature=related*.

SECTION 3: REFUTATIONAL WRITING

# MISCONCEPTION ABOUT THE METHODS OF SCIENTIFIC INVESTIGATIONS (NATURE OF SCIENCE)   24

Many people think that there is one method to conduct scientific research. In other words, some people think that the process for doing all scientific research is the same and that there is a specific ordered set of steps (or a method) that all scientists follow during a scientific investigation.

Write a one- to two-page refutational essay to convince someone who thinks that there is only one scientific method that this idea is a misconception. As you write your paper, remember to

- clearly state the misconception that you are trying to refute;
- include several specific facts, details, reasons, and/or examples that show or demonstrate why the misconception is inaccurate;
- explain how scientists conduct many different types of investigations and scientists engage in many different types of activities and usually do not engage in sequence of activities in all the investigations that they plan and carry out;
- include several specific facts, details, reasons, and/or examples that show or demonstrate how there are many different methods of science;
- present your ideas in a clear and logical order, including an introduction, body, and conclusion;
- use a variety of words and well-constructed sentences to create tone and voice;
- include at least five references in your essay, and be sure to cite your references correctly; and
- correct errors in capitalization, punctuation, sentence structure, spelling, and grammar.

You will have _____ to research this topic, and then plan, write, edit, and produce a final product. You must use at least five different reference materials (e.g., your textbook, online resources, and so on) during this process. Be sure to cite all your references in the text of your essay and include a reference section.

You must complete your **research** by _____.

Your **prewrite** (an outline, a concept map, etc.) is due on _____.

Your **initial draft** of your essay is due on _____.

Your **final draft** of your essay is due on _____.

SECTION 3: REFUTATIONAL WRITING

# 24 MISCONCEPTION ABOUT THE METHODS OF SCIENTIFIC INVESTIGATIONS TEACHER NOTES

## Purpose

The purpose of this activity is to help students understand that scientists plan and carry out many different types of scientific investigations, and as a result, there is not one single scientific method that all scientists must follow. This activity is also designed to address many of the *Common Core State Standards for English Language Arts and Literacy*, which have a strong emphasis on literacy in science. These standards include writing arguments focused on discipline-specific content, writing in a clear and coherent manner, and developing and strengthening their papers through a process of planning, revising, editing, and rewriting. The Common Core State Standards for English Language Arts and Literacy (NGA and CCSSO 2010) also calls for students to be able to conduct a short research project, gather relevant information from multiple print and digital sources, assess the credibility and accuracy of each source, and quote or paraphrase the data and conclusions of others while avoiding plagiarism. This writing assignment provides an opportunity for students to develop these skills in the context of science.

## The Content and Related Concepts

Scientists conduct investigations in order to answer a wide range of questions. Scientists, therefore, plan and carry out different types of investigations based on the question they are attempting to answer. For example, scientists conduct literature reviews, analyze existing data sets, conduct systematic observations in the field or in the lab, develop models to test ideas, and design experiments. The traditional way of teaching science is by showing students a classic "scientific method that involves only a very general and incomplete version of the work of scientists" (NRC 2012, p. 24). There is some version of this classic method in the beginning of textbooks, usually coupled with a model that teachers expect students to follow (see Figure 24.1). Although each book or resource usually describes the scientific method and provides a model, it seems that often textbooks neglect to mention that the method is not absolute. Additionally, readers don't seem to notice that the model is different in each resource. Yet the teaching of the scientific method often includes the idea that all scientists agree on what that method is, when the traditional idea of the scientific method is really more like a representation of how scientists write up the results of their studies rather than how they are building knowledge (University of California Berkeley 2011).

The order of stages in each investigation also differs based on what is being investigated, what is known about what is being investigated, and what prior experiences and unique background the researcher has. Differ-

# SECTION 3: REFUTATIONAL WRITING

## MISCONCEPTION ABOUT THE METHODS OF SCIENTIFIC INVESTIGATIONS
### TEACHER NOTES 24

*Figure 24.1. Two Inaccurate Depictions of the Nature of Scientific Inquiry*

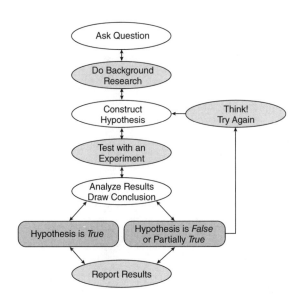

ent scientific disciplines also employ different methods to collect or analyze data, rely on different core theories to frame their work, and use different criteria to evaluate and support scientific claims. There is no single universal step-by-step scientific method that all scientists follow. Traditional scientific methods have emphasized the asking of a question to solve a problem, and the models are designed to provide steps to be able to solve those problems (BSCS 2005). The use of these models in science instruction can lead teachers into teaching science as a cookbook lab (University of California Berkeley 2011) and can create a misconception that there is a single correct and agreed upon method for doing science. Not all science is about solving problems, and often the explorations don't start with a question that seeks to find an answer. Often the questions are seeking to identify causal relationships or eliminate possibilities. Inquiry for different science disciplines looks different depending on the topic that is being explored, the way the question was initiated, and the tools that are available for understanding the topic. Because of this, scientists in different disciplines engage in scientific inquiry in different ways (Schleigh and Manda, forthcoming). It is therefore more important to introduce students to the various methods of science or describe the different practices of scientists.

In addition, many discoveries are made by accident. They may be made when we are exploring a different question. The phrase *The Principle of Limited Sloppiness* has been associated with science to describe fortuitous or acci-

## SECTION 3: REFUTATIONAL WRITING

## 24 MISCONCEPTION ABOUT THE METHODS OF SCIENTIFIC INVESTIGATIONS
### TEACHER NOTES

dental discoveries. It suggests that researchers should be sloppy enough that unexpected things can happen, but not so sloppy that they can't find or can't recognize the event. Rather than teaching students to follow a strict procedure, it would be important to emphasize that if there are any changes in their procedure, that they log those changes in a journal and document any reasons those changes took place.

## Curriculum and Instructional Considerations

This activity can be used at any point in a traditional biology curriculum. We recommend, however, that it is used after teachers introduce students to the methods of science or the various practices of science. Teachers can then have students conduct research about this topic by gathering relevant information from multiple print and digital sources. The students, however, should be taught how to assess the credibility and accuracy of each source they use, how to quote or paraphrase the information they find, and how to use a standard format for citation in their paper. Students also need to be taught to avoid plagiarism. Students must have an opportunity to conduct research and write during this activity in order for it to address the Common Core State Standards for English Language Arts and Literacy.

## Middle School

Students are being introduced to the processes of science as a scientific method for an experimental design. Traditional labs focus on following directions to confirm a known outcome. Students are not likely to understand the value and purpose of revising an experiment, and they are also highly focused on getting a correct answer for a grade. Although following procedures can be important for some experiments, it can also lead to misconceptions about how science is done and what the nature of science is. Students are likely to think that there is only one way of doing science, which includes following a specific procedure in order to get a right answer. This removes the sense of creativity and opportunities for discovery and invention (AAAS 2009). With the elimination of creativity in science, students are likely to lose interest in science (Schleigh, Messer, and Miles, forthcoming).

## High School

In the high school classroom, many of the labs in which students engage are designed by the instructor or the textbook. Often this is to eliminate safety concerns since many high school labs involve chemicals and sophisticated equipment. In addition, it may be assumed that students have already learned how to design an experiment. Therefore, the focus of the labs in the classroom is to observe a known outcome to confirm a science concept. Students are then graded on their ability to complete the labs with the expected outcome, receiving a correct answer. Students are not often permitted to use their creativity or to deviate from the procedures that have been given to them. The misconception that science is procedural is substantiated in the cookbook labs, and students may continue to lose interest in doing sci-

# SECTION 3: REFUTATIONAL WRITING

## MISCONCEPTION ABOUT THE METHODS OF SCIENTIFIC INVESTIGATIONS
### TEACHER NOTES 24

ence and in pursuing science careers (Schleigh, Messer, and Miles, forthcoming).

## Recommendations for Implementing the Activity

This activity takes between 100 and 400 minutes of instructional time to complete, depending on how a teacher decides to spend time in class. See Appendix F (pp. 370–371) for more information on how to implement this activity.

## Assessment

This activity can be used to identify prior knowledge and knowledge development if it is implemented as a preactivity to a science unit (or course) and then as a postactivity of the same unit (or course). The rubric in Appendix D (p. 368) can be used to assess the students' essays and to compare previous work to determine changes in their ideas, writing skills, and skills in developing a scientific argument. The rubric focuses on the content, the structure of the argument, and mechanics of the essay. The Mechanics section of the rubric is well aligned with the Common Core State Standards for English Language Arts and Literacy. We strongly recommend that teachers use the Comments or Suggestions sections to give students detailed feedback so they will understand what they did wrong, why it is wrong, and ways they can improve their performance next time.

The following is an excellent example of a sixth-grade group essay, with the rubric scores (Figure 24.2, p. 282) to illustrate how the essays can be assessed.

(a) If you think that you know the scientific method and you think that the scientific method has 6 steps or even 8 steps, you probably have a misconception. (b) Science does not have just one method. (c) This can be really confusing because in our book, right in the front, there is a picture of the steps that we are supposed to take to do science correctly. Why do they put that there? (d) We looked online and found out that there are LOTS of different models for the scientific method and we wondered who was wrong. We didn't think they could all be right. (e) What we found out was that we also had a misconception. (f) The scientific method does not have only one way of doing it. (g) That is because there are lots of ways of doing science. (h) For example, when a biologist does an experiment, sometimes she will do an experiment where she is testing a certain question. Like if she wanted to know if rats will be allergic to something. She might really do some steps like ask questions and make a hypothesis and make an experiment to test her hypothesis and analyze her test results and then make a conclusion. (i) But sometimes the biologist might do a different kind of steps. Like maybe she will look at coral reefs and see that they look different in one place. And she might ask other people what they saw in their places and that means that she is comparing observations to do her science

## SECTION 3: REFUTATIONAL WRITING

### 24 MISCONCEPTION ABOUT THE METHODS OF SCIENTIFIC INVESTIGATIONS
### TEACHER NOTES

*Figure 24.2. Activity 24 Student Sample Scored Rubric*

| Aspect of the Essay | Point Value 0 | Point Value 1 | Comments or Suggestions |
|---|---|---|---|
| **The claim** | | | |
| The claim that is being advanced is clear | | 1 | b, f |
| The claim that is being refuted (the misconception) is clear | | 1 | a |
| **The evidence** | | | |
| Describes the evidence that supports the claim that is being advanced | | 1 | h, i, m |
| Describes the evidence as examples, applications, observations, etc. rather than provided as sets of facts | | 1 | h, i, |
| Multiple sources used to support the argument | 0 | | c, d but not listed or cited |
| Interpretation of the literature is correct | | 1 | p, t, |
| **The justification** | | | |
| Explains why the evidence supporting the claim that is being advanced is important or relevant | | 1 | k, q |
| Links the evidence to important concepts or principles | | 1 | m, r, s, t |
| **The challenge** | | | |
| Explains why the claim being refuted is inaccurate | | 1 | g, h, i, k |
| Explains how or why the misconception may have been developed | | 1 | c |
| **Language of science** | | | |
| Use of scientific terms is correct | | 1 | |
| Does not use rhetorical references that misrepresent the nature of science or scientific inquiry | | 1 | w |
| **Mechanics** | | | |
| The order and arrangement of the sentences enhances the development of the main idea (organization) | 0 | | Paragraphs with structure to organize the points, examples, and counterarguments would have been helpful. |
| The author used complete sentences, proper subject-verb agreement, and kept the tense constant (grammar) | | 1 | |
| The author used appropriate spelling, punctuation, and capitalization (conventions) | | 1 | |
| **Word choice and voice** | | | |
| The author employs a broad range of words and uses the right word (e.g., *affect* vs. *effect*, *their* vs. *there*, etc.). | | 1 | |
| The sentences are written in an active (rather than passive) voice. | | 1 | |
| **Total score** | | 15 / 17 | |

## SECTION 3: REFUTATIONAL WRITING

## MISCONCEPTION ABOUT THE METHODS OF SCIENTIFIC INVESTIGATIONS
### TEACHER NOTES 24

instead of actually testing something. **(j)** That is still doing science. **(k)** Maybe then that will make her want to do a special experiment on the coral by her (but that might be hard because you can't really change the coral since it is under the water and hard to get to). **(l)** If you think about astronomers they would probably do science different too because they have to only look at light and they don't get to go places to actually test things. **(m)** So astronomers would use models that they make. **(n)** Sometimes those would be like computer things and sometimes it would be like real hard objects in a model. **(o)** They would do their science by watching the models and then deciding what they think that means. **(p)** They still have questions but they don't answer the questions in the same way like in a lab with test tubes and stuff. **(q)** So even though our book tells us that there is one way to do science, we know that there is lots of ways to do it and we know that different kinds of scientists do science methods in different ways. **(r)** That's because they don't have the same kinds of tools and they don't get to touch everything in the same way. **(s)** Something that they should all be doing is asking questions and discussing what they think. **(t)** That means that they have to argue so that they can convince other people how their experiment (since it could be different) is still important and how it means something is correct. **(u)** That is what we are doing now. **(v)** We are trying to convince you that you are not right about science having one method and we are giving you examples to support our argument. **(w)** So really we are doing science and we didn't follow the scientific method in our book.

## Standards Addressed in This Activity

This activity can be used to address the following dimensions outlined in *A Framework for K–12 Science Education* (NRC 2012):

## Scientific Practices

- Engaging in argument from evidence
- Obtaining, evaluating, and communicating information

This activity can be used to address the following standards for literacy in science from the *Common Core State Standards for English Language Arts* (NGA and CCSSO 2010):

## Writing

- Text types and purposes
- Production and distribution of writing
- Research to build and present knowledge
- Range of writing

## References

American Association for the Advancement of Science (AAAS). 2009. *Benchmarks for science literacy*. Washington, DC: AAAS.

Biological Sciences Curriculum Study (BSCS). 2005. *Doing Science: The process of scientific inquiry*. National Institutes of Health (NIH).

## SECTION 3: REFUTATIONAL WRITING

## 24 MISCONCEPTION ABOUT THE METHODS OF SCIENTIFIC INVESTIGATIONS
### TEACHER NOTES

*http://science.education.nih.gov/supplements/nih6/inquiry/default.htm*.

Gedney, L. 1985. *Unexpected scientific discoveries are often the most important, article #741.* Alaska Science Forum. Geophysical Institute, University of Alaska Fairbanks. *www2.gi.alaska.edu/ScienceForum/ASF7/741.html*.

National Governors Association Center (NGA) for Best Practices, and Council of Chief State School Officers (CCSSO). 2010. *Common core state standards for English language arts and literacy.* Washington, DC: National Governors Association for Best Practices, Council of Chief State School.

National Research Council (NRC). 2012. *A framework for K–12 science education: Practices, crosscutting concepts, and core ideas*. Washington, DC: National Academies Press.

Schleigh, S.P., and A. Manda. Forthcoming. *What science teachers and scientists know about science.*

Schleigh, S. P., T. Messer, and R. Miles. Forthcoming. *The relationship between creativity and science: A progression of interest.*

University of California Berkeley. 2011. *Understanding how science really works.* *http://undsci.berkeley.edu/teaching/misconceptions.php*.

## SECTION 3: REFUTATIONAL WRITING

# MISCONCEPTION ABOUT LIFE ON EARTH (EVOLUTION) 25

Many people think that species do not evolve over time. In other words, some people think that a change in the gene frequency in a population from one generation to the next (microevolution) does not happen, and the formation of a new species from an existing species (macroevolution) has never occurred.

Write a one- to two-page refutational essay to convince someone who thinks that species do *not* evolve over time that this idea is a misconception (from a scientific perspective). As you write your paper, remember to

- clearly state the misconception that you are trying to refute;
- include several specific facts, details, reasons, and/or examples that show or demonstrate why the misconception is inaccurate;
- explain how evolution is descent with modification;
- include several specific facts, details, reasons, and/or examples that show or demonstrate the difference between macroevolution and microevolution;
- present your ideas in a clear and logical order, including an introduction, body, and conclusion;
- use a variety of words and well-constructed sentences to create tone and voice;
- include at least five references in your essay, and be sure to cite your references correctly; and
- correct errors in capitalization, punctuation, sentence structure, spelling, and grammar.

You will have _____ to research this topic, and then plan, write, edit, and produce a final product. You must use at least five different reference materials (e.g., your textbook, online resources, and so on) during this process. Be sure to cite all your references in the text of your essay and include a reference section.

You must complete your **research** by _____.

Your **prewrite** (an outline, a concept map, and so on) is due on _____.

Your **initial draft** of your essay is due on _____.

Your **final draft** of your essay is due on _____.

SCIENTIFIC ARGUMENTATION **IN BIOLOGY**: 30 CLASSROOM ACTIVITIES

SECTION 3: REFUTATIONAL WRITING

# 25 MISCONCEPTION ABOUT LIFE ON EARTH
## TEACHER NOTES

## Purpose

The purpose of this activity is to help students understand biological evolution and the mechanisms that drive it. This activity is also designed to address many of the Common Core State Standards for English Language Arts and Literacy, which have a strong emphasis on literacy in science. These standards include writing arguments focused on discipline-specific content, writing in a clear and coherent manner, and developing and strengthening their papers they write through a process of planning, revising, editing, and rewriting. The *Common Core State Standards for English Language Arts and Literacy* (NGA and CCSSO 2010) also calls for students to be able to conduct a short research project, gather relevant information from multiple print and digital sources, assess the credibility and accuracy of each source, and quote or paraphrase the data and conclusions of others while avoiding plagiarism. This writing assignment provides an opportunity for students to develop these skills in the context of science.

## The Content and Related Concepts

Biological evolution is defined as descent with modification. This definition includes both small-scale evolution and large-scale evolution. Small-scale evolution, or microevolution, is a change in the gene frequency of a population from one generation to the next. Large-scale evolution, or macroevolution, in contrast, refers to the descent of different species from a common ancestor over many generations. The central idea of biological evolution is that all life on Earth shares a common ancestor, and through the process of descent with modification, the common ancestor of all life on Earth gave rise to the wide range of diversity that we see all around us today. Biological evolution is driven by processes such as genetic mutation, gene flow, genetic drift, and natural selection. To learn more about biological evolution, see the websites under Resources, which are excellent for both teachers and students.

## Curriculum and Instructional Considerations

This activity should be used during a unit on biological evolution. Teachers can have students conduct research about this topic by gathering relevant information from multiple print and digital sources. The students, however, should be taught how to assess the credibility and accuracy of each source they use, how to quote or paraphrase the information they find, and how to use a standard format for citation in their paper. Students also need to be taught to avoid plagiarism. Students must have an opportunity to conduct research and write during this activity in order for it to address the Common Core State Standards for English Language Arts and Literacy.

## SECTION 3: REFUTATIONAL WRITING

## MISCONCEPTION ABOUT LIFE ON EARTH
### TEACHER NOTES 25

### Middle School

Students should be able to use classification to group organisms hierarchically (Leach et al. 1992); however, they are likely to have a more restrained definition for *animal* that may not include humans, insects, or fish (Mintzes et al. 1991). They are also likely to have difficulty in understanding that classifications are not static, that biologists may not agree with the placement of all organisms in a classification, and that some organisms can be classified as both a bird and an animal (Bell 1981).

### High School

Students should be able to describe hierarchical organizations of organisms using taxonomies (Leach et al. 1992) and should have a stronger sense for the definition of animal. They are likely to still hold the misconception that traits are inherited by one parent, that some characteristics are always inherited by the mother while others are always inherited by the father, or that inheritance is related to the interaction between parent and offspring (Deadman and Kelly 1978; Kargbo, Hobbs, and Erickson 1980; Clough and Wood-Robinson 1985). While they may have some understanding of the genetic material and the role of parents in carrying and transferring traits and characteristics, the students may still hold the misconception that environmentally produced characteristics can be inherited, especially over several generations (Clough and Wood-Robinson 1985).

## Recommendations for Implementing the Activity

This activity takes 100–400 minutes of instructional time to complete depending on how a teacher decides to spend time in class. See Appendix F (pp. 370–371) for more information on how to implement this activity.

## Assessment

This activity can be used to identify prior knowledge and knowledge development if it is implemented as a preactivity to a science unit (or course) and then as a postactivity of the same unit (or course). The rubric in Appendix D (p. 368) can be used to assess the students' essays and to compare previous work to determine changes in their ideas, writing skills, and skills in developing a scientific argument. The rubric focuses on the content, the structure of the argument, and mechanics of the essay. The Mechanics section of the rubric is well aligned with the Common Core State Standards for English Language Arts and Literacy. We strongly recommend that teachers use the Comments or Suggestions sections to give students detailed feedback so they will understand what they did wrong, why it is wrong, and ways they can improve their performance next time.

The following is an example of an essay that was written by a 10th-grade student followed by a rubric (Figure 25.1, p. 288) to illustrate how to assess a student's response.

> **(a)** In the news and at school people are always talking about the difference between evolution and creationism. **(b)**

# SECTION 3: REFUTATIONAL WRITING

## 25 MISCONCEPTION ABOUT LIFE ON EARTH
### TEACHER NOTES

*Figure 25.1. Activity 25 Student Sample Scored Rubric*

| Aspect of the Essay | Point Value | | Comments or Suggestions |
|---|---|---|---|
| | 0 | 1 | |
| **The claim** | | | |
| The claim that is being advanced is clear | 0 | | No claim |
| The claim that is being refuted (the misconception) is clear | 0 | | No clear counterargument |
| **The evidence** | | | |
| Describes the evidence that supports the claim that is being advanced | 0 | | attempts to support (f) but confusing |
| Describes the evidence as examples, applications, observations, etc. rather than provided as sets of facts | | 1 | e, q, |
| Multiple sources used to support the argument | 0 | | None provided |
| Interpretation of the literature is correct | 0 | | e, r, s, t |
| **The justification** | | | |
| Explains why the evidence supporting the claim that is being advanced is important or relevant | | 1 | s |
| Links the evidence to important concepts or principles | | 1 | p, u |
| **The challenge** | | | |
| Explains why the claim being refuted is inaccurate | | 1 | u |
| Explains how or why the misconception may have been developed | | 1 | m |
| **Language of science** | | | |
| Use of scientific terms is correct | 0 | | k, t |
| Does not use rhetorical references that misrepresent the nature of science or scientific inquiry | 0 | | b (bolded section), m |
| **Mechanics** | | | |
| The order and arrangement of the sentences enhances the development of the main idea (organization) | 0 | | Not focused and needs improvement in fluency |
| The author used complete sentences, proper subject-verb agreement, and kept the tense constant (grammar) | 0 | | Underlined sections |
| The author used appropriate spelling, punctuation, and capitalization (conventions) | | 1 | |
| **Word choice and voice** | | | |
| The author employs a broad range of words and uses the right word (e.g., *affect* vs. *effect*, *their* vs. *there*, etc.). | | 1 | |
| The sentences are written in an active (rather than passive) voice. | 0 | | |
| **Total score** | 7 / 17 | | |

# SECTION 3: REFUTATIONAL WRITING

## MISCONCEPTION ABOUT LIFE ON EARTH
### TEACHER NOTES 25

There are lots of <u>people argument</u> about which one is <u>right and if they</u> took time to learn about the theories in evolution they might be able to know **which one is right**. (c) There are lots of things you should know about evolution but most important is microevolution and macroevolution. (d) If you learned about these two really important terms you would know a lot about evolution.

(e) Microevolution is when small changes happen in a group of animals. Like in some birds that Darwin studied he learned that the birds didn't have as many that had a long beak and so he could tell that the group of birds were changing. (f) Since evolution is actually a word for change, even for living things and nonliving things, <u>seeing these changes shows in the bird groups or populations shows that they</u> are evolving. (g) Evolution really happens. Most people don't really disagree with seeing these kinds of changes. (h) In fact, we see changes like this with humans. Have you ever noticed how Chinese people might have more slanted eyes than other people like Germans? (i) Or that there are different shades of brown in Spanish groups? (j) Like Mexicans and people from Spain don't really have the same skin color. (k) It might be possible to count how many times some of these characteristics shows up and that is called population frequency. (l) So this can be used to have evidence of evolution on a micro scale – or microevolution.

(m) Most of the time Creationists and other people that don't believe in evolution, it is because they don't like to think that people can come from other species or that any species can come from other species. (n) They say that it is like saying god didn't create living things and they don't think that there is real evidence for species changing. (o) Species changing is called macroevolution. (p) It is true that we don't have the same evidence for macroevolution as microevolution but we still have evidence. (q) For example, scientists have found lots of fossils that are of different kinds of skulls. (r) It is like a puzzle where you can put pieces together and see how the shapes and colors seem to fit. (s) Scientists did the same thing and that is how they know that different species have changed and how they are related to each other. (t) There is a tree that can be used to help with a model for knowing what things are related and the characteristics are used to show which things are more related than other things. (u) In macroevolution scientist can also use things like DNA that shows how we get our characteristics. (v) The hardest thing about the macroevolution is that we didn't get to see the evolution so it is called inference instead of real observations or real evidence. (w) But

## SECTION 3: REFUTATIONAL WRITING

### 25 MISCONCEPTION ABOUT LIFE ON EARTH
TEACHER NOTES

that still counts because inferences are always used for evidence in science.

## Standards Addressed in This Activity

This activity can be used to address the following dimensions outlined in *A Framework for K–12 Science Education* (NRC 2012):

### Scientific Practices

- Engaging in argument from evidence
- Obtaining, evaluating, and communicating information

### Crosscutting Concepts

- Cause and effect: Mechanisms and explanations
- Stability and change

### Life Sciences Core Ideas

- Biological evolution: Unity and diversity

This activity can be used to address the following standards for literacy in science from the *Common Core State Standards for English Language Arts and Literacy* (NGA and CCSSO 2010):

### Writing

- Text types and purposes
- Production and distribution of writing
- Research to build and present knowledge
- Range of writing

## References

Bell, B. 1981. *When is an animal, not an animal?* Journal of Biological Education 15: 213–218.

Clough, E. E., and C. Wood-Robinson. 1985. *Children's understanding of inheritance.* Journal of Biological Education 19: 304–310.

Deadman, J., and P. Kelly. 1978. *What do secondary school boys understand about evolution and heredity before they are taught the topics?* Journal of Biological Education 12: 7–15.

Kargbo, D., E. Hobbs, and G. Erickson. 1980. *Children's beliefs about inherited characteristics.* Journal of Biological Education 14: 137–146.

Leach, J., R. Driver, P. Scott, and C. Wood-Robinson. 1992. *Progression in understanding of ecological concepts by pupils aged 5 to 16.* Leeds, UK: The University of Leeds, Centre for Studies in Science and Mathematics Education.

Mintzes, J., J. Trowbridge, M. Arnaudin, and J. Wandersee. 1991. Children's biology: Studies on conceptual development in the life sciences. In *The psychology of learning science*, ed. S. Glynn, R. Yeany, and B. Britton, 179–202. Hillsdale, NJ: Lawrence Erlbaum Associates.

National Governors Association Center (NGA) for Best Practices, and Council of Chief State School Officers (CCSSO). 2010. *Common core state standards for English language arts and literacy.* Washington, DC: National Governors Association for Best Practices, Council of Chief State School.

## SECTION 3: REFUTATIONAL WRITING

### MISCONCEPTION ABOUT LIFE ON EARTH
TEACHER NOTES 25

National Research Council (NRC). 2012. *A framework for K–12 science education: Practices, crosscutting concepts, and core ideas*. Washington, DC: National Academies Press.

## Resources

Understanding Evolution
 *http://evolution.berkeley.edu*

Evolution
 *www.pbs.org/wgbh/evolution*

SECTION 3: REFUTATIONAL WRITING

# MISCONCEPTION ABOUT BACTERIA (MICROBIOLOGY) 26

Many people think that all bacteria cause disease. In other words, some people think that all bacteria are bad for humans.

Write a one- to two-page refutational essay to convince someone who thinks all bacteria cause disease that this idea is a misconception (from a scientific perspective). As you write your paper, remember to

- clearly state the misconception that you are trying to refute;
- include several specific facts, details, reasons, and/or examples that show or demonstrate why the misconception is inaccurate;
- explain the important functions that bacteria serve in in living systems;
- include several specific facts, details, reasons, and/or examples that show or demonstrate how bacteria can be beneficial in living systems;
- present your ideas in a clear and logical order, including an introduction, body, and conclusion;
- use a variety of words and well-constructed sentences to create tone and voice;
- include at least five references in your essay, and be sure to cite your references correctly; and
- correct errors in capitalization, punctuation, sentence structure, spelling, and grammar.

You will have _____ to research this topic, and then plan, write, edit, and produce a final product. You must use at least five different reference materials (e.g., your textbook, online resources, and so on) during this process. Be sure to cite all your references in the text of your essay and include a reference section.

You must complete your **research** by _____.

Your **prewrite** (an outline, a concept map, and so on) is due on _____.

Your **initial draft** of your essay is due on _____.

Your **final draft** of your essay is due on _____.

SECTION 3: REFUTATIONAL WRITING

# 26 MISCONCEPTION ABOUT BACTERIA
## TEACHER NOTES

## Purpose

The purpose of this activity is to help students learn about the wide diversity of bacteria found on Earth and some of the roles bacteria play in biological systems. This activity is also designed to address many of the Common Core State Standards for English Language Arts and Literacy, which have a strong emphasis on literacy in science. These standards include writing arguments focused on discipline-specific content, writing in a clear and coherent manner, and developing and strengthening their papers through a process of planning, revising, editing, and rewriting. The *Common Core State Standards for English Language Arts and Literacy* (NGA and CCSSO 2010) also calls for students to be able to conduct a short research project, gather relevant information from multiple print and digital sources, assess the credibility and accuracy of each source, and quote or paraphrase the data and conclusions of others while avoiding plagiarism. This writing assignment provides an opportunity for students to develop these skills in the context of science.

## The Content and Related Concepts

Bacteria are ubiquitous. Bacteria are microscopic, single-celled, living organisms that do not contain a nucleus and are therefore usually classified as prokaryotes. They reproduce by fission or by forming spores and are thought to be the first forms of life on Earth. For that reason, bacteria might be considered the oldest (or most enduring) living organisms on Earth. They display a wide diversity of shapes, sizes, and colors (morphologies). For example, some bacteria are so small that they are only 0.3 micrometers long and invisible to the naked eye; while others are up to a 0.5 millimeters long and are visible to the naked eye. In part, because of the variety and the difficulty with which to observe these rather small living organisms, the concept of species for bacteria is a continued debate. The reason that there is some discourse as to the ability to organize bacteria into species is that bacteria are so closely related and because bacteria, when combined with different strains, seem to have the ability to recombine alleles, making the description of a species muddled for bacteria (Hanage, Fraser, and Spratt 2005).

Although the concept and definition of a bacterial species is a recurrent problem of taxonomists (Staley 2006), it is recognized that there are many different kinds of bacteria requiring a wide variety of supportive environments in order to thrive. In fact, there is some form of bacterial "species" in every known ecosystem including the bodies of other living organisms. They can be found in soil, lakes, streams, hot springs, and even deep inside the Earth's crust. In addition, they live on and in the bodies of other organisms. Bacteria

# SECTION 3: REFUTATIONAL WRITING

## MISCONCEPTION ABOUT BACTERIA
### TEACHER NOTES 26

are vital in recycling nutrients, with many steps in various nutrients cycles depending on these organisms (e.g., the fixation of nitrogen from the atmosphere or converting sulfur compounds around hydrothermal vents located at the bottom of the ocean). There are approximately 10 times as many bacterial cells in the human body as there are human cells. The vast majority of the bacteria found on and in the human body, however, are harmless and a few are even beneficial. Therefore, it is a misconception to think that all bacteria cause disease, because only a small fraction of all the species of bacteria found on the Earth cause a disease in humans.

## Curriculum and Instructional Considerations

This activity can be used as part of a unit about biodiversity, a unit about microbiology, or an ecology unit.

### Middle School

Health topics are common curriculum units and serve as a means of introducing microbes. This activity would serve as an introduction to microbes and the classification of living things, beyond the obvious physical characteristics, following a unit on health, body functions, and systems. Prior to engaging in this activity, the teacher should help students to identify what they know about the classification of living things and basic characteristics of living things.

The middle-level students may have a basic idea for what is considered living and nonliving; however, they may continue to limit their criteria to breathing, movement, reproduction and death. This would interfere with their ability to identify bacteria as living organisms. Introducing viruses could further confuse them in identifying criteria for living and nonliving. Middle-level students may believe that fire, clouds, and the Sun are possibly classified as living while bacteria, plants, fungi, and certain animals would be classified as nonliving (Bell and Freyberg 1985; Leach et al. 1992).

### High School

This activity would follow a unit on DNA and disease, birth defects, and illnesses that are hereditary versus nonhereditary. This may be a unit within a biology course or an anatomy course and would help students to develop a stronger understanding for the classification of living things as well as for how organisms in an ecosystem impact humans and other living things.

High school students are able to develop hierarchical classifications for living organisms but may continue to recognize bacteria as nonliving organisms. They may continue to distinguish between living and nonliving, rarely mentioning structural criteria (cells) or

## SECTION 3: REFUTATIONAL WRITING

## 26 MISCONCEPTION ABOUT BACTERIA
TEACHER NOTES

biochemical characteristics (DNA) (Brumby 1982; Leach et al. 1992)

Teachers can have students conduct research about this topic by gathering relevant information from multiple print and digital sources. The students, however, should be taught how to assess the credibility and accuracy of each source they use, how to quote or paraphrase the information they find, and how to use a standard format for citation in their paper. Students also need to be taught to avoid plagiarism. Students must have an opportunity to conduct research and write during this activity in order for it to address the Common Core State Standards for English Language Arts and Literacy.

## Recommendations for Implementing the Activity

This activity takes between 100 and 400 minutes of instructional time to complete depending on how a teacher decides to spend time in class. See Appendix F (pp. 370–371) for more information on how to implement this activity.

## Assessment

This activity can be used to identify prior knowledge and knowledge development if it is implemented as a preactivity to a science unit (or course) and then as a postactivity of the same unit (or course). The rubric in Appendix D (p. 368) can be used to assess the students'

essays and to compare previous work to determine changes in their ideas, writing skills, and skills in developing a scientific argument. The rubric focuses on the content, the structure of the argument, and mechanics of the essay. The Mechanics section of the rubric is well aligned with the Common Core State Standards for English Language Arts and Literacy. We strongly recommend that teachers use the Comments or Suggestions sections to give students detailed feedback so they will understand what they did wrong, why it is wrong, and ways they can improve their performance next time.

The following is a sample written by an 11th-grade student followed by a rubric (Figure 26.1) used to score the sample.

(a) When a person gets sick they usually hear that the cause is from either bacteria or a virus. (b) This leads many people to believe that bacteria and viruses are the same thing and all are bad for humans. (c) This is a misconception that can be proven wrong with a little more information.

(d) First off viruses and bacteria are not the same at all. (e) Bacteria are larger than viruses by about a hundred times and are much more complex. (f) Bacteria are alive and contain all they need to replicate themselves whereas a virus is not alive and must have

## SECTION 3: REFUTATIONAL WRITING

*Figure 26.1. Activity 26 Student Sample Scored Rubric*

| Aspect of the Essay | Point Value 0 | Point Value 1 | Comments or Suggestions |
|---|---|---|---|
| **The claim** | | | |
| The claim that is being advanced is clear | 0 | | d? argument is about harmfulness of bacteria not about similarities of bacteria and viruses |
| The claim that is being refuted (the misconception) is clear | | 1 | b, c |
| **The evidence** | | | |
| Describes the evidence that supports the claim that is being advanced | | 1 | e, f, g, i, n, o |
| Describes the evidence as examples, applications, observations, etc. rather than provided as sets of facts | | 1 | p, q |
| Multiple sources used to support the argument | 0 | | None provided |
| Interpretation of the literature is correct | | 1 | Although vague, the information is correct |
| **The justification** | | | |
| Explains why the evidence supporting the claim that is being advanced is important or relevant | 0 | | |
| Links the evidence to important concepts or principles | 0 | | |
| **The challenge** | | | |
| Explains why the claim being refuted is inaccurate | 0 | | |
| Explains how or why the misconception may have been developed | | 1 | a, b, l |
| **Language of science** | | | |
| Use of scientific terms is correct | | 1 | |
| Does not use rhetorical references that misrepresent the nature of science or scientific inquiry | 0 | | Presents as list of facts rather than as observations, theories, etc. |
| **Mechanics** | | | |
| The order and arrangement of the sentences enhances the development of the main idea (organization) | 0 | | No structure in paragraphs and comparisons are difficult to follow |
| The author used complete sentences, proper subject-verb agreement, and kept the tense constant (grammar) | | 1 | |
| The author used appropriate spelling, punctuation, and capitalization (conventions) | | 1 | |
| **Word choice and voice** | | | |
| The author employs a broad range of words and uses the right word (e.g., *affect* vs. *effect*, *their* vs. *there*, etc.). | | 1 | |
| The sentences are written in an active (rather than passive) voice. | | 1 | |
| **Total score** | 10 / 17 | | |

# SECTION 3: REFUTATIONAL WRITING

## 26 MISCONCEPTION ABOUT BACTERIA
### TEACHER NOTES

help from another living organism to reproduce. **(g)** A virus actually invades other cells and uses those cells to reproduce more copies of itself.

**(h)** Bacteria are found everywhere, from the deepest oceans to the tallest mountain tops and everywhere in between, in fact bacteria can be found in the atmosphere and in clouds. **(i)** Most bacteria (about 99%) are harmless or even helpful to humans; disease is caused by only a few of them.

**(j)** Harmful bacteria can cause such illnesses as tetanus, pneumonia, syphilis, strep throat, tuberculosis and cholera. **(k)** Bacteria can cause food poisoning such as botulism and it was bacteria-carrying fleas found on rats and mice that caused the bubonic plague and the deaths of millions of people. **(l)** With statements such as these it is easy to understand how the misconception came about, but not all bacteria are harmful.

**(m)** Helpful bacteria are currently residents on and in every human being. **(n)** Some bacteria live in our intestines and help in digestion and destroying harmful organisms while others help break down lactose and help to release beneficial vitamins such as vitamin B and K. **(o)** Other bacteria live inside the mouth, nose, throat and intestines of humans and keep out other harmful microbes from living there.

**(p)** One such example is the probiotic bacteria found in various dairy and soy products. **(q)** These bacteria have been shown to help to prevent and relieve infections and problems such as diarrhea and evidence suggests they may help to prevent colon cancer as well.

**(r)** Bacteria can survive without humans but humans cannot survive with bacteria.

## Standards Addressed in This Activity

This activity can be used to address the following dimensions outlined in *A Framework for K–12 Science Education* (NRC 2012):

## Scientific Practices

- Engaging in argument from evidence
- Obtaining, evaluating, and communicating information

## Crosscutting Concepts

- Cause and effect: Mechanisms and explanations
- Energy and matter: Flows, cycles, and conservation

# SECTION 3: REFUTATIONAL WRITING

## MISCONCEPTION ABOUT BACTERIA
### TEACHER NOTES 26

### Life Sciences Core Ideas

- Ecosystems: Interactions, energy, and dynamics

This activity can be used to address the following standards for literacy in science from the *Common Core State Standards for English Language Arts and Literacy* (NGA and CCSSO 2010):

### Writing

- Text types and purposes
- Production and distribution of writing
- Research to build and present knowledge
- Range of writing

## References

Bell, B., and P. Freyberg. 1985. Language in the science classroom. In *Learning in science*, ed. R. Osborne and P. Freyberg, 29–40. Auckland, NZ: Heinemann.

Brumby, M. 1982. Students' perceptions of the concept of life. *Science Education* 66: 613–622.

Hanage, W., C. Fraser, and B. Spratt. 2005. Fuzzy species among recombinogenic bacteria. BioMed Central Biology. *www.biomedcentral.com/1741-7007/3/6*.

Leach, J., R. Driver, P. Scott, and C. Wood-Robinson. 1992. *Progression in understanding of ecological concepts by pupils aged 5 to 16*. Leeds, UK: The University of Leeds, Centre for Studies in Science and Mathematics Education.

National Governors Association Center (NGA) for Best Practices, and Council of Chief State School Officers (CCSSO). 2010. *Common core state standards for English language arts and literacy*. Washington, DC: National Governors Association for Best Practices, Council of Chief State School.

National Research Council (NRC). 2012. *A framework for K–12 science education: Practices, crosscutting concepts, and core ideas*. Washington, DC: National Academies Press.

Staley, J. 2006. The bacterial species dilemma and the genomic-phylogenetic species concept. *Philosophical Transactions of the Royal Society of London* 47 (3): 177–182.

Venkataraman, S,. V. Reddy, S. P. Reddy, N. Idamakanti, P. Hallenbeck, and J. Loo. 2006. Structure of Seneca Valley Virus-001, an oncolytic picornavirus representing a new genus. *Structure* (16): 1555–1561.

SECTION 3: REFUTATIONAL WRITING

# MISCONCEPTION ABOUT INTERACTIONS THAT TAKE PLACE BETWEEN ORGANISMS (ECOLOGY)

## 27

Many people think that all interactions between organisms are competitive in nature. In other words, some people think that two or more organisms interact with one another, one will benefit and one will not.

Write a one- to two-page refutational essay to convince someone who thinks that all interactions that take place between animals are competitive in nature that this idea is a misconception. As you write your paper, remember to

- clearly state the misconception that you are trying to refute;
- include several specific facts, details, reasons, and/or examples that show or demonstrate why the misconception is inaccurate;
- discuss several different types of interactions that can happen between organisms;
- include several specific facts, details, reasons, and/or examples that show or demonstrate how some interactions are not competitive;
- present your ideas in a clear and logical order, including an introduction, body, and conclusion;
- use a variety of words and well-constructed sentences to create tone and voice;
- include at least five references in your essay, and be sure to cite your references correctly; and
- correct errors in capitalization, punctuation, sentence structure, spelling, and grammar.

You will have _____ to research this topic, and then plan, write, edit, and produce a final product. You must use at least five different reference materials (e.g., your textbook, online resources, and so on) during this process. Be sure to cite all your references in the text of your essay and include a reference section.

You must complete your **research** by _____.

Your **prewrite** (an outline, a concept map, etc.) is due on _____.

Your **initial draft** of your essay is due on _____.

Your **final draft** of your essay is due on _____.

SECTION 3: REFUTATIONAL WRITING

# 27 MISCONCEPTION ABOUT INTERACTIONS THAT TAKE PLACE BETWEEN ORGANISMS
## TEACHER NOTES

## Purpose

The purpose of this activity is to help students understand some of the different types of interactions that can, and do, take place between organisms within an ecosystem. This activity is also designed to address many of the Common Core State Standards for English Language Arts and Literacy, which have a strong emphasis on literacy in science. These standards include writing arguments focused on discipline-specific content, writing in a clear and coherent manner, and developing and strengthening their papers they write through a process of planning, revising, editing, and rewriting. The *Common Core State Standards for English Language Arts and Literacy* (NGA and CCSSO 2010) also calls for students to be able to conduct a short research project, gather relevant information from multiple print and digital sources, assess the credibility and accuracy of each source, and quote or paraphrase the data and conclusions of others while avoiding plagiarism. This writing assignment provides an opportunity for students to develop these skills in the context of science.

## The Content and Related Concepts

An ecosystem consists of a community that involves interactions between biotic and abiotic factors. These interactions work together as a system with various relationships that help keep the ecosystem healthy and virile. Changes in the ecosystem (e.g., an introduction of a new species) can impact the balance and the health of the initial ecosystem. Such effects may include the reduction of one species and the increase of another.

The relationships organisms and species have on one another vary by the type of interaction, and that interaction can sometimes have a positive effect, a negative effect, or no effect at all. The interactions are defined by the mechanism of the interaction and by the effect of the interaction. The effect of the interaction can further be classified by the strength, duration, and direction of the interaction.

The interactions that take place between organisms within an ecosystem can be grouped into three broad categories. The first category is competitive interactions. These interactions refer to any situation in which two or more organisms compete for the same resource. This type of interaction can occur within a species or between two or more different species. In this type of interaction, some individuals are better than others at obtaining some type of limited resource. The second broad category is interactions that are cooperative in nature. This category includes symbiotic relationships such as mutualism and communalism. In mutualism, the close, long-term relationship is beneficial to both species. In commensalism, the relationship benefits one species but not the other yet

# SECTION 3: REFUTATIONAL WRITING

## MISCONCEPTION ABOUT INTERACTIONS THAT TAKE PLACE BETWEEN ORGANISMS 27
### TEACHER NOTES

the other is not harmed by the relationship. The third category is exploitative interactions. This category includes parasitism. In a parasitic relationship, one species benefits from the relationship, while the other is harmed.

## Curriculum and Instructional Considerations

This activity should be used during a unit on ecosystems. Teachers can have students conduct research about this topic by gathering relevant information from multiple print and digital sources. The students, however, should be taught how to assess the credibility and accuracy of each source they use, how to quote or paraphrase the information they find, and how to use a standard format for citation in their paper. Students also need to be taught to avoid plagiarism. Students must have an opportunity to conduct research and write during this activity in order for it to address the Common Core State Standards for English Language Arts and Literacy (NGA and CCSSO 2010).

### Middle School

Students are introduced to food webs and the trophic levels that interact with one another. They may have had some instruction on the ecosystem and the effect on an ecosystem when an organism is added to or removed from the ecosystem. Most discussions in the classroom focus on the predator and prey relationship with the use of a food web. Students are able to identify groups of organisms by hierarchy (Leach et al. 1992) and can identify direct interactions that are related to predator and prey relationships but may still have difficulty with understanding that animals are indirectly impacted by the resources in an ecosystem. Students may still have difficulty understanding that a food source can be scarce in an ecosystem, and they are likely to think that an organism can change its diet according to the availability of a particular source (Leach et al. 1992). This adds to students' confusion about the theory of natural selection as middle grade students are likely to think that organisms can adapt individually and deliberately, rather than as a population over generations (Brumby 1979; Clough and Wood-Robinson 1985).

### High School

In the high school classroom, the instruction includes a focus on the environment and the ecosystem, and there is an emphasis on the theory of evolution and natural selection. However, high school students may continue to have difficulty accepting that resources have limitations in an ecosystem when the resource is a population that is needed to support another population (Leach et al. 1992). This adds to their misunderstanding of the kinds of interactions that animals have and the development of an understanding for the carrying capacity as a factor of the limitations of a population. The high school student is also likely to continue having difficulty understanding natural selection and that the environment itself is directly related to evolution rather than the relationships and interactions between and among species (Bishop and Anderson 1990; Brumby 1979).

## SECTION 3: REFUTATIONAL WRITING

## 27 MISCONCEPTION ABOUT INTERACTIONS THAT TAKE PLACE BETWEEN ORGANISMS
### TEACHER NOTES

## Recommendations for Implementing the Activity

This activity takes 100–400 minutes of instructional time to complete depending on how a teacher decides to spend time in class. See Appendix F (pp. 370–371) for more information on implementing this activity.

## Assessment

This activity can be used to identify prior knowledge and knowledge development if it is implemented as a pre/post activity prior to the unit on ecology and then following the unit on ecology. The rubric in Appendix D (p. 368) can be used to assess the students' essays and to compare previous work to determine changes in their ideas, writing skills, and skills in developing a scientific argument. The rubric focuses on the content, the structure of the argument, and mechanics of the essay. The mechanics section of rubric is well aligned with the Common Core State Standards for English Language Arts and Literacy. We strongly recommend that teachers use the Comments or Suggestions sections to give students detailed feedback so they will understand what they did wrong, why it is wrong, and ways they can improve their performance next time.

The following is an example of an essay written by an eleventh-grade student, and it has been scored to help identify how the rubric (Figure 27.1) would be applied.

(a) The misconception that all interactions between animals are competitive is wrong and may be accepted as being true due to common statements like "survival of the fittest". (b) I think the misconception is about how competition is defined between different animals based on the animal interactions among a number of competitive species because their interactions may all be competitive, however not all species are competitive. (c) According to the website About.com there are four basic groups of species interactivity. (d) These are:

- Competitive interactions
- Consumer-resource interactions
- Detritivore-detritis interactions
- Mutualistic interactions

(e) Competitive interactions are when more than one species is competing for the same resource like food, water, or shelter. (f) For example a lion and a hyena both fighting over a gazelle is a competitive interaction. (g) In my example the food source is the gazelle and since there is only one gazelle it is a limited resource. (h) Both the lion and the hyena would be negatively impacted since the food source will have been used up. (i) But not all species are after

# SECTION 3: REFUTATIONAL WRITING

## MISCONCEPTION ABOUT INTERACTIONS THAT TAKE PLACE BETWEEN ORGANISMS — 27
### TEACHER NOTES

*Figure 27.1. Activity 27 Student Sample Scored Rubric*

| Aspect of the Essay | Point Value 0 | Point Value 1 | Comments or Suggestions |
|---|---|---|---|
| **The claim** | | | |
| The claim that is being advanced is clear | 0 | | |
| The claim that is being refuted (the misconception) is clear | | 1 | a |
| **The evidence** | | | |
| Describes the evidence that supports the claim that is being advanced | | 1 | d |
| Describes the evidence as examples, applications, observations, etc. rather than provided as sets of facts | | 1 | e, f, g, k, l, etc. |
| Multiple sources used to support the argument | 0 | | About.com is not a valid resource |
| Interpretation of the literature is correct | | 1 | |
| **The justification** | | | |
| Explains why the evidence supporting the claim that is being advanced is important or relevant | | 1 | v, w |
| Links the evidence to important concepts or principles | 0 | | Ecosystems; survival of the fittest? |
| **The challenge** | | | |
| Explains why the claim being refuted is inaccurate | 0 | | Gives examples of how it is inaccurate but doesn't specifically explain why it is inaccurate |
| Explains how or why the misconception may have been developed | | 1 | a |
| **Language of science** | | | |
| Use of scientific terms is correct | | 1 | |
| Does not use rhetorical references that misrepresent the nature of science or scientific inquiry | 0 | | States ideas as truths and facts |
| **Mechanics** | | | |
| The order and arrangement of the sentences enhances the development of the main idea (organization) | | 1 | |
| The author used complete sentences, proper subject-verb agreement, and kept the tense constant (grammar) | 0 | | b: run-on sentences |
| The author used appropriate spelling, punctuation, and capitalization (conventions) | 0 | | c, b: run-on sentences commas semicolons |
| **Word choice and voice** | | | |
| The author employs a broad range of words and uses the right word (e.g., *affect* vs. *effect*, *their* vs. *there*, etc.). | | 1 | |
| The sentences are written in an active (rather than passive) voice. | | 1 | |
| **Total score** | 10 / 17 | | |

# SECTION 3: REFUTATIONAL WRITING

## 27 MISCONCEPTION ABOUT INTERACTIONS THAT TAKE PLACE BETWEEN ORGANISMS
### TEACHER NOTES

the same resources and therefore not all interactions are competitive. **(j)** That is because the animals in the interactions may have different relationships. **(k)** The lion and the gazelle are not competing for the same resources. **(l)** Instead they are a relationship that is a consumer-resource interaction. **(m)** Every time a gazelle is eaten by a lion (or hyena) it is a negative thing for the negative impact for the gazelle because there are now fewer gazelles and it is also a positive impact because the gazelle does not have to compete with other gazelles for food or water resources.

**(n)** Using the same example as above there is also a consumer –resource interaction between the Lion and gazelle. **(o)** The lion is the consumer and the gazelle is the resource and in this interaction the gazelle species is negatively impacted because there are now fewer numbers of them and the lion species is positively impacted as they have been fed for another day.

**(p)** Another interaction that is not competitive is the detritivore-detritis interaction and using the gazelle carcass again but this time with a vulture picking away at the rotting flesh. **(q)** In this interaction it is positive impact is for the vulture (who would be the consumer this time) but has no impact on the gazelle species in this relationship because the gazelle is already dead.

**(r)** In a mutualistic interaction both the consumer and resource species benefit. **(s)** In the gazelle and lion scenario if the gazelle was weak or carried bad genes then it would be beneficial to the gazelle species to be rid of the weaker animal as well as being beneficial to the lion. **(t)** This scenario leads back to the saying "survival of the fittest" where the stronger animals within a species survive and the weaker become a resource for other consumer species.

**(u)** Another example that isn't directly in this scenario is a bird picking lice off the lion's mane. **(v)** The bird is interacting with the lion but is not in competition with the lion or any of the other animals. **(w)** Therefore not all animal interactions are competitive.

## Standards Addressed in This Activity

This activity can be used to address the following dimensions outlined in *A Framework for K–12 Science Education* (NRC 2012):

# SECTION 3: REFUTATIONAL WRITING

## MISCONCEPTION ABOUT INTERACTIONS THAT TAKE PLACE BETWEEN ORGANISMS — 27
### TEACHER NOTES

## Scientific Practices

- Engaging in argument from evidence
- Obtaining, evaluating, and communicating information

## Crosscutting Concepts

- Patterns
- Cause and effect: Mechanisms and explanations
- Energy and matter: Flows, cycles, and conservation

## Life Sciences Core Ideas

- Ecosystems: Interactions, energy, and dynamics

This activity can be used to address the following standards for literacy in science from the *Common Core State Standards for English Language Arts and Literacy* (NGA and CCSSO 2010):

## Writing

- Text types and purposes
- Production and distribution of writing
- Research to build and produce knowledge
- Range of writing

## References

Bishop, B., and C. Anderson. 1990. Student conceptions of natural selection and its role in evolution. *Journal of Research in Science Teaching* 27: 415–427.

Brumby, M. 1979. Problems in learning the concept of natural selection. *Journal of Biological Education* 13 (2): 119–122.

Clough, E. E., and C. Wood-Robinson. 1985. How secondary students interpret instances of biological adaptation. *Journal of Biological Education* 19: 125–130.

Leach, J., R. Driver, P. Scott, and C. Wood-Robinson. 1992. *Progression in understanding of ecological concepts by pupils aged 5 to 16*. Leeds, UK: The University of Leeds, Centre for Studies in Science and Mathematics Education.

National Governors Association Center (NGA) for Best Practices, and Council of Chief State School Officers (CCSSO). 2010. *Common core state standards for English language arts and literacy*. Washington, DC: National Governors Association for Best Practices, Council of Chief State School.

National Research Council (NRC). 2012. *A framework for K–12 science education: Practices, crosscutting concepts, and core ideas*. Washington, DC: National Academies Press.

SECTION 3: REFUTATIONAL WRITING

# MISCONCEPTION ABOUT PLANT REPRODUCTION (BOTANY) 28

Many people think that plants do not sexually reproduce. In other words, some people think that sexual reproduction is something that only animals do and that plants do not do.

Write a one- to two-page refutational essay to convince someone who thinks that plants do not reproduce sexually that this idea is a misconception. As you write your paper, remember to

- clearly state the misconception that you are trying to refute;
- include several specific facts, details, reasons, and/or examples that show or demonstrate why the misconception is inaccurate;
- explain how plants reproduce;
- include several specific facts, details, reasons, and/or examples that show or demonstrate plants reproduce;
- present your ideas in a clear and logical order, including an introduction, body, and conclusion;
- use a variety of words and well-constructed sentences to create tone and voice;
- include at least five references in your essay, and be sure to cite your references correctly; and
- correct errors in capitalization, punctuation, sentence structure, spelling, and grammar.

You will have _____ to research this topic, and then plan, write, edit, and produce a final product. You must use at least five different reference materials (e.g., your textbook, online resources, and so on) during this process. Be sure to cite all your references in the text of your essay and include a reference section.

You must complete your **research** by _____.

Your **prewrite** (an outline, a concept map, and so on) is due on _____.

Your **initial draft** of your essay is due on _____.

Your **final draft** of your essay is due on _____.

SECTION 3: REFUTATIONAL WRITING

# 28 MISCONCEPTION ABOUT PLANT REPRODUCTION
## TEACHER NOTES

## Purpose

The purpose of this activity is to help students understand plant reproduction and the difference between sexual and asexual forms of reproduction. This activity is also designed to address many of the Common Core State Standards for English Language Arts and Literacy, which have a strong emphasis on literacy in science. These standards include writing arguments focused on discipline-specific content, writing in a clear and coherent manner, and developing and strengthening papers through a process of planning, revising, editing, and rewriting. The *Common Core State Standards for English Language Arts and Literacy* (NGA and CCSSO 2010) also calls for students to be able to conduct a short research project, gather relevant information from multiple print and digital sources, assess the credibility and accuracy of each source, and quote or paraphrase the data and conclusions of others while avoiding plagiarism. This writing assignment provides an opportunity for students to develop these skills in the context of science.

## The Content and Related Concepts

Plants can reproduce by asexual or sexual means. Asexual reproduction produces new individuals without the fusion of gametes, so the offspring are genetically identical to the parent plant and one another (except when mutations occur). Sexual reproduction, in contrast, produces offspring through the fusion of gametes. This process results in offspring that are genetically different from the parent plants and one another. Often the term *sex* is confused with or discussed in association of intercourse, which would indicate a direct contact between a male and female and their sexual parts. This has become an inaccurately contextualized word that has led to misconceptions about the ability of plants to sexually reproduce.

Angiosperms are an example of plants that can reproduce through asexual or sexual means. Angiosperms are the most diverse type of plants on land, but they can be easily distinguished from other types of plants because of their reproductive organs (flowers). Sexual reproduction in angiosperms is a complex process. It begins with the production of both male (pollen) and female (ovules) gametes. Pollen is transferred to the ovules through a process called pollination. Plants are immobile, so pollination often requires the pollen to be dispersed by wind, water, or an animal. Once the process of pollination is complete, the pollen can fertilize the ovules. The ovules then grow into seeds within a fruit. Once the seeds are mature, the fruit ripens and can be dispersed (usually with the aid of an animal). The seeds are then freed from the fruit, and then under the right conditions, the seeds will germinate and grow into new plants.

# SECTION 3: REFUTATIONAL WRITING

## MISCONCEPTION ABOUT PLANT REPRODUCTION
### TEACHER NOTES 28

## Curriculum and Instructional Considerations

This activity can be used during a unit about genetics or during a unit on plant structure and function. It can also be used during a unit about ecosystems. Teachers can have students conduct research about this topic by gathering relevant information from multiple print and digital sources. The students, however, should be taught how to assess the credibility and accuracy of each source they use, how to quote or paraphrase the information they find, and how to use a standard format for citation in their paper. Students also need to be taught to avoid plagiarism. Students must have an opportunity to conduct research and write during this activity in order for it to address the Common Core State Standards for English Language Arts and Literacy.

## Middle School

In the middle-level classroom, instruction usually encompasses basic comparisons between plants and animals. Some instruction may include the comparison at a cellular level; however, most instruction in the middle-level focuses on the comparison in terms of food sources, trophic levels, and energy and matter. This neglect of comparison in terms of reproduction could add to the confusion about the transfer of genetic material and the role of natural selection that occurs in both plants and animals that middle-level students carry with them into high school (Bishop and Anderson 1990; Brumby 1979). Students are also likely to consider plants and animals as having different types of materials (Stavy, Eisen, and Yaakobi 1987), which adds to their difficulty in understanding and comparing the reproductive processes in living organisms.

## High School

In the high school classroom, the instruction focuses more on the cellular differences and processes of plants and animals. However, the students continue to carry a misunderstanding of the relationship of sexual reproduction in terms of natural selection, and they may have difficulty comparing plants and animals in sexual interactions (Bishop and Anderson 1990; Brumby 1979). In addition, because the focus on evolution and natural selection often emphasizes animals rather than plants, students continue to carry a misunderstanding of plants that they are living organisms that share similar materials and similar reproductive processes (Stavy, Eisen, and Yaakobi 1987).

## Recommendations for Implementing the Activity

This activity takes 100–400 minutes of instructional time to complete depending on how a teacher decides to spend time in class. See Appendix F (pp. 370–371) for more information on how to implement this activity.

## Assessment

This activity can be used to identify prior knowledge and knowledge development if it is implemented as a preactivity to a science unit (or course) and then as a postactivity of the same unit (or course). The rubric in Appendix

## SECTION 3: REFUTATIONAL WRITING

## 28 MISCONCEPTION ABOUT PLANT REPRODUCTION
### TEACHER NOTES

D (p. 368) can be used to assess the students' essays and to compare previous work to determine changes in their ideas, writing skills, and skills in developing a scientific argument. The rubric focuses on the content, the structure of the argument, and mechanics of the essay. The Mechanics section of the rubric is well aligned with the Common Core State Standards for English Language Arts and Literacy. We strongly recommend that teachers use the Comments or Suggestions sections to give students detailed feedback so they will understand what they did wrong, why it is wrong, and ways they can improve their performance next time.

The following is an example of an essay written by a ninth-grade student and the scoring of the essay using the rubric (Figure 28.1).

**(a)** If you asked someone if a plant could have sex like humans, they would probably say no. **(b)** And they would be both right and wrong. **(c)** Plants are not animals, and therefore they are a little different than animals. **(d)** For example, they do not have appendages in the same way that we do. **(e)** However, it is a misconception that plants don't have sex. **(f)** Actually, some plants do have sex and that is because they have parts that get fertilized (a seed or egg) thorough pollination and parts that do the fertilizing (the stamen). **(g)** These are the reproductive parts of a plant. **(h)** It is considered sex because the plants share their DNA with each other and they pass on their characteristics to their offspring (children). **(i)** It is kind of like when Mendel was testing the pea plants. **(j)** He could have a parent breed with another parent and they would have offspring with similar characteristics as both parents. **(k)** "Sexual reproduction is of great significance in that, because of the fusion of two separate parental nuclei, the offspring inherit endlessly varied combinations of characteristics that provide a vast testing ground for new variations that may not only improve the species but ensure its survival. This probably explains the predominance of sexual reproduction among higher forms." (reproduction: the free dictionary)

## Standards Addressed in This Activity

This activity can be used to address the following dimensions outlined in *A Framework for K–12 Science Education* (NRC 2012):

### Scientific Practices

- Engaging in argument from evidence
- Obtaining, evaluating, and communicating information

### Crosscutting Concepts

- Patterns
- Cause and effect: Mechanisms and explanations
- Structure and function

# SECTION 3: REFUTATIONAL WRITING

## MISCONCEPTION ABOUT PLANT REPRODUCTION
### TEACHER NOTES 28

*Figure 28.1. Activity 28 Student Sample Scored Rubric*

| Aspect of the Essay | Point Value 0 | Point Value 1 | Comments or Suggestions |
|---|---|---|---|
| **The claim** | | | |
| The claim that is being advanced is clear | 0 | | f (buried in 6th line) |
| The claim that is being refuted (the misconception) is clear | 0 | | e (buried in 5th line) |
| **The evidence** | | | |
| Describes the evidence that supports the claim that is being advanced | | 1 | f, g, h |
| Describes the evidence as examples, applications, observations, etc. rather than provided as sets of facts | | 1 | i, j |
| Multiple sources used to support the argument | 0 | | "the free dictionary" isn't valid |
| Interpretation of the literature is correct | | 1 | |
| **The justification** | | | |
| Explains why the evidence supporting the claim that is being advanced is important or relevant | | 1 | h |
| Links the evidence to important concepts or principles | | 1 | k |
| **The challenge** | | | |
| Explains why the claim being refuted is inaccurate | 0 | | |
| Explains how or why the misconception may have been developed | 0 | | c, d? |
| **Language of science** | | | |
| Use of scientific terms is correct | | 1 | d, f, g, h |
| Does not use rhetorical references that misrepresent the nature of science or scientific inquiry | | 1 | b |
| **Mechanics** | | | |
| The order and arrangement of the sentences enhances the development of the main idea (organization) | | 1 | |
| The author used complete sentences, proper subject-verb agreement, and kept the tense constant (grammar) | | 1 | |
| The author used appropriate spelling, punctuation, and capitalization (conventions) | | 1 | |
| **Word choice and voice** | | | |
| The author employs a broad range of words and uses the right word (e.g., *affect* vs. *effect*, *their* vs. *there*, etc.). | | 1 | |
| The sentences are written in an active (rather than passive) voice. | | 1 | |
| **Total score** | 12 / 17 | | |

## SECTION 3: REFUTATIONAL WRITING

## 28 MISCONCEPTION ABOUT PLANT REPRODUCTION
TEACHER NOTES

### Life Sciences Core Ideas

- From molecules to organisms: Structures and processes
- Heredity: Inheritance and variation of traits

This activity can be used to address the following standards for literacy in science from the *Common Core State Standards for English Language Arts and Literacy* (NGA and CCSSO 2010):

### Writing

- Text types and purposes
- Production and distribution of writing
- Research to build and present knowledge
- Range of writing

### References

Bishop, B., and C. Anderson. 1990. *Student conceptions of natural selection and its role in evolution. Journal of Research in Science Teaching* 27: 415–427.

Brumby, M. 1979. Problems in learning the concept of natural selection. *Journal of Biological Education* 13 (2): 119–122.

National Governors Association Center (NGA) for Best Practices, and Council of Chief State School Officers (CCSSO). 2010. *Common core state standards for English language arts and literacy*. Washington, DC: National Governors Association for Best Practices, Council of Chief State School.

National Research Council (NRC). 2012. *A framework for K–12 science education: Practices, crosscutting concepts, and core ideas*. Washington, DC: National Academies Press.

Stavy, R., Y. Eisen, and D. Yaakobi. 1987. How students aged 13–15 understand photosynthesis. *International Journal of Science Education* 9: 105–115.

SECTION 3: REFUTATIONAL WRITING

# MISCONCEPTION ABOUT INHERITANCE OF TRAITS (GENETICS) 29

Many people think that all human traits are inherited. In other words, some people think that artistic skills learned through practice or large muscles developed through countless hours in the gym, or other acquired characteristics can be inherited in the same way that hair color, eye shape, and baldness are inherited. They think that acquired characteristics are transferred to our offspring in the same way that inherited traits are transferred—through genes.

Write a one- to two-page refutational essay to convince someone who thinks the traits that people acquired over their lifetime can be passed down to their children that this idea is a misconception. As you write your paper, remember to

- clearly state the misconception that you are trying to refute;
- include several specific facts, details, reasons, and/or examples that show or demonstrate why the misconception is inaccurate;
- explain how scientists conduct many different types of investigations and scientists engage in many different types of activities and usually do not engage in sequence of activities in all the investigations that they plan and carry out;
- include several specific facts, details, reasons, and/or examples that show or demonstrate how there are many different methods of science;
- present your ideas in a clear and logical order, including an introduction, body, and conclusion;
- use a variety of words and well-constructed sentences to create tone and voice;
- include at least five references in your essay, and be sure to cite your references correctly; and
- correct errors in capitalization, punctuation, sentence structure, spelling, and grammar.

You will have _____ to research this topic, and then plan, write, edit, and produce a final product. You must use at least five different reference materials (e.g., your textbook, online resources, and so on) during this process. Be sure to cite all your references in the text of your essay and include a reference section.

You must complete your **research** by _____.

Your **prewrite** (an outline, a concept map, and so on) is due on _____.

Your **initial draft** of your essay is due on _____.

Your **final draft** of your essay is due on _____.

SECTION 3: REFUTATIONAL WRITING

# 29 MISCONCEPTION ABOUT INHERITANCE OF TRAITS
## TEACHER NOTES

### Purpose

The purpose of this activity is to help students understand that acquired traits, such as injuries and specific skills, are not passed down from parent to offspring. This activity is also designed to address many of the Common Core State Standards for English Language Arts and Literacy, which have a strong emphasis on literacy in science. These standards include writing arguments focused on discipline-specific content, writing in a clear and coherent manner, and developing and strengthening papers through a process of planning, revising, editing, and rewriting. The *Common Core State Standards for English Language Arts and Literacy* (NGA and CCSSO 2010) also calls for students to be able to conduct a short research project, gather relevant information from multiple print and digital sources, assess the credibility and accuracy of each source, and quote or paraphrase the data and conclusions of others while avoiding plagiarism. This writing assignment provides an opportunity for students to develop these skills in the context of science.

### The Content and Related Concepts

The debate about nature versus nurture has a long history, and it is related to questions about the inheritance of acquired characteristics. Acquired characteristics refer to any modification in an organism that results from an injury, disease, use or disuse of a body part, and influence of other organisms. Examples of acquired characteristics are changes in physical appearance caused by a communicable disease, muscle atrophy, and the ability to play a musical instrument after taking lessons for several years. Many people thought that these types of acquired characteristics could be passed down from parent to offspring before the development of the gene theory of inheritance. Jean Baptiste Lamarck (1744–1829) took this idea one step further and proposed that species evolve when individuals adapt to their environment and transmit those acquired traits to their offspring. Lamarck, for example, proposed that individual giraffes developed longer necks over their lifetimes as they stretched to reach the leaves of high trees. This acquired trait gave them an advantage that they then passed down to their offspring. The gene theory of inheritance, however, does not support Lamarck's ideas. This theory suggests that the gene is the unit of heredity in organisms. The gene is a stretch of DNA that codes for a protein that has a specific function within a cell. Genes, therefore, hold the information needed to build and maintain an organism's cells. This information determines the traits of an organism, although the expression of these traits is often influenced by environmental factors (such as diet).

Complicating the identification of characteristics that are inherited is in part the variety

## SECTION 3: REFUTATIONAL WRITING

## MISCONCEPTION ABOUT INHERITANCE OF TRAITS
### TEACHER NOTES 29

of phenotypes that are the result of incomplete dominance and codominance. In addition, many characteristics may be influenced by both inheritance and acquisition. For example, weight has been shown to be related to genetics in terms of how a body metabolizes food. However, weight is also something that is influenced by the environment and is an acquired characteristic, since the foods we eat, the amounts we eat, and the balance of exercise are choices that affect our weight. This means that an overweight parent is not necessarily going to have an overweight child.

## Curriculum and Instructional Considerations

This activity is best used as part of a unit on genetics. Teachers can have students conduct research about this topic by gathering relevant information from multiple print and digital sources. The students, however, should be taught how to assess the credibility and accuracy of each source they use, how to quote or paraphrase the information they find, and how to use a standard format for citation in their paper. Students also need to be taught to avoid plagiarism. Students must have an opportunity to conduct research and write during this activity in order for it to address the Common Core State Standards for English Language Arts and Literacy.

## Recommendations for Implementing the Activity

This activity takes 100–400 minutes of instructional time to complete depending on how a teacher decides to spend time in class. See Appendix F (pp. 370–371) for more information on how to implement this activity.

## Assessment

This activity can be used to identify prior knowledge and knowledge development if it is implemented as a preactivity to a science unit (or course) and then as a postactivity of the same unit (or course). The rubric in Appendix D (p. 368) can be used to assess the students' essays and to compare previous work to determine changes in their ideas, writing skills, and skills in developing a scientific argument. The rubric focuses on the content, the structure of the argument, and mechanics of the essay. The Mechanics section of the rubric is well aligned with the Common Core State Standards for English Language Arts and Literacy. We strongly recommend that teachers use the Comments or Suggestions sections to give students detailed feedback so they will understand what they did wrong, why it is wrong, and ways they can improve their performance next time.

The following is an example of an essay written by a 10th-grade student. This student completed the activity as an introduction to a lesson on genetics in a high school biology course. (See Figure 29.1 on page 318 for the scored rubric.)

(a) Many people wrongly believe that all human traits are passed on through genetics or inherited. (b) They believe that a <u>persons</u> food preference, musical taste, artistic ability or other acquired characteristics are inherited the same way in which a <u>persons</u> hair

## SECTION 3: REFUTATIONAL WRITING

### 29 MISCONCEPTION ABOUT INHERITANCE OF TRAITS
### TEACHER NOTES

*Figure 29.1. Activity 29 Student Sample Scored Rubric*

| Aspect of the Essay | Point Value 0 | Point Value 1 | Comments or Suggestions |
|---|---|---|---|
| **The claim** | | | |
| The claim that is being advanced is clear | | 1 | c |
| The claim that is being refuted (the misconception) is clear | | 1 | a, b |
| **The evidence** | | | |
| Describes the evidence that supports the claim that is being advanced | | 1 | k, l |
| Describes the evidence as examples, applications, observations, etc. rather than provided as sets of facts | | 1 | k, l |
| Multiple sources used to support the argument | 0 | | PBS; blogs not valid? |
| Interpretation of the literature is correct | | 1 | d, g, m, n, o, p |
| **The justification** | | | |
| Explains why the evidence supporting the claim that is being advanced is important or relevant | 0 | | |
| Links the evidence to important concepts or principles | 0 | | Vaguely genetics? |
| **The challenge** | | | |
| Explains why the claim being refuted is inaccurate | | | h, i |
| Explains how or why the misconception may have been developed | 0 | | e: not really how misconception is developed |
| **Language of science** | | | |
| Use of scientific terms is correct | | 1 | |
| Does not use rhetorical references that misrepresent the nature of science or scientific inquiry | | 1 | |
| **Mechanics** | | | |
| The order and arrangement of the sentences enhances the development of the main idea (organization) | 0 | | Seems like cut and paste and some sentence structure (e) |
| The author used complete sentences, proper subject-verb agreement, and kept the tense constant (grammar) | | 1 | |
| The author used appropriate spelling, punctuation, and capitalization (conventions) | | 1 | Small punctuation errors |
| **Word choice and voice** | | | |
| The author employs a broad range of words and uses the right word (e.g., *affect* vs. *effect*, *their* vs. *there*, etc.). | | 1 | |
| The sentences are written in an active (rather than passive) voice. | | 1 | |
| **Total score** | | 11 / 17 | |

# SECTION 3: REFUTATIONAL WRITING

## MISCONCEPTION ABOUT INHERITANCE OF TRAITS
### TEACHER NOTES 29

color, eye shape, or height is inherited; through genetics.

**(c)** Humans are made up of mix of acquired and inherited traits. **(d)** Acquired traits cannot be passed on genetically whereas inherited traits are passed down from generation to generation.

**(e)** Searching through the internet there are a number of blogs that debate <u>this issue which leads to more people with the misconception</u>. **(f)** The best way to straighten out the misconception is through examples.

**(g)** Acquired traits are those that are picked up after a birth and cannot be transferred through DNA or RNA. **(h)** If it was possible then if a person loses a limb through an accident then their offspring would be born missing that limb. **(i)** Or, a fashion designer would be able to genetically transfer their sense of fashion. **(j)** It is true that their sense of fashion may very well be passed on, but through learned means not genetic coding. **(k)** Food preference is acquired and can be proven as young children adopted to culturally different families grow up with their adopted family's preferences toward food which would not occur if it was genetically passed down. **(l)** The same thing with musical taste, if it was genetically passed down and not culturally biased then I would not be a fan of heavy metal.

**(m)** The appearance of an organism is due to both inherited and acquired traits. **(n)** Charles Darwin's theory holds that inheritance is due to DNA passed on by parents. **(o)** Acquired traits are learned or developed, and do not affect the genes, according to PBS.org.

**(p)** While the misconception that ALL traits are passed on genetically is false there is still a lot of debate as to exactly what can and cannot be inherited or acquired.

## Standards Addressed in This Activity

This activity can be used to address the following dimensions outlined in *A Framework for K–12 Science Education* (NRC 2012):

### Scientific Practices
- Engaging in argument from evidence
- Obtaining, evaluating, and communicating information

### Crosscutting Concepts
- Patterns
- Structure and function

### Life Sciences Core Ideas
- Heredity: Inheritance and variation of traits

This activity can be used to address the following standards for literacy in science from

## SECTION 3: REFUTATIONAL WRITING

### 29 MISCONCEPTION ABOUT INHERITANCE OF TRAITS
TEACHER NOTES

the *Common Core State Standards for English Language Arts and Literacy* (NGA and CCSSO 2010):

- Writing text types and purposes
- Production and distribution of writing
- Research to build and present knowledge
- Range of writing

## References

National Governors Association Center (NGA) for Best Practices, and Council of Chief State School Officers (CCSSO). 2010. *Common core state standards for English language arts and literacy.* Washington, DC: National Governors Association for Best Practices, Council of Chief State School.

National Research Council (NRC). 2012. *A framework for K–12 science education: Practices, crosscutting concepts, and core ideas.* Washington, DC: National Academies Press.

SECTION 3: REFUTATIONAL WRITING

# MISCONCEPTION ABOUT INSECTS (ECOLOGY) 30

Many people think that some organisms do not play an important role in an ecosystem. For example, some people think that termites are nothing but pests because they damage our property. These people, as result, often want to get rid of all termites. In other words, some people think that insects, such as termites, are not important and that eliminating all of them would not have a negative impact an ecosystem.

Write a one- to two-page refutational essay to convince someone who thinks that it would be a good idea to get rid of all termites on Earth that this idea would have a negative impact on an ecosystem. As you write your paper, remember to

- clearly state the misconception that you are trying to refute;
- include several specific facts, details, reasons, and/or examples that show or demonstrate why the misconception is inaccurate;
- explain the important functions insects, such as termites, play in an ecosystem;
- include several specific facts, details, reasons, and/or examples that show the important role insects play in an ecosystem;
- present your ideas in a clear and logical order, including an introduction, body, and conclusion;
- use a variety of words and well-constructed sentences to create tone and voice;
- include at least five references in your essay, and be sure to cite your references correctly; and
- correct errors in capitalization, punctuation, sentence structure, spelling, and grammar.

You will have _____ to research this topic, and then plan, write, edit, and produce a final product. You must use at least five different reference materials (e.g., your textbook, online resources, and so on) during this process. Be sure to cite all your references in the text of your essay and include a reference section.

You must complete your **research** by _____.

Your **prewrite** (an outline, a concept map, and so on) is due on _____.

Your **initial draft** of your essay is due on _____.

Your **final draft** of your essay is due on _____.

SECTION 3: REFUTATIONAL WRITING

# 30 MISCONCEPTION ABOUT INSECTS
## TEACHER NOTES

## Purpose

The purpose of this activity is to help students learn about the important role that insects play in ecosystems and the specific role that termites, which are often viewed as a nuisance, play in specific ecosystem. It is specifically meant to help students learn about the role of the decomposers and their interactions in an ecosystem. This activity is also designed to address many of the Common Core State Standards for English Language Arts and Literacy, which have a strong emphasis on literacy in science. These standards include writing arguments focused on discipline-specific content, writing in a clear and coherent manner, and developing and strengthening papers through a process of planning, revising, editing, and rewriting. The *Common Core State Standards for English Language Arts and Literacy* (NGA and CCSSO 2010) also calls for students to be able to conduct a short research project, gather relevant information from multiple print and digital sources, assess the credibility and accuracy of each source, and quote or paraphrase the data and conclusions of others while avoiding plagiarism. This writing assignment provides an opportunity for students to develop these skills in the context of science.

## The Content and Related Concepts

Termites are a very important component of an ecosystem. They are found most abundantly in forests, specifically tropical forests, and there are thousands of species that play a vital role in the recycling of nitrogen back into the atmosphere. Termites eat wood, and they help to break down decaying tree trunks in the forest. Many people view termites as a nuisance, because they will eat wood found in buildings. However, without termites, fallen trees would not decompose as quickly, and their nutrients would not be returned to the soils as quickly. Because termites help with the process of decomposing and adding nutrients to the soil, they help with plant growth. Termites are also an important food source for some animals. Like other living organisms in the ecosystem, they interact with every other element in their local environment. Although they are not primary food sources on the trophic level of the food chain (plants are), they are important for secondary and tertiary organisms since they are plentiful and provide the nutrition for many animals in the secondary level. Even more importantly, they help the living organisms in the primary level by adding nutrients back into the soil. In addition, because they are in the category of decomposers in the trophic levels of a food chain, they directly interact with abiotic elements in the ecosystem. They manage to keep forest floors clean and help rotting tress to decay faster.

## Curriculum and Instructional Considerations

This activity should be used during an ecology unit. Students should be provided with opportunities to review multiple perspectives of the termite. For example, students should be given articles that indicate the value of the termites in the ecosystem (see Resource, p. 327), the negative interactions of termites in an ecosystem, and the food webs that include the termites. Teachers can have students conduct research about this topic by gathering relevant information from multiple print and digital sources. The students, however, should be taught how to assess the credibility and accuracy of each source they use, how to quote or paraphrase the information they find, and how to use a standard format for citation in their paper. Students also need to be taught to avoid plagiarism. Students must have an opportunity to conduct research and write during this activity in order for it to address the Common Core State Standards for English Language Arts and Literacy.

## Middle School

Students are likely to have the misconception that organisms at the top of a trophic level will have more energy, misunderstanding the storage of and the use of energy in living organisms. They may think that some populations of organisms are larger than others in order to meet the demands of food for other populations (Leach et al. 1992), and they may think that small organisms are unimportant in the ecosystem overall. They are likely to have misconceptions about how plants get their nutrients (Bell and Brook 1994), and they are likely to think that the food plants get is from the soil rather than from the processes of photosynthesis, which requires water and air (Anderson et al. 1990). They are also likely to think that the nutrients that animals and plants have are all different entities and that different kinds of living organisms need completely different kinds of nutrients. They are most likely going to confuse energy with food, force, and temperature since these concepts are often taught alongside energy but separate in terms of conceptual coherence.

## High School

Students will focus more of their content on biology and environmental studies than the middle-level students and are therefore more likely to be able to identify connections to the food chains more easily and without much prompting. However, they are not likely to understand the interactions between organisms in a food web that are indirect causal interactions. This is usually because the curriculum focuses on isolated ecosystems and food chains that are directly interactive. In addition, students are likely to have negative perceptions about some organisms in an ecosystem, misunderstanding the value of all living organisms in a healthy system. This will lead to their support of eradicating pests. In addition, high school students may not recognize the concept of matter that is trans-

# SECTION 3: REFUTATIONAL WRITING

## 30 MISCONCEPTION ABOUT INSECTS
### TEACHER NOTES

ferred through the chains and are likely to see it as being created and destroyed rather than transferred and conserved, in the same way as energy is transferred and conserved.

## Recommendations for Implementing the Activity

This activity takes 100–400 minutes of instructional time to complete depending on how a teacher decides to spend time in class. For more information on how to implement this activity, see Appendix F (pp. 370–371).

## Assessment

This activity can be used to identify prior knowledge and knowledge development if it is implemented as a preactivity to a science unit (or course) and then as a postactivity of the same unit (or course). The rubric in Appendix D (p. 368) can be used to assess the students' essays and to compare previous work to determine changes in their ideas, writing skills, and skills in developing a scientific argument. The rubric focuses on the content, the structure of the argument, and mechanics of the essay. The Mechanics section of the rubric is well aligned with the Common Core State Standards for English Language Arts and Literacy. We strongly recommend that teachers use the Comments or Suggestions sections to give students detailed feedback so they will understand what they did wrong, why it is wrong, and ways they can improve their performance next time.

To illustrate how to score the students' arguments and counterarguments, consider the following example written by a 10th-grade student and the following rubric (Figure 30.1).

> We are constantly hearing that termites are pests from people on the news and even on commercials. Everyone must have seen a pest control commercial that says we can spray for termites and get rid of them for up to a year. **(a)** They claim on these commercials that termites are extremely dangerous pests that cause considerable damage to people's homes and lives. Most people should be knowledgeable in the fact that termites can get into your home start <u>breading</u> and destroy the foundation of your home. As humans if anything destroys our home, we automatically are going to think the worst of it. We are going to think this insect is damaging my home that I live in, so it must be destroyed. **(b)** So, people take every precaution to destroy these so-called "pests". However, does anybody ever think of the benefits that termites could be for us and our world at large? **(c)** Normally, we do not take the time to think about what good things an insect like this can do. As humans, we always

## SECTION 3: REFUTATIONAL WRITING

### MISCONCEPTION ABOUT INSECTS
### TEACHER NOTES 30

*Figure 30.1. Activity 30 Student Sample Scored Rubric*

| Aspect of the Essay | Point Value | | Comments or Suggestions |
|---|---|---|---|
| | 0 | 1 | |
| **The claim** | | | |
| The claim that is being advanced is clear | 0 | | h (vague and buried) |
| The claim that is being refuted (the misconception) is clear | | 1 | a |
| **The evidence** | | | |
| Describes the evidence that supports the claim that is being advanced | | 1 | d, e, f, g |
| Describes the evidence as examples, applications, observations, etc. rather than provided as sets of facts | | 1 | d, e, f, g |
| Multiple sources used to support the argument | 0 | | None provided |
| Interpretation of the literature is correct | | 1 | Concepts and applications are correct |
| **The justification** | | | |
| Explains why the evidence supporting the claim that is being advanced is important or relevant | | 1 | i |
| Links the evidence to important concepts or principles | 0 | | Indirectly implies ecosystem relationships |
| **The challenge** | | | |
| Explains why the claim being refuted is inaccurate | 0 | | c (weak) |
| Explains how or why the misconception may have been developed | | 1 | a, b |
| **Language of science** | | | |
| Use of scientific terms is correct | 0 | | Terms expected were not used |
| Does not use rhetorical references that misrepresent the nature of science or scientific inquiry | | 1 | |
| **Mechanics** | | | |
| The order and arrangement of the sentences enhances the development of the main idea (organization) | | 1 | |
| The author used complete sentences, proper subject-verb agreement, and kept the tense constant (grammar) | | 1 | |
| The author used appropriate spelling, punctuation, and capitalization (conventions) | | 1 | Breeding vs breading? |
| **Word choice and voice** | | | |
| The author employs a broad range of words and uses the right word (e.g., *affect* vs. *effect*, *their* vs. *there*, etc.). | | 1 | |
| The sentences are written in an active (rather than passive) voice. | | 1 | |
| **Total score** | 12 / 17 | | |

## SECTION 3: REFUTATIONAL WRITING

### 30 MISCONCEPTION ABOUT INSECTS
### TEACHER NOTES

think of the negative aspect, because our first inclination is to think of the negative part. We need to change this misconception that termites are only pests, and see what good they can bring to us. Are termite's really just pests or can they be beneficial? Let's think about it. What can termites do to help out humans? For example, how many times do we see rotting trees that nobody will take the time or money to cut down? **(d)** Termites could be a great asset in destroying these rotting trees. **(e)** Termites are supposed to like eating wood that is not living like rotting trees or the wood that builds our homes. Basically if the termites will eat the rotting trees in our yards instead of our homes, this will help clean up our yards and forest by naturally destroying something that was taken up unnecessary room. So, if the termite breaks down the old wood, it is feeding the termites, so they do not have to search for food in other places like the foundation of your home. **(f)** Also, if termites are breaking down old trees or logs, they must be adding something back into the ground. So, they must be adding nutrients back into the soil which also helps plants grow better. We probably have all heard people complain that their plants won't grow in the yard, or the growth of their plants has been stunted. Termites might be a major factor in helping plants grow because of added nutrients in the soil which will help the plants in their growing process. **(g)** Next, termites also go through the soil, so this penetrates the soil which can help aerate the soil too. So, this will also help in plant growth, because the roots of plants will have an easier time growing in the soil as well. **(h)** <u>Basically termites can help in so many different ways, they are not just pests, but they are helping our ecological systems too.</u> **(i)** They are breaking down unwanted materials such as rotting trees. They are also helping add nutrients to the soil, and to aerate the soil. So, basically we can say that termites are not all bad. They are helping humans. So, we should not consider termites to be all bad. **(j)** Now we can put this misconception that termites are harmful to humans and their homes to rest.

## Standards Addressed in This Activity

This activity can be used to address the following dimensions outlined *A Framework for K–12 Science Education* (NRC 2012):

## Scientific Practices

- Engaging in argument from evidence
- Obtaining, evaluating, and communicating information

## Crosscutting Concepts

- Cause and effect: Mechanisms and explanations

## SECTION 3: REFUTATIONAL WRITING

### MISCONCEPTION ABOUT INSECTS
#### TEACHER NOTES 30

- Energy and matter: Flows, cycles, and conservation

## Life Sciences Core Ideas

- Ecosystems: Interactions, energy, and dynamics

This activity can be used to address the following standards for literacy in science from the *Common Core State Standards for English Language Arts and Literacy* (NGA and CCSSO 2010):

## Writing

- Text types and purposes
- Production and distribution of writing
- Research to build and present knowledge
- Range of writing

## References

Anderson, C., T. Sheldon, and J. Dubay. 1990. The effects of instruction on college nonmajors' conceptions of respiration and photosynthesis. *Journal of Research in Science Teaching* 27: 761–776.

Bell, B., and A. Brook. 1984. *Aspects of secondary students understanding of plant nutrition*. Leeds, UK: University of Leeds.

Leach, J., R. Driver, P. Scott, and C. Wood-Robinson. 1992. *Progression in understanding of ecological concepts by pupils aged 5 to 16*. Leeds, UK: The University of Leeds.

National Governors Association Center (NGA) for Best Practices, and Council of Chief State School Officers (CCSSO). 2010. *Common core state standards for English language arts and literacy*. Washington, DC: National Governors Association for Best Practices, Council of Chief State School.

National Research Council (NRC). 2012. *A framework for K–12 science education: Practices, crosscutting concepts, and core ideas*. Washington, DC: National Academies Press.

## Resource

HubPages. 2012. The beneficial role of termites in ecosystems. *http://vines.hubpages.com/hub/The-beneficial-role-of-termites-in-ecosystems*

# ASSESSMENTS & STUDENT SAMPLES

| | |
|---|---|
| **Generate an Argument Samples** | 333 |
| **Evaluate Alternatives Samples** | 343 |
| **Refutational Writing Samples** | 351 |

# ASSESSMENTS AND STUDENT SAMPLES

In this book, we provide an Assessments chapter that is meant to help teachers identify the components of an argument in students' work and how to use the rubric to provide feedback and grading.

## Difficulty in Assessing

Like most assessments beyond the multiple-choice format, scoring a student's written work can be difficult and time-consuming. The use of rubrics can help make scoring more reliable; however, even a rubric that is too general will lead to unreliable evaluations and one that is too specific will return the teacher to a time-intensive effort. A simplified rubric with multiple criteria serves as a tool to identify both multiple and specific qualities of a written scientific argument.

In this book, we have provided a student sample followed by a scored rubric and descriptions of scoring for each activity. The samples provided are identified as high, medium, and low based on quality of the work rather than on grade or age level. It should be noted that the samples have been altered by adding bolded letters in order to help the reader identify what part of the argument from the student's work is being scored or referred to in the rubric and the feedback. It is not suggested that teachers have students use this structuring in their writing. The samples' scores identify how well and in what ways the student successfully met the criteria for demonstration of content knowledge, of skills in developing a scientific argument, and of written communication skills. We recognize that it's unreasonable to expect a teacher to score multiple students with each rubric every time the students engaged in a scientific discussion. We therefore suggest that teachers focus on specific sections for scoring while requiring students to complete the assignment as a whole. For example, a teacher may choose to first identify how well the students can

- identify claims (critical thinking and argumentation component),
- use terms (science content knowledge), and
- write appropriate sentence structures (English mechanics).

The teacher would only look for and score those specific components on the rubric. Feedback would be provided for just those components as well. The next time, the teacher might focus again on those components to evaluate student growth, or the teacher might focus on another subset within the rubric to continue building such as evidence (critical thinking and argumentation and science content) or word choice (English mechanics).

This would help to reduce the amount of time that teachers spend on the assessments and it would help students to focus on specific components of their work for improvement. It is strongly suggested that teachers use the Comments and Suggestions sections to indicate where the students have successfully met the criteria. Teachers should also provide critical suggestive feedback in the Comments and Suggestions section of the rubric to help the students continue to improve.

# ASSESSMENTS AND STUDENT SAMPLES

## Assessments in Terms of Evaluating a Learning Progression

This book intentionally provides activities that can fit naturally at many points within the middle-level and high school curriculum. As such, the activities—assessed with the rubrics—naturally serve as a multilevel evaluation tool. Teachers may use the activities

- at the beginning of a unit (or course) as a way to introduce the science content (diagnostic assessment for prior knowledge),
- embedded in a unit (or course) as a way to identify how well students are developing an understanding for science content within a larger unit (formative assessment), and/or
- at the end of instruction of a unit (or course) as a way of identifying what students know after the unit has been taught (summative assessment).

A teacher using these activities as assessments could identify student growth and gains in science content knowledge, scientific argumentation skills, and written communication skills by implementing and assessing pre/post student work. It is not our intent to provide examples of student gains in this book but to offer a variety of student voices by providing individual samples of student work at various levels of quality. We do this to allow the teacher to get a sense for what strong, quality work might look like and how the rubric can be used to measure that quality. The following sections provide individual student's work for a single activity from each of the three instructional approaches: Generate an Argument, Evaluate Alternatives, and Refutational Writing. These student samples were collected from different students, in different classes, at different schools, engaging in the activities at different points within their curriculum.

The student samples in each of the activities do not reflect any specific gains or attempt to specifically represent exemplary (or poor) qualities of student work. They have each been randomly collected from various test classrooms, which used the activities in their curriculum at different points for different purposes, to indicate how the rubric can be used to evaluate student knowledge and skills.

# GENERATE AN ARGUMENT SAMPLES

# STUDENT SAMPLE 1: LOW
## ACTIVITY 1: CLASSIFYING BIRDS IN THE UNITED STATES

## QUESTION
**(a)** How many species do these 10 different birds represent?

## CLAIM
**(b)** There are 6 different species of bird represented: **(c)**
- A+B + F
- G + I
- D + J
- C
- E
- H

## EVIDENCE
**(d)** Interactions matched among pairs.
- These interactions were mating interactions.
- The birds in those groups also have the same shared things:
  » Clutch size
  » Habitats matched

## RATIONALE
**(e)** Species can only mate with the same species. That is the definition for species according to **(f)** Biology Online.org is:

"An individual belonging to a group of organisms (or the entire group itself) having common characteristics and (usually) are capable of mating with one another to produce **fertile** offspring. Failing that (for example the Liger) It has to be ecologically and recognisably the same."

**(g)** Thus the interactions are important interactions. Also, since they have other life interactions the same (the clutch and habitat) they are even more likely the same species.

## TEACHER FEEDBACK
Although you used some important terms, such as *interactions*, *species*, *organism* and *characteristics*, I am not convinced that you understand them since there was no elaboration. Also, I am curious how you would have answered other possible combinations or number of species. How are you sure that the species are in those groups and that there are only 6 species in this set of 10?

Please remember that when you write the arguments, you need to demonstrate your skills in written communication. This would be a great outline for an essay, but it is not a complete essay. To improve on this assignment, you could elaborate and extend the discussions and ideas.

# GENERATE AN ARGUMENT SAMPLES

*Generate an Argument Student Sample 1 Scored Rubric*

| Aspect of the Argument | Point Value 0 | Point Value 1 | Comments or Suggestions |
|---|---|---|---|
| **The claim** | | | |
| The claim is sufficient | | 1 | b |
| The claim is accurate | | 1 | |
| **The evidence** | | | |
| Includes data | 0 | | |
| Includes an analysis of the data | 0 | | |
| Includes an interpretation of the analysis | | 1 | d (very weak) |
| **The justification of the evidence** | | | |
| Explains why the evidence is important or why it is relevant | | 1 | E (incomplete) |
| Links the evidence to an important concept or principle | 0 | | |
| **Language of science** | | | |
| Appropriate use of scientific terms | | 1 | |
| Used phrases that are consistent with the nature of science | 0 | | In the quote from *BiologyOnline.org*, the word usually appears. What does that mean for science? |
| **Mechanics** | | | |
| The order and arrangement of the sentences enhances the development of the main idea (organization) | | 1 | Focused and straight forward but lacking depth |
| The author used complete sentences, proper subject-verb agreement, and kept the tense constant (grammar) | 0 | | Elaboration is missing. |
| The author used appropriate spelling, punctuation, and capitalization (conventions) | 0 | | This is really an outline more than an essay or a complete argument. |
| **Total score** | 6 / 12 | | |

Overall, Sample 1 Low is a weak argument as it is lacking in depth and content. The author provided information but did not provide evidence that clearly supports the ideas. In addition, the author did not explain clearly how that evidence was important for the argument. It is interesting that the author copied a definition from a resource but didn't address the use of the word usually in the argument.

# GENERATE AN ARGUMENT SAMPLES

# STUDENT SAMPLE 2: MEDIUM
## ACTIVITY 1: CLASSIFYING BIRDS IN THE UNITED STATES

**(a)** Based on the information shown our group claims that there are five different species of birds. **(b)** Although all of the <u>birds has</u> very similar body shape and coloration, <u>there</u> characteristics make them unique to <u>one another's species</u>. **(c)** We decided to compare the information given for each of the birds and categorized the birds into five groups. We focused mainly on the clutch size, habitat and size of the <u>birds itself</u>.

**(d)** Our first group of species consists of bird A, B, C, and F. **(e)** This <u>groups</u> clutch size contains 4–6 grayish eggs; they live in habitats of deciduous woodlands and shade trees, and are between 16–21 cm in length. **(f)** Our second group of species consists of birds D, and J. **(g)** This group's clutch size contains 3–5 white eggs with dark brown and purple splotches; they live in tree plantations, city parks, and suburban areas with palm or eucalyptus trees and shrubbery, and are between 18–20 cm in length. **(h)** Our third group of species consists of bird E. **(i)** This group's clutch size is 2–4 white eggs with purple streaks, lives in the forest and scattered groves of trees that are near water, and is 23–25 cm in length. **(j)** Our fourth group of species consist of birds G and I. This group's clutch size is 3–5 bluish white eggs, lives in woodlands in semi-desert areas, yucca trees or palms in deserts, and sycamores or cottonwoods in canyons, and is about 18–21 cm in length. **(k)** Our fifth group of species consists of bird H. This group's clutch size contains 4 whitish eggs with black streaks, lives in open country with scattered trees, orchards, or gardens and is 20 cm in length.

**(l)** The reason that we used the characteristics that we used for <u>the groups was because</u> we wanted to see what the birds had in common. **(m)** <u>Species has things</u> in common besides just how they look. **(n)** They also have to have other things the <u>same like where</u> they live, what kind of offspring (or eggs they have) and what they eat. **(o)** Also, they should have mating the same. **(p)** All of these birds in the groups mated with each <u>other so that also what</u> makes them a species. **(q)** Even though they don't always mate in those groups we <u>thought that maybe that is</u> because the birds were males trying to mate with <u>males or something like that</u>. **(r)** So, the mating wasn't as important for finding out who was related to each other.

**(s)** In conclusion, our group categorized these ten birds into five groups based on the information provided. **(t)** By focusing on <u>the birds</u> clutch sizes, habitats and the size of bird we were able to decide which birds belonged in the same <u>species with one another</u>.

# GENERATE AN ARGUMENT SAMPLES

### Generate an Argument Student Sample 2 Scored Rubric

| Aspect of the Argument | Point Value 0 | Point Value 1 | Comments or Suggestions |
|---|---|---|---|
| **The claim** | | | |
| The claim is sufficient | | 1 | a |
| The claim is accurate | 0 | | Errors in species |
| **The evidence** | | | |
| Includes data | | 1 | d–k |
| Includes an analysis of the data | 0 | | c but weak |
| Includes an interpretation of the analysis | | 1 | p, q, r |
| **The justification of the evidence** | | | |
| Explains why the evidence is important or why it is relevant | | 1 | l, m, n, o (weak) |
| Links the evidence to an important concept or principle | | 1 | Attempts with m and n? |
| **Language of science** | | | |
| Appropriate use of scientific terms | | 1 | m |
| Used phrases that are consistent with the nature of science | 0 | | Only makes claims and observations |
| **Mechanics** | | | |
| The order and arrangement of the sentences enhances the development of the main idea (organization) | | 1 | Focused and organized; issue with some sentence structures |
| The author used complete sentences, proper subject-verb agreement, and kept the tense constant (grammar) | 0 | | Several errors: b, l, m, q, etc. |
| The author used appropriate spelling, punctuation, and capitalization (conventions) | 0 | | Possessive and plural confusions occasionally |
| **Total score** | 7 / 12 | | |

# GENERATE AN ARGUMENT SAMPLES

## TEACHER FEEDBACK

You did a nice job of organizing your thoughts so that the argument was easy for me to read and understand. However, there are some errors in the use of plurals, possessive, and tense that weaken my understanding.

Your argument starts with a clear claim, and you provided the information you used to make your claim. I have the sense that you are not sure what the definition of *species* is and when there might be examples that stray from that definition. I think if you do some more work on learning about species and the other things that can determine species (besides the specific characteristics), you may find that some of your argument would be challenged. You might then include that information to anticipate the challenge, which would strengthen your argument.

> Although the overall score is not very high in this sample, most of errors were in the writing mechanics. The argument itself is incomplete and has errors; however, it is stronger than Student Sample 1 Low. This group also has some misunderstanding or incomplete understanding of the concepts of species. There are several errors in their claims and content, and therefore, the line of reasoning for the evidence is weak. For example, they focused on mostly the physical characteristics and the behaviors and mating practices or opportunities.

## GENERATE AN ARGUMENT SAMPLES

# STUDENT SAMPLE 3: HIGH
## ACTIVITY 1: CLASSIFYING BIRDS IN THE UNITED STATES

**(a)** We think that there are five different species represented in this sample of birds that you have given us. **(b)** We started to figure out the number of species by looking at the mating practices of the birds. **(c)** A simple definition of species is when the individuals in a group will mate with the individuals in another group to share genetic material (our textbook) and they produce fertile offspring (our textbook). Sometimes species will breed but their offspring can't reproduce (like a liger or a donkey). When that happens, it means that the parents are willing to mate but are actually different species. **(d)** We don't know from the information that you gave us if any of these birds mate and produce fertile offspring, but we are going to assume that they do and the mating means that they would fit the simple definition of species. **(e)** Since E and H do not breed outside of their own group, by simple definition of a species, they would be each one species. **(f)** Some people might say that E and H are variations of the same species because they have eggs with streaks (although different colors) and that they have not mated only because they are in different states (Tx and Fl). **(g)** So we looked at the other characteristics to see if they could have been "transplants" that just didn't get an opportunity to mate. **(h)** But all of the other characteristics are different like their sizes, the number of eggs, and the kinds of nests. **(i)** E is in TX and has an opportunity to mate with many other birds but doesn't. And H is in FL and can breed with F and C, but doesn't. **(j)** Based on the breeding, the opportunity and the other characteristics, these two are separate species from all others. They may have a common ancestor though since their eggs have streaks.

**(k)** The same kind of reasoning is how we decided that D and J were one species with two variations and that G and I were one species with two variations. **(l)** To help see the interactions between the different birds we plotted where they lived on the map that was given and how they traveled during the winter (or not). **(m)** D and J are in all the same locations and don't migrate and G and I are in the same locations and don't migrate.

**(n)** A B C and F are the hardest to decide because there is some mating between them but not all. **(o)** A will mate with B and F but F won't mate with B, and C will mate with F but not A. If we look at only the mating definition of species then there could be four different possibilities: **(p)**

one species (A + B + C + F)
two species (A + B and F + C) or (A + B + F and C)
three species (A + B  and A + F  and F + C )

**(q)** We used the same mapping plan to see if the birds in these four groups were in the same locations and found that A and B are mostly on the western side of the states and F and C are mostly on the eastern (or middle longitudes) of the states. **(r)** There are only three states where they have an opportunity to interact and mate and that is ND, SD and TX. **(s)** Since they are found in these three states together, then we think there must be a reason that they don't all mate and that could be related to their migration patterns. **(t)** Although all of them migrate south (either to the tropics or S. Atlantic region) we think that it may be that they are migrating at slightly different times and that has something to do with their mating. **(u)** If so, then we think that these are really all one species with four different variations and their migration patterns are why they haven't mated with each other. They have similar nests, the same eggs and eat the same kinds of foods. **(v)** Nests might not be considered a characteristic that helps to know what the species is if they are living in different places since there might be different materials to make a nest. **(w)** But since they do live in different places and the nests are the same structure, we think this shows that it is a characteristic of the species. **(x)** The same thing for the food, and they eat the same kinds of food. **(y)** The only other differences seem to be song but that could be a variation since B and F both don't have song and mate with A but not with each other.

**(z)** One important characteristic we noticed was the eggs. **(aa)** If D and J and G and I are the same species, and they have the same kind of eggs (color and clutch size) in their species groups, then that could mean that the eggs would be a way of knowing if they are related species. **(bb)** A, B, C and F all have the same color of eggs and clutch size. We did some research online and could not find that the characteristics of the eggs (color and shape) would change as a variation in the species of birds. **(cc)** It might be a variation but we could not find anything about that. **(dd)** We did see that the geographic location could make a difference in when the eggs are laid. **(ee)** A group of scientists (P. Olsen and T.G. Marples) in Australia looked at birds and saw that they were laying eggs a few days apart based on latitude and longitude. **(ff)** If geographic location matters, then maybe these four groups are not mating because their period of doing so is slightly off based on their locations. **(gg)** This could mean that these four groups are interacting but they have slightly different periods to lay eggs and bred and that might be why they aren't mating. **(hh)** But since they have the same eggs, they are still the same species. **(ii)** This is more evidence that they are most likely the same species even if they are not

# GENERATE AN ARGUMENT SAMPLES

Our map

mating. **(jj)** We would like to know more about when they migrate and when their breeding period is and until then, we will call these one species.

**(kk)** Our final conclusion is that there are a total of 5 species in this population of birds that you gave us: (E), (H), (D+J), (G+I) and (A+B+C+F).

**Reference**

1993, P Olsen and TG Marples: <u>Geographic-Variation in Egg size, Clutch Size and Date of Laying of Australian Raptors (Falconiformes and Strigiformes)</u>. EMU Austral Ornithology found online at *http://www.publish.csiro.au/paper/MU9930167.htm*

*Generate an Argument Student Sample 3 Scored Rubric*

| Aspect of the Argument | Point Value 0 | Point Value 1 | Comments or Suggestions |
|---|---|---|---|
| **The claim** | | | |
| The claim is sufficient | | 1 | a |
| The claim is accurate | | 1 | |
| **The evidence** | | | |
| Includes data | | 1 | (l) Map, e, i, y, etc. |
| Includes an analysis of the data | | 1 | r, v, dd, |
| Includes an interpretation of the analysis | | 1 | n, s, t, w, aa, ff, gg, hh, ii |
| **The justification of the evidence** | | | |
| Explains why the evidence is important or why it is relevant | | 1 | c, d |
| Links the evidence to an important concept or principle | | 1 | j, s, u |
| **Language of science** | | | |
| Appropriate use of scientific terms | | 1 | |
| Used phrases that are consistent with the nature of science | | 1 | "…more evidence that they are most likely…" |
| **Mechanics** | | | |
| The order and arrangement of the sentences enhances the development of the main idea (organization) | 0 | | Sentences are good, but could use more structure for paragraphs and overall argument |
| The author used complete sentences, proper subject-verb agreement, and kept the tense constant (grammar) | | 1 | |
| The author used appropriate spelling, punctuation, and capitalization (conventions) | | 1 | |
| **Total score** | | 11 / 12 | |

## GENERATE AN ARGUMENT SAMPLES

### TEACHER FEEDBACK

This was a very nicely written argument. You certainly used the evidence provided to identify interactions and patterns well. It was great how you plotted the locations of the birds on the map to help identify possible interactions. Because there was so much information and you are examining both your claim and other possible alternative claims, at times it was difficult to follow your logic. It might have been easier to follow if you used headings or restructured your paragraphs.

> Although there was an expected answer for six species, the students used the evidence provided, they explained the analysis of their evidence, and they used good rationale to argue their point. For this, they received credit for the accuracy of their claim. They did not consider the concept of prezygotic barriers in defining species. If they had added this to their accepted definition, they may have recognized that there were six rather than five species in this group of birds. The students address this concept (without using the vocabulary) by identifying interactions but do not recognize the concept as a part of the definition of *species*. This is a very strong scientific argument, because the group provides alternative possibilities as well as their own claim and provides information to rule out each of the other possibilities.

EVALUATE ALTERNATIVES SAMPLES

# STUDENT SAMPLE 1: LOW
## ACTIVITY 20: TERMITE TRAILS

**(a)** Why do termites follow lines? **(b)** Actually they don't follow all lines. **(c)** Some lines they will follow and other lines they won't. **(d)** So the question is really why do they follow some lines. **(e)** We think that they are following some lines because they are attracted to a wavelength like bees. **(f)** They see a special wavelength of color and that is all they can see. **(g)** The color of wavelength they see is blue. **(h)** We tested our idea by making lines with other colors of pens like red, green, yellow, orange, and violet. **(i)** The termites only followed the lines with the blue colored pen. **(j)** And they seemed afraid of the red lines because they stayed far away from that color. **(k)** Since the blue wavelength has the most energy and the red wavelengths have the smallest energy the termites need to have a lot of energy to see. **(l)** We proved with our experiment that we were correct and they see certain colors.

## TEACHER FEEDBACK

It sounds like you had some good ideas about how to test your claims, and you did a good job of telling your "story" in an order that I could easily follow it. However, your argument could be improved by finding a way to test the alternative ideas. Also, your writing could be improved by organizing the argument into sections or paragraphs. To do so, you would need to expand on the details of your experiment by including what you did, how you did it, why you did it that way, and what you think the results mean.

> This is a very weak argument for several reasons. The authors do not describe their evidence very clearly, and they do not address the alternative arguments at all. They make conclusions based on obvious misconceptions about the color and light concepts and attempt to use that as a rational for their evidence.

# EVALUATE ALTERNATIVES SAMPLES

*Evaluate Alternatives Student Sample 1 Scored Rubric*

| Aspect of the Argument | Point Value 0 | Point Value 1 | Comments or Suggestions |
|---|---|---|---|
| **The claim** | | | |
| The claim is sufficient | 0 | | e (not specific or clear) |
| The claim is accurate | 0 | | |
| **The evidence** | | | |
| Includes data | 0 | | h (attempted but weak) |
| Includes an analysis of the data | | 1 | i (weak) |
| Includes an interpretation of the analysis | | 1 | j, k |
| **The justification of the evidence** | | | |
| Explains why the evidence is important or why it is relevant | 0 | | |
| Links the evidence to an important concept or principle | 0 | | |
| **The challenge** | | | |
| The alternative explanation(s) being challenged is explicit | 0 | | |
| Explains why the alternative explanation is inaccurate | 0 | | |
| **Language of science** | | | |
| Appropriate use of scientific terms | 0 | | Misconceptions about light and color |
| Used phrases that are consistent with the nature of science | 0 | | l ("…proved…. we were correct….") inconsistent with NOS |
| **Mechanics** | | | |
| The order and arrangement of the sentences enhances the development of the main idea (organization) | 0 | | Focused but not structured into paragraphs. |
| The author used complete sentences, proper subject-verb agreement, and kept the tense constant (grammar) | | 1 | |
| The author used appropriate spelling, punctuation, and capitalization (conventions) | 0 | | d, j |
| **Total score** | | 3/14 | |

# STUDENT SAMPLE 2: MEDIUM
## ACTIVITY 20: TERMITE TRAILS

## Question

(a) Why do termites follow the lines?

## Our Claim

(b) Me and my partner think that explanation #3 is the correct answer: **Termites are cooperative and navigate by smell. The ink in the pens contains a chemical that smells the same as the pheromones that are secreted by the termites. As a result, the termites follow the line because it smells like a trail left by another termite.**

## Our Method

(c) We used blue ink from lots of different kinds of pens including a ball point, a watercolor pen, a sharpie, a crayon and a higher. (d) The color did not seem to be a factor. (e) But when we used the same kind of pen but in a different color the termites followed the circles. (f) We also tried lots of different colors in all of these pens and the termites followed only the yellow and pink highlighter.

## Our Results

(g) The termites followed the lines from the yellow and pink high lighter and from the blue and black felt pens. (h) They did not follow any other pens.

## Relevant Evidence

(i) The scent should matter more than the color in real life because termites live in dark places. (j) They have to smell more than they have to see so they should have very strong smelling.

## Alternative Explanations

(k) Explanation #1 can't be right because we gave them different colors and they didn't follow them. (l) Also they live in dark places.

(m) Explanation #2 is not right because even though the termites followed some of our lines they obviously didn't follow all of our lines.

## Conclusion

(n) Termites can travel with pens when the pens have special chemicals and the termites are cooperative. (o) Termites don't use their sight but they use their smell to travel.

## EVALUATE ALTERNATIVES SAMPLES

*Evaluate Alternatives Student Sample 2 Scored Rubric*

| Aspect of the Argument | Point Value 0 | Point Value 1 | Comments or Suggestions |
|---|---|---|---|
| **The claim** | | | |
| The claim is sufficient | | 1 | b |
| The claim is accurate | | 1 | |
| **The evidence** | | | |
| Includes data | 0 | | c, e, f, g, h (but weak) |
| Includes an analysis of the data | 0 | | None given |
| Includes an interpretation of the analysis | | 1 | d |
| **The justification of the evidence** | | | |
| Explain why the evidence is important or why it is relevant | 0 | | Attempts with i, j, but incorrect |
| Links the evidence to an important concept or principle | 0 | | Attempts with n, o, but too weak |
| **The challenge** | | | |
| The alternative explanation to be challenged is explicit | | 1 | k, m |
| Explains why the alternative explanation is inaccurate | 0 | | k, m (weak; disconnected; incomplete) |
| **Language of science** | | | |
| Appropriate use of scientific terms | 0 | | None used |
| Used phrases that are consistent with the nature of science | 0 | | d (vague) |
| **Mechanics** | | | |
| The order and arrangement of the sentences enhances the development of the main idea (organization) | 0 | | More detail and elaboration needed |
| The author used complete sentences, proper subject-verb agreement, and kept the tense constant (grammar) | | 1 | |
| The author used appropriate spelling, punctuation, and capitalization (conventions) | | 1 | |
| **Total score** | | 6 / 14 | |

# EVALUATE ALTERNATIVES SAMPLES

## TEACHER FEEDBACK

The use of headings was an effective way to help readers know what you were thinking and seemed to help you stay focused on the argument. I would have liked to see more elaboration on your observations and explanations, however. That might have helped to make your argument sound more like an essay than a short bulleted list.

You could improve the argument itself by giving more detail on the methods and by describing the results in more detail. Also, remember that the relevance of the evidence is how you are explaining that your evidence "counts" as good evidence, not just explaining why the result makes sense.

> Although the argument does not have a lot of depth and elaboration, the argument is fairly strong. The students clearly labeled the components and focused their argument.

EVALUATE ALTERNATIVES SAMPLES

# STUDENT SAMPLE 3: HIGH
ACTIVITY 20: TERMITE TRAILS

(a) We did an experiment that involved termites and pen lines and we <u>determined that the termites were able to confirm</u> that the termites were attracted to some pen lines but not to all pen lines. (b) We think that the reason they were attracted to some of the pen lines is because termites are attracted to something in the ink. (c) We think that Explanation # 3 is probably true and that explanation # 1 and explanation #2 are probably not true.

(d) We first tried drawing the pen lines in the figure eight with our pens. (e) We noticed that the termites followed the pen lines. (f) When then drew different figures and straight lines. (g) We noticed that the termites followed those lines as well. (h) We started to draw lines with other colors. (i) We used a black pen and a red pen and the termites followed those lines. (j) We also drew lines with orange pens, yellow pens, green pens and purple pens. (k) The termites did not follow those lines. (l) We noticed that some of the pens smelled bad so we thought that maybe the termites smelled them too. (m) So we tried pens of all one kind first and then pens of all another kind. (n) The termites followed the pens that had the Bic name on them. (o) They also followed the pens that were from the Hilton hotel. (p) But they didn't follow the other pens. (q) We put the termites in a plastic container that was see through and put them on top of the pen marks and they didn't seem to be following any of the lines. (r) They just moved randomly around the container.

(s) All of these <u>observations that we made is what makes</u> us think that the termites don't navigate by sight. (t) By showing how the termites don't follow all lines we are showing that they are not relying on site to navigate. (u) By showing that they like pens that were from the same company we can show that the termites are following something that is in those pens or something that isn't in those pens. (v) Since explanation # 1 and #2 are both about how the termites see by sight we don't think that those explanations are correct. (w) We do think they navigate by smell so Explanation # 3 is probably right. (x) (y) But we didn't actually find out if pheromones were in the ink. (z) If they do navigate by pheromones or by smell of some kind, that is important for their survival because they live mostly in dark places. (aa) Navigation with some other sense besides sight would be helpful for their survival to find food and to find mates. (bb) The reason we aren't sure about the pheromones is because we didn't actually test if there were pheromones in the ink. (cc) Also, if these termites are sterile then why would they be attracted to pheromones since pheromones are chemicals that

# EVALUATE ALTERNATIVES SAMPLES

*Evaluate Alternatives Student Sample 3 Scored Rubric*

| Aspect of the Argument | Point Value 0 | Point Value 1 | Comments or Suggestions |
|---|---|---|---|
| **The claim** | | | |
| The claim is sufficient | | 1 | b, c |
| The claim is accurate | | 1 | |
| **The evidence** | | | |
| Includes data | | 1 | e, g, k, n, o, p, r |
| Includes an analysis of the data | 0 | | Observations and conclusions but not comparisons to develop patterns for analysis |
| Includes an interpretation of the analysis | 0 | | Vague; not connected to data/analysis |
| **The justification of the evidence** | | | |
| Explains why the evidence is important or why it is relevant | | 1 | t, u |
| Links the evidence to an important concept or principle | | 1 | aa, cc |
| **The challenge** | | | |
| The alternative explanation(s) being challenged is (are) explicit | | 1 | c |
| Explains why the alternative explanation is inaccurate | | 1 | t |
| **Language of science** | | | |
| Appropriate use of scientific terms | 0 | | None used |
| Used phrases that are consistent with the nature of science | | 1 | w, x, bb, dd, ee "…probably right…" "… didn't actually find out.." |
| **Mechanics** | | | |
| The order and arrangement of the sentences enhances the development of the main idea (organization) | 0 | | d-r disconnected from purpose and reasoning so hard to follow |
| The author used complete sentences, proper subject-verb agreement, and kept the tense constant (grammar) | | 1 | Minimal; see underlined |
| The author used appropriate spelling, punctuation, and capitalization (conventions) | | 1 | |
| **Total score** | 10 /14 | | |

## EVALUATE ALTERNATIVES SAMPLES

are used for attacking mates? **(dd)** Because we don't think we have all the information yet we decided that we don't really know if explanation #3 is right either. **(ee)** But it is the most right for now.

### TEACHER FEEDBACK

You did an excellent job thinking about the different explanations and the questions you might ask or the problems you might try to solve to eliminate the explanations. However, your argument would have been easier to follow and stronger to evaluate if you put the data into a table and provided specific comparisons of variables that you felt were being tested. Following this with the reasons for those variables would have helped to make stronger connections between your decisions, your observations, and your interpretations.

> This is a fairly strong example of an argument. The student evaluates ways to eliminate the alternative arguments and clearly considers the tentativeness of the knowledge based on limited observations.

# REFUTATIONAL WRITING SAMPLES

# STUDENT SAMPLE 1: LOW
## ACTIVITY 23: MISCONCEPTION ABOUT THE WORK OF SCIENTISTS

(a) The idea <u>that Science</u> does not involve imagination and creativity is a huge misconception. (b) As with every subject, hearing lectures and <u>learning about the past of science</u> may seem like it lacks creativity, but when applying science, and creating experiments it involves a lot of imagination and creativeness to make it work and worth looking at. (c) Experiments, research and inventions all involve a very <u>imaginative person</u> to do. (d) Think about it <u>this way, you</u> wouldn't want to go to a science fair and look at the inventions made by boring people would you? (e) Not to mention all of the things we use today that were invented by <u>imaginative</u> scientists may not exist. (f) Every day we send people up into the sky to travel in a plane, now that took some imagination.

(g) Sometimes in science, we follow someone else's steps to see how <u>something works, we follow</u> a procedure. (h) This is simply to gain an understanding of how something <u>works, in</u> order for you to learn and <u>gain insight on what</u> things are used for and why. (i) When creating your own experiments which you will do a lot in science you must use your imagination to make it work. (j) Science is all about having a starting step and using a series of steps to come out with the final result. (k) The stepping stones in between must be figured out by you, which involves more creativity then anything.

(l) As you can see, science involves a lot of creativity and imagination. (m) If science did not involve the creativity it does, we would not have the information and technology that we have today. (n) For example, it is creativity that led scientists to try new things and do research to discover and create cures and medicines that help almost any disease out there. (o) A creative mind can only let our technology evolve that much more.

# REFUTATIONAL WRITING SAMPLES

*Refutational Writing Student Sample 1 Scored Rubric*

| Aspect of the Essay | Point Value 0 | Point Value 1 | Comments or Suggestions |
|---|---|---|---|
| **The claim** | | | |
| The claim that is being advanced is clear | | 1 | a, l |
| The claim that is being refuted (the misconception) is clear | 0 | | Implies via claim |
| **The evidence** | | | |
| Describes the evidence that supports the claim that is being advanced | 0 | | |
| Describes the evidence as examples, applications, observations, etc. rather than provided as sets of facts | 0 | | c, d, e, f not clear; gives examples of when creativity is important but not clear how this is related to what scientists do |
| Multiple sources used to support the argument | 0 | | None provided |
| Interpretation of the literature is correct | 0 | | None provided |
| **The justification** | | | |
| Explains why the evidence supporting the claim that is being advanced is important or relevant | 0 | | d (irrelevant) |
| Links the evidence to important concepts or principles | 0 | | |
| **The challenge** | | | |
| Explains why the claim being refuted is inaccurate | 0 | | h, i (weak) |
| Explains how or why the misconception may have been developed | 0 | | b? |
| **Language of science** | | | |
| Use of scientific terms is correct | 0 | | j, k |
| Does not use rhetorical references that misrepresent the nature of science or scientific inquiry | 0 | | j, h |
| **Mechanics** | | | |
| The order and arrangement of the sentences enhances the development of the main idea (organization) | 0 | | Underlined |
| The author used complete sentences, proper subject-verb agreement, and kept the tense constant (grammar) | | 1 | |
| The author used appropriate spelling, punctuation, and capitalization (conventions) | 0 | | Punctuation is lacking |
| **Word Choice and voice** | | | |
| The author employs a broad range of words and uses the right word (e.g., *affect* vs. *effect*, *their* vs. *there*, etc.). | 0 | | Several word choices |
| The sentences are written in an active (rather than passive) voice. | 0 | | d |
| **Total score** | 2 / 17 | | |

# REFUTATIONAL WRITING SAMPLES

## TEACHER FEEDBACK

To improve the argument, you will need to be more specific about how creativity is important in the development of science or how scientists themselves use creativity to do science. You have some good ideas of events and examples, but I am not sure what they (scientists or science) are doing during those examples that are creative. Adding some specific behaviors or thinking processes would help you develop your evidence more clearly. Also, think about why people might have the misconception. What is it that scientists do or why is it that people perceive scientists as not being creative?

To improve the writing, focus first on organizing the points you want to make. The writing contains a voice and tone in that you are casual when you write, but readers will have some trouble following your ideas. I would suggest that you outline your ideas first and develop complete sentences and paragraphs second. Start with your claim, then have a point that you think supports your claim (the evidence), and then explain that evidence. This is a structure that would work for one full paragraph.

> Overall, this is a very weak scientific argument. The author's statements are repetitive and attempt to make claims without actual evidence to support the claims. The author seems to assume that by giving examples of science events that the examples are, in themselves, evidence. It is unclear, however, how those examples are evidence and whether they are actually relevant to the argument.
>
> The writing is distracting and hard to follow with serious errors in sentence structure, organization, word choice, and punctuation. Although paragraphs have been used, the paragraphs themselves are not focused.

# REFUTATIONAL WRITING SAMPLES

# STUDENT SAMPLE 2: MEDIUM
## ACTIVITY 23: MISCONCEPTION ABOUT THE WORK OF SCIENTISTS

Dear 8th grader:

**(a)** I would like to explain to you about the misconception that science is not creative. **(b)** I am hoping that I can change your mind like my mind was changed. **(c)** Science is very creative to do and scientists are very creative people. **(d)** There are a lot of people that think that this isn't true but that is probably because they have not really done science right.

**(e)** When we first started to do our astronomy projects our teacher asked us to draw a picture of a scientist. **(f)** I was really surprised when I saw that my scientist was just like a lot of other kids in my class. **(g)** We mostly drew nerdy science guys with crazy hair and in a lab by themselves. **(h)** This is because people think that is really what a scientist does and what a scientist looks like. **(i)** This showed us that a lot of people think the same way about scientists and about science. **(j)** Also if you see the movies and cartoons and you think about the famous scientists you will probably notice that they also seem to be crazy guys doing crazy things and they seem pretty nerdy too. **(k)** We are doing science in class with fun projects and that helps us to see that our science guys are not really what scientists look like or what science is really like.

**(l)** My teacher wanted us to change our minds so we learned about how science has to use images from telescopes. **(m)** We got an image from the astronomy banks and we got to go talk about how to put the pictures to color. **(n)** This is how science can be creative. **(o)** We don't actually get to see how things in the far away sky look because we don't see that kind of wavelength. **(p)** But with our tools and technology we can be creative to see the different ways that the astronomy things will look when we color them. **(q)** All of our pictures were different even when the pictures were of the same objects in the sky. **(r)** We were trying to be the most creative by making them different and really colorful.

**(s)** Another example of creativity that we learned about is when we got to build things to test and to learn about science. **(t)** Our teacher asked us to look up Rube Goldberg Apparatus and we saw videos of them. **(u)** She let us work in our teams and we <u>builded</u> a invention ourselves that had to show how we were using energy to make something happen. **(v)** This was fun and everyone's was different. We got to be creative and we got to test our ideas when we were creative and we got to see who's ideas were the most creative. **(w)** Even though this doesn't seem like science it is. **(x)** Just because we didn't do exactly what other groups did and exactly what the teacher said for every experiment we were learning how to do science.

# REFUTATIONAL WRITING SAMPLES

## *Refutational Writing Student Sample 2 Scored Rubric*

| Aspect of the Essay | Point Value 0 | Point Value 1 | Comments or Suggestions |
|---|---|---|---|
| **The claim** | | | |
| The claim that is being advanced is clear | | 1 | c |
| The claim that is being refuted (the misconception) is clear | | 1 | a, d |
| **The evidence** | | | |
| Describes the evidence that supports the claim that is being advanced | | 1 | o, p, q, r |
| Describes the evidence as examples, applications, observations, etc. rather than provided as sets of facts | | 1 | v |
| Multiple sources used to support the argument | 0 | | Personal experiences |
| Interpretation of the literature is correct | 0 | | None provided |
| **The justification** | | | |
| Explains why the evidence supporting the claim that is being advanced is important or relevant | | 1 | n, p, q, r |
| Links the evidence to important concepts or principles | | 1 | h, i, j, k |
| **The challenge** | | | |
| Explains why the claim being refuted is inaccurate | | 1 | aa, w |
| Explains how or why the misconception may have been developed | | 1 | j, x, y, z |
| **Language of science** | | | |
| Use of scientific terms is correct | 0 | | |
| Does not use rhetorical references that misrepresent the nature of science or scientific inquiry | | 1 | |
| **Mechanics** | | | |
| The order and arrangement of the sentences enhances the development of the main idea (organization) | | 1 | |
| The author used complete sentences, proper subject-verb agreement, and kept the tense constant (grammar) | | 1 | |
| The author used appropriate spelling, punctuation, and capitalization (conventions) | | 1 | |
| **Word choice and voice** | | | |
| The author employs a broad range of words and uses the right word (e.g., *affect* vs. *effect*, *their* vs. *there*, etc.). | 0 | | Simple vocabulary and sentence structures |
| The sentences are written in an active (rather than passive) voice. | 0 | | |
| **Total score** | 12 / 17 | | |

**SCIENTIFIC ARGUMENTATION IN BIOLOGY: 30 CLASSROOM ACTIVITIES**

# REFUTATIONAL WRITING SAMPLES

**(y)** Sometimes science is going to be boring and following steps and finding the answer you are supposed to find. **(z)** And that is why it seems like science is not really creative. **(aa)** But it is or we wouldn't have new things because you have to think of new things to make new things. **(bb)** And that is what scientists do. **(cc)** And they try to find a different way to solve problems to help get the discoveries. **(dd)** It would not be bad to be a scientist if you liked to be creative.

## TEACHER FEEDBACK

Very nicely written! You did a great job with remembering some of the activities that we have done so far to help you prove that science is creative. I also like that you organized your paragraphs. Each one started with a topic sentence and developed the point from there with supporting details. To improve your argument, you could find some resources that explain what creativity is and what it means to be creative. Then you could use that as evidence when you describe the experiences you have had. Or you could describe creativity by using a known scientist as an example. To improve your writing, use more science terms and vocabulary. We have covered many terms this year so far that could fit in your descriptions and explanations.

> Overall, this is a strong argument that is well organized and easy to read. The student clearly states the claim and then provides a rationale for why the argument is important. The student then provides examples of the creative activities that he knows scientists use based on his own experiences. However, authentic examples of evidence would be stronger than merely the personal experiences of the author. The paragraphs are well structured, but word choice and use of vocabulary is lacking.

# REFUTATIONAL WRITING SAMPLES

# STUDENT SAMPLE 3: HIGH
## ACTIVITY 23: MISCONCEPTION ABOUT THE WORK OF SCIENTISTS

**(a)** Our Claim: To be a scientist means to be creative and to do science means to think creatively

**(b)** The Misconception: People think that being a scientist means that you have to follow a procedure and there is no deviation from that procedure. They think scientists are boring and uncreative and so is science.

**(c)** Our Evidence: Being a scientist is actually pretty difficult because not only is there a lot of information that you have to know, but you also have to be creative; and that is hard! Our evidence is the examples of creativity that have led to new science discoveries. **(d)** These are a few that we know of:

Alfred Wegner used the outline shapes of the continents to show that there was contentital drift and the continents were connected before. Nobody else thought of this and they didn't believe him when he first showed them. **(e)** *http://academic.emporia.edu/aberjame/histgeol/wegener/wegener.htm*

Alexander graham bell invented the phone by using a new idea to talk to people from far away. Martin Cooper invented the first cell phone or mobile phone by trying to think of a new way to communicate after he was inspired by the ideas of Star trek. Even though he didn't come up with the idea, how he made it work was new and original. **(f)** *http://en.wikipedia.org/wiki/Communicator_(Star_Trek)*

Thermometers have been made in different ways. The scientists that made them found new ways to make them so that they could measure temperature in different ways. Galileo made one with water in 1593. Santorio Santorio used numbers on the thermometer to make it a scale in 1612 so you could measure temperature in a person's mouth. He added to the thermometer and he used it in a new way. Ferdinand the II used alchol instead of water in the thermometer in 1654. Daniel farhenheit use mercury in 1714 and he set a standard scale for the movement of mercury in the glass tube. Anders Celcius changed the scale so that it compared to water instead of to mercury in 1742. Lord Kelvin changed the scale to compare it to energy in matter in 1848. **(g)** The way we measure temperature today is based on what each of these scientists did and how they added to the information before and made it new and different. Each of them was creative in a different way. **(h)** *http://inventors.about.com/od/tstartinventions/a/History-Of-The-Thermometer.htm*

**(i)** Our Rationale: We think that these are good evidence that science is creative and that to be a good scientist

# REFUTATIONAL WRITING SAMPLES

you have think creatively because of the definition of creativity and because of other uses of creativity that we know are creative. Sometimes being creative means to have a completely new idea and sometimes it means you have to modify the ideas you already know just a little bit and sometimes it means you have to think of a way to make your imagination come true. We looked up the definition of creative and we found that **(j)** R. Standler (1998) defines creativity as something that has "never been done before. Particularly important instances of creativity include discoveries…" the evidences we have are examples of new ways of thinking about observations that the scientists made and new ways of doing things that helped them to understand better.

**(k)** Also, if you think about when people would all agree that creativity is used it would be in art. Art is creative when it is something new or different. **(l)** So, this makes our points about how the scientists were creative. **(m)** They wouldn't have been able to do something new if they did the same thing that everyone else did or if they followed a known procedure. That is not new by definition.

Conclusion: **(n)** One of the most difficult things about being a scientist is to be able to think creatively. Some people say that science is boring and hard because it has lots of things to remember and lots of formulas. That is partly true. But even more true is the part about being creative. You have to think differently from everyone else. Coming up with new ideas after millions of years of people coming up with ideas is really mind-blowing! **(o)** Science Daily in 2011 "Why we crave creativity but reject creative ideas" says that we crave creativity but when we have it we are nervous because it means uncertainty. If science is about asking a question, it also means that we would be uncertain. This is another direct connection. **(p)** *http://www.sciencedaily.com/releases/2011/09/110903142411.htm* for science being creative.

**(q)** Scientists have to see things a little differently than other people. Maybe that is why scientists are usually antisocial. If you knew someone that could think of things differently than you and everyone you knew, you would think that person was weird. **(r)** That is probably why we always think of scientists as being crazy too. They have lots of new ideas that the rest of us are not used to.

**(s)** Our Suggestions: We think that if science is supposed to be creative that we should be able to have more freedom in science class so that we can be real scientist. **(t)** Mostly we learn about facts and we memorize stuff and we do labs

## REFUTATIONAL WRITING SAMPLES

*Refutational Writing Student Sample 3 Scored Rubric*

| Aspect of the Essay | Point Value 0 | Point Value 1 | Comments or Suggestions |
|---|---|---|---|
| **The claim** | | | |
| The claim that is being advanced is clear | | 1 | a |
| The claim that is being refuted (the misconception) is clear | | 1 | b |
| **The evidence** | | | |
| Describes the evidence that supports the claim that is being advanced | | 1 | c |
| Describes the evidence as examples, applications, and observations, rather than provided as sets of facts | | 1 | d |
| Multiple sources are used to support the argument | | 1 | e, f, h, j, p |
| Interpretation of the literature is correct | | 1 | |
| **The justification** | | | |
| Explains why the evidence supporting the claim that is being advanced is important or relevant | | 1 | i, j, k, l, m |
| Links the evidence to important concepts or principles | | 1 | g, m, n, q, s |
| **The challenge** | | | |
| Explains why the claim being refuted is inaccurate | 0 | | |
| Explains how or why the misconception might have been developed | | 1 | r, t, v |
| **Language of science** | | | |
| Use of scientific terms is correct | | 1 | |
| Does not use rhetorical references that misrepresent the nature of science or scientific inquiry | | 1 | |
| **Mechanics** | | | |
| The order and arrangement of the sentences enhances the development of the main idea (organization) | | 1 | |
| The author used complete sentences, proper subject-verb agreement and consistent tense (grammar) | 0 | | |
| The author used appropriate spelling, punctuation, and capitalization (conventions) | 0 | | |
| **Word choice and voice** | | | |
| The author employs a broad range of words and uses the right word (e.g., *affect* vs. *effect*, *their* vs. *there*, etc.) | | 1 | |
| The sentences are written in active (rather than passive) voice | | 1 | |
| **Total score** | | 14/17 | |

that other people have already done. We learn how to follow the procedures. **(u)** We think that is why people have a misconception. **(v)** They learn science with procedures and they think science is hard and boring. **(w)** It is hard because we have to be creative but it isn't boring if we get to actually discover something and do some new things that we think is interesting.

## TEACHER FEEDBACK

Your group did an excellent job of organizing each paragraph and providing a clear label for the key components in your argument. You had some good examples that definitely show how creativity is important in science. I would like to know if you think that science is always creative or if the counterargument can sometimes be correct? This would have made your argument stronger.

The writing was also well done. There were only a few mistakes with your word choice and grammar. Please remember that resources such as Wikipedia and Yahoo Answers are not reliable and should not be used for citations. Also, you should look again at the rules for how to cite your resources in the text.

> Overall, this is a very strong scientific argument. The authors addressed most of the components of the argument and labeled them nicely. The writing was also strong as it was well organized and focused. The authors have obvious skills in the mechanics of writing but have some errors that may have been due to oversight.

# APPENDIX

**Appendix A:** School Pilot Sites and Contributors to the Book .... 363

**Appendix B:** Rubric for Generate an Argument .................. 366

**Appendix C:** Rubric for Evaluate Alternatives .................. 367

**Appendix D:** Rubric for Refutational Writing .................. 368

**Appendix E:** Two Options for Implementing the Generate an Argument Activities .................. 369

**Appendix F:** Two Options for Implementing the Refutational Writing Activities .................. 370

# APPENDIX A:
## SCHOOL PILOT SITES AND CONTRIBUTORS TO THE BOOK

## Pilot Sites

### Beth Shields Middle School (6–8)
15732 Beth Shields Way
Ruskin, Florida 33573

- Rural
- Low socioeconomic status (SES) (85% of students receive a free or reduced-price lunch)
- Public

### Chapel Hill High School (9–12)
1709 High School Road
Chapel Hill, North Carolina 27516

- Urban
- High SES (25% of students receive a free or reduced-price lunch)
- Public

### D. H. Conley (9–12)
2006 Worthington Road
Greenville, North Carolina 27858

- Suburban
- Middle SES (58% of students receive a free or reduced-price lunch)
- Public

### E. B. Aycock (6–8)
1325 Red Banks Road
Greenville, North Carolina 27858

- Suburban
- Middle SES (58% of students receive a free or reduced-price lunch)
- Public

### Fairview Middle School (6–8)
3415 Zillah Street
Tallahassee, Florida 32305

- Urban
- Low SES (80% of students receive a free or reduced-price lunch)
- Public

### Florida State University Schools (K–12)
3000 School House Road
Tallahassee, Florida 32311

- Suburban
- Middle SES (40% of students receive a free or reduced-price lunch)
- Charter

### Forest Hills Middle School (6–8)
1210 Forest Hills
Wilson, North Carolina 27893

- Suburban
- Middle SES (62% of students receive a free or reduced-price lunch)
- Public

# APPENDIX A

### Lake Asbury Junior High School (7–9)
2851 Sandridge Road
Green Cove Springs, Florida 32043
- Suburban
- Middle SES (35% of students receive a free or reduced price lunch)
- Public

### Lincoln High School (9–12)
3838 Trojan Trail
Tallahassee, Florida 32311
- Urban
- Middle SES (35% of students receive a free or reduced price lunch)
- Public

### New Bern High School (9–12)
4200 Academic Drive
New Bern, North Carolina 28562
- Suburban
- Middle SES (54% of students receive a free or reduced price lunch)
- Public

### Palm Springs Middle School (6–8)
1025 West 56 Street
Hialeah, Florida 33012
- Suburban
- Low SES (63% of students receive a free or reduced price lunch)
- Public

### Swift Creek Middle School (6–8)
2100 Pedrick Road
Tallahassee, Florida 32317
- Suburban
- Middle SES (40% of students receive a free or reduced price lunch)
- Public

### The Meadow School (K–12)
8601 Scholar Lane
Las Vegas, Nevada 89128
- Suburban
- High SES (15% of students receive a free or reduced price lunch)
- Private

### The Oakwood School (K–12)
4000 MacGregor Downs Road
Greenville, North Carolina 27834
- Rural
- High SES (12% of students receive a free or reduced price lunch)
- Private

### Turrentine Middle School (6–8)
1710 Edgewood Avenue
Burlington, North Carolina 27215
- Rural
- Middle SES (64% of students receive a free or reduced price lunch)
- Public

## Washington High School (9–12)

400 Slatestone Road
Washington, North Carolina 27889

- Rural
- Low SES (84% of students receive a free or reduced-price lunch)
- Public

## Contributors to the Book

Katherine Alligood
Stephanie Buck
Kathleen Casulli
Brandon Coltraine
Rebecca Jordan
Tim Messer
Amanda Parfitt
Kelly Riley
Brian Schleigh
Thomas Townsend
Jessica White

# APPENDIX B:
## RUBRIC FOR GENERATE AN ARGUMENT

| Aspect of the Argument | Point Value 0 | Point Value 1 | Comments or Suggestions |
|---|---|---|---|
| **The claim** | | | |
| The claim is sufficient | | | |
| The claim is accurate | | | |
| **The evidence** | | | |
| Includes data | | | |
| Includes an analysis of the data | | | |
| Includes an interpretation of the analysis | | | |
| **The justification of the evidence** | | | |
| Explains why the evidence is important or why it is relevant | | | |
| Links the evidence to an important concept or principle | | | |
| **Language of science** | | | |
| Appropriate use of scientific terms | | | |
| Used phrases that are consistent with the nature of science | | | |
| **Mechanics** | | | |
| The order and arrangement of the sentences enhances the development of the main idea (organization) | | | |
| The author used complete sentences, proper subject-verb agreement, and kept the tense constant (grammar) | | | |
| The author used appropriate spelling, punctuation, and capitalization (conventions) | | | |
| **Total score** | /12 | | |

# APPENDIX C:
## RUBRIC FOR EVALUATE ALTERNATIVES

| Aspect of the Argument | Point Value | | Comments or Suggestions |
|---|---|---|---|
| | 0 | 1 | |
| **The claim** | | | |
| The claim is sufficient | | | |
| The claim is accurate | | | |
| **The evidence** | | | |
| Includes data | | | |
| Includes an analysis of the data | | | |
| Includes an interpretation of the analysis | | | |
| **The justification of the evidence** | | | |
| Explains why the evidence is important or why it is relevant | | | |
| Links the evidence to an important concept or principle | | | |
| **The challenge** | | | |
| The alternative explanation(s) being challenged is (are) explicit | | | |
| Explains why the alternative explanation is inaccurate | | | |
| **Language of science** | | | |
| Appropriate use of scientific terms | | | |
| Used phrases that are consistent with the nature of science | | | |
| **Mechanics** | | | |
| The order and arrangement of the sentences enhances the development of the main idea (organization) | | | |
| The author used complete sentences, proper subject-verb agreement, and kept the tense constant (grammar) | | | |
| The author used appropriate spelling, punctuation, and capitalization (conventions) | | | |
| **Total score** | /14 | | |

# APPENDIX D:
## RUBRIC FOR REFUTATIONAL WRITING

| Aspect of the Essay | Point Value 0 | Point Value 1 | Comments or Suggestions |
|---|---|---|---|
| **The claim** | | | |
| The claim that is being advanced is clear | | | |
| The claim that is being refuted (the misconception) is clear | | | |
| **The evidence** | | | |
| Describes the evidence that supports the claim that is being advanced | | | |
| Describes the evidence as examples, applications, observations, etc. rather than provided as sets of facts | | | |
| Multiple sources used to support the argument | | | |
| Interpretation of the literature is correct | | | |
| **The justification** | | | |
| Explains why the evidence supporting the claim that is being advanced is important or relevant | | | |
| Links the evidence to important concepts or principles | | | |
| **The challenge** | | | |
| Explains why the claim being refuted is inaccurate | | | |
| Explains how or why the misconception may have been developed | | | |
| **Language of science** | | | |
| Use of scientific terms is correct | | | |
| Does not use rhetorical references that misrepresent the nature of science or scientific inquiry | | | |
| **Mechanics** | | | |
| The order and arrangement of the sentences enhances the development of the main idea (organization) | | | |
| The author used complete sentences, proper subject-verb agreement, and kept the tense constant (grammar) | | | |
| The author used appropriate spelling, punctuation, and capitalization (conventions) | | | |
| **Word choice and voice** | | | |
| The author employs a broad range of words and uses the right word (e.g., *affect* vs. *effect*, *their* vs. *there*, etc.). | | | |
| The sentences are written in an active (rather than passive) voice. | | | |
| **Total score** | / 17 | | |

# APPENDIX E:
# TWO OPTIONS FOR IMPLEMENTING THE GENERATE AN ARGUMENT ACTIVITIES

Option A

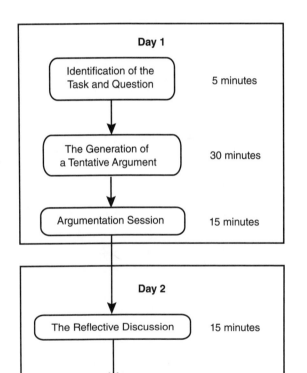

Option B

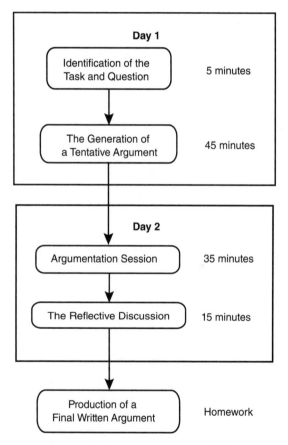

In Option A, the students are given time to complete all five stages of the lesson during class. Stages 1–3 are completed on day 1 and Stages 4 and 5 are completed on day 2. This option for implementing the activity works best in schools where students are not expected to complete much homework or if students need to be encouraged to write more during the school day.

In Option B, students complete Stages 1 and 2 of the lesson during class on day 1. The students then complete the argumentation session and the reflective discussion during day 2 of the lesson. The final written argument is then assigned as homework and returned the next day.

# APPENDIX F:
## TWO OPTIONS FOR IMPLEMENTING THE REFUTATIONAL WRITING ACTIVITIES

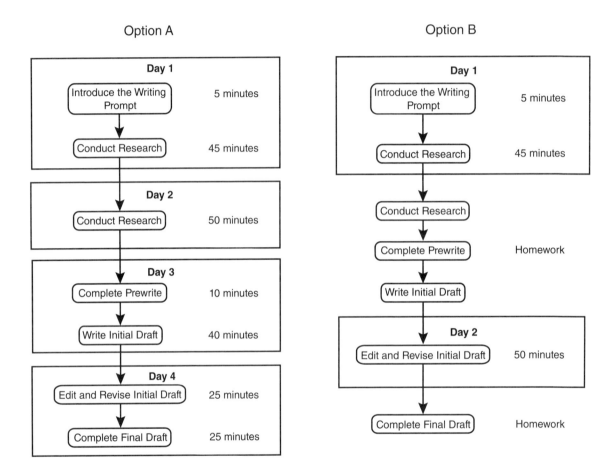

In Option A, the students are given time to complete all five stages of the lesson during class. The teacher introduces the activity and gives students time to begin conducting their research during day 1. Students are given another day to conduct their research during day 2. This is also an appropriate time for a teacher to show students how to gather information from several sources and how to access the credibility and accuracy of the source of information. The teacher gives students time to complete their prewrite and initial draft on day 3. This is also an appropriate time for the teacher to show students how to use appropriate citations in a paper. On day 4, students edit their own papers or the papers of their classmates. The students then complete the final draft. This option for implementing the activity works best in schools where students are not expected to complete much homework or if students need to be encouraged to write more during the school day and need to learn how to write a formal paper.

In Option B, the teacher introduces the students to the activity and then gives the students the rest of the period for research. The students then complete the research, prewrite, and initial draft of the paper as homework. On day 2 of the activity (which may be several school days later), the students have an opportunity to revise and edit their initial draft. We recommend that students review the papers of their classmates and provide each other with feedback about how to improve the content and writing mechanics. The students can then complete the final draft as homework. This option works well in schools where students are expected to complete homework outside of class and have many of the prerequisite skills they need to write a formal paper.

# INDEX

*Page numbers printed in **boldface** type refer to figures or tables.*

**A**

*A Framework for K–12 Science Education: Practices, Crosscutting Concepts, and Core Ideas,* xii, xv–xvi, xxxiii, xxxix
   standards addressed in Evaluate Alternatives activities, 134–135
      Cell Size and Diffusion, 189–190
      Environmental Influence on Genotypes and Phenotypes, 200–201
      Healthy Diet and Weight, 237
      Hominid Evolution, 217
      Liver and Hydrogen Peroxide, 179
      Movement of Molecules in or out of Cells, 169
      Plant Biomass, 158
      Plants and Energy, 228
      Spontaneous Generation, 146
      Termite Trails, 247–248
   standards addressed in Generate an Argument activities, 2–3
      Characteristics of Viruses, 130–131
      Classifying Birds in the United States, 16–17
      Color Variation in Venezuelan Guppies, 26
      Decline in Saltwater Fish Populations, 101
      Desert Snakes, 42–43
      DNA Family Relationship Analysis, 64–65
      Evolutionary Relationships in Mammals, 78–79
      Fruit Fly Traits, 54
      History of Life on Earth, 110
      Surviving Winter in the Dust Bowl, 122
   standards addressed in Refutational Writing activities, 250–251
      Misconception About Bacteria, 298–299
      Misconception About Inheritance of Traits, 319
      Misconception About Insects, 327–328
      Misconception About Interactions That Take Place Between Organisms, 306–307
      Misconception About Life on Earth, 290
      Misconception About Plant Reproduction, 312, 314
      Misconception About the Methods of Scientific Investigations, 283
      Misconception About the Nature of Scientific Knowledge, 268
      Misconception About the Work of Scientists, 274
      Misconception About Theories and Laws, 259
Activities, xxxii–xxxix. *See also specific activities*
   assessments of, xxxv, xxxviii, xxxix, 331–360 (*See also* Assessment(s); Rubrics)
   for Evaluate Alternatives, xxxiii, 133 (*See also* Evaluate Alternatives instructional model)
   field testing of, xiii, 363–365
   flexibility of, xxxiii
   *Framework* matrices for, xxxix, 2–3, 134–135, 250–251
   for Generate an Argument, xxxii–xxxiii, 1 (*See also* Generate an Argument instructional model)
   how to use, xxxiii–xxxiv
   purpose of, xxxviii
   safety concerns for, xxxiii, 143, 150, 160, 172, 182, 188, 192, 220, 239, 272, 280
   small-group format for, xviii, xix, **xix, xxvii,** xxviii, xxx, xxxiv
   Teacher Notes for, xiii, xxxiii, xxxviii–xxxix
   teacher's role during, xxxiv–xxxv, **xxxvi–xxxvii**
   time required for, xxxiv
Alternative explanations, xviii, xxvi. *See also* Evaluate Alternatives instructional model
   introduction of, xxvi–xxviii
   tentative arguments and counterarguments for, xxviii–xxix, **xxix**
Animals
   Classifying Birds in the United States, 5–17
   Color Variations in Venezuelan Guppies, 19–27
   Decline in Saltwater Fish Populations, 81–101
   Desert Snakes, 29–43
   Evolutionary Relationships in Mammals, 67–79
   Fruit Fly Traits, 45–54
   Hominid Evolution, 203–218
   Misconceptions About Insects, 321–327
   Termite Trails, 239–248
Argumentation session
   for Evaluate Alternatives, xxix–xxx
   for Generate an Argument, **xxii,** xxii–xxiv
   round-robin format for, xxiii, **xxiii–xxiv,** xxx, 6, 19, 30, 46, 55, 68, 83, 104, 114, 124, 139, 151, 161, 173, 182, 193, 208, 221, 231, 240
   for specific activities
      Cell Size and Diffusion, 182–183, **183**
      Environmental Influence on Genotypes and Phenotypes, 192–193, **193**
      Healthy Diet and Weight, 231, **231**
      Hominid Evolution, 208–209, **209**
      Liver and Hydrogen Peroxide, 173, **173**
      Movement of Molecules in or out of Cells, **160,** 161–162
      Plant Biomass, 151, **151**
      Plants and Energy, 221, **221**
      Spontaneous Generation, **139,** 139–140
      Termite Trails, 240–241, **241**
Assessment(s), xxxix, xxxv, 331–360. *See also* Rubrics
   difficulty of, 331
   section of Teacher Notes, xxxix
   and student samples for specific activities, 333–360
      Cell Size and Diffusion, 188
      Characteristics of Viruses, 129–130
      Classifying Birds in the United States, 15–16, 333–342, **334, 336, 341**
      Color Variations in Venezuelan Guppies activity, 26–27
      Decline in Saltwater Fish Populations, 99–101

# INDEX

Desert Snakes, 41–42
DNA Family Relationship Analysis, 63–64
Environmental Influence on Genotypes and Phenotypes, 199–200
Evolutionary Relationships in Mammals, 77–78
examples of high, medium, and low quality, 333–360
Fruit Fly Traits, 53–54
Healthy Diet and Weight, 234–237
History of Life on Earth, 109–110
Hominid Evolution, 216–217
Liver and Hydrogen Peroxide, 178–179
Misconception About Bacteria, 296–298, **297**
Misconception About Inheritance of Traits, 317–319, **318**
Misconception About Insects, 324–326, **325**
Misconception About Interactions That Take Place Between Organisms, 304–306, **305**
Misconception About Life on Earth, 287–290, **288**
Misconception About Plant Reproduction, 311–312, **313**
Misconception About the Methods of Scientific Investigations, 281–283, **282**
Misconception About the Nature of Scientific Knowledge, 265–268, **267**
Misconception About the Work of Scientists, 272–274, **273**, 351–360, **352, 355, 359**
Misconception About Theories and Laws, 256–259, **257**
Movement of Molecules in or out of Cells activity, 168–169
Plant Biomass, 155–158
Plants and Energy, 226–228
Spontaneous Generation, 145–146
Surviving Winter in the Dust Bowl, 121–122
Termite Trails, 246–247, 343–350, **344, 346, 349**
in terms of evaluating a learning progression, 332

## B
Benedict's solution, 160, 164, **167**, 168
Benedict's test, 160, 168
Botany
    Environmental Influence on Genotypes and Phenotypes, 191–201
    Misconception About Plant Reproduction, 309–314
    Plant Biomass, 149–158
    Plants and Energy, 219–228

## C
Cell Size and Diffusion activity, 181–190
    argumentation session for, 182–183, **183**
    getting started on, 181–182
    introduction to, 181
    materials for, 181, 187, **188**
        recipe for phenolphthalein agar, 188–189
    purpose of, 185
    recording your method and observations for, 182
    standards addressed in, 134–135, 189–190
    Teacher Notes for, 185–190
        assessment, 188
        content and related concepts, 185–186
        curriculum and instructional considerations, 186
        options for implementation, 186–187, **187**
    time required for, 186
    writing an argument for, 184
Cell theory: Spontaneous Generation, 137
Characteristics of Viruses activity, 123–131
    developing a claim for, 123–124, **124**
    information about viruses and other objects found on Earth for, **125–126**
    introduction to, 123, **123**
    materials for, 129, **130**
    purpose of, 128
    research question for, 123
    standards addressed in, 2–3, 130–131
    Teacher Notes for, 128–131
        assessment, 129–130
        content and related concepts, 128–129
        curriculum and instructional considerations, 129
        recommendations for implementation, 129
    time required for, 129
    writing an argument for, 127
Chemical reactions: Liver and Hydrogen Peroxide, 171
Cladogram, **71,** 71–72
Claims, ix–x
    criteria for evaluation of, x–xi
    definition of, ix
    development of, xviii (*See also specific activities*)
    evaluation of, ix, xv
    learning to articulate, xvi
    predictive power of, xi
    reasons for, ix
    revision of, xvi, xviii
    supporting evidence for, ix, xvi
Classifying Birds in the United States activity, 5–17
    developing a claim for, 5–6, **6**
    information about the 10 birds for, **7–10**
    introduction to, 5, **5**
    map of United States for, **10**
    materials for, 15, **15**
    purpose of, 12
    research question for, 5
    standards addressed in, 2–3, 16–17
    Teacher Notes for, 12–17
        assessment, 15–16, 333–342
        classification of the 10 birds, **13**
        content and related concepts, 12–14
        curriculum and instructional considerations, 14–15
        names of the 10 birds, **14**
        recommendations for implementation, 14–15
        student sample scored rubrics, **334, 336, 342**
    time required for, 15
    writing an argument for, 11
Classroom discussions, xviii, xxxiv
    argumentation session, xxii–xxiv, **xxii–xxiv,** xxix–xxx (*See also* Argumentation session)
    within and between groups, xxxv

# INDEX

reflective, xxiv, xxx
round-robin format for, xxiii, **xxiii–xxiv,** xxx, 6, 19, 30, 46, 55, 68, 83, 104, 114, 124, 139, 151, 161, 173, 182, 193, 208, 221, 231, 240
Color Variations in Venezuelan Guppies activity, 19–27
   developing a claim for, 19–20, **20**
   information about pools where guppies were found for, **21**
   information about theory of natural selection for, 22
   introduction to, 19, **19**
   map of pool locations for, **22**
   materials for, 26, **26**
   purpose of, 24
   research question for, 19
   standards addressed in, 2–3, 26
   Teacher Notes for, 24–27
      assessment, 26–27
      content and related concepts, 24–25
      curriculum and instructional considerations, 25
      recommendations for implementation, 25–26
   time required for, 25–26
   writing an argument for, 23
*Common Core State Standards for English Language Arts and Literacy,* xii, xxxiii, xxxix
   standards addressed in Evaluate Alternatives activities, 135
      Cell Size and Diffusion, 190
      Environmental Influence on Genotypes and Phenotypes, 201
      Healthy Diet and Weight, 237
      Hominid Evolution, 217–218
      Liver and Hydrogen Peroxide, 179–180
      Movement of Molecules in or out of Cells, 169
      Plant Biomass, 158
      Plants and Energy, 228
      Spontaneous Generation, 146–147
      Termite Trails, 248
   standards addressed in Generate an Argument activities, 3
      Characteristics of Viruses, 131
      Classifying Birds in the United States, 17
      Color Variations in Venezuelan Guppies, 27
      Decline in Saltwater Fish Populations, 101
      Desert Snakes, 43
      DNA Family Relationship Analysis, 65
      Evolutionary Relationships in Mammals, 79
      Fruit Fly Traits, 54
      History of Life on Earth, 110
      Surviving Winter in the Dust Bowl, 122
   standards addressed in Refutational Writing activities, 251
      Misconception About Bacteria, 294, 299
      Misconception About Inheritance of Traits, 316, 320
      Misconception About Insects, 322, 327
      Misconception About Interactions That Take Place Between Organisms, 302, 307
      Misconception About Life on Earth, 286, 290
      Misconception About Plant Reproduction, 310, 314
      Misconception About the Methods of Scientific Investigations, 278, 283
      Misconception About the Nature of Scientific Knowledge, 262, 268
      Misconception About the Work of Scientists, 270, 274
      Misconception About Theories and Laws, 254, 259
Critical-thinking skills, xxii, xxvi, xxxiii, xxxiv, 98, 331

## D

Darwin, Charles, ix, 67, 68, 289, 319
Data generation, xxviii
Decline in Saltwater Fish Populations activity, 81–101
   developing a claim for, 83, **83**
   information about selected fish populations found around Florida coast for, **89–95**
   information on annual observations of young of the year of select fish along Florida Atlantic coast for, **87–88**
   information on annual standardized commercial catch rates along Florida Atlantic coast for, **85–86**
   introduction to, 81–83
   materials for, 99, **99**
   purpose of, 96
   research question for, 82
   standards addressed in, 2–3, 101
   Teacher Notes for, 96–101
      assessment, 99–101
      content and related concepts, 96–97
      curriculum and instructional considerations, 97–98
      recommendations for implementation, 98–99
   time required for, 99
   writing an argument for, 84
Desert Snakes activity, 29–43
   developing a claim for, 30, **30**
   information about the desert snakes for, **31–32**
   information on primary snake predators for, 34–36
      badgers, 36, **36**
      long-tailed weasel, 35, **35**
      raptors, 34, **34**
   information on theory of natural selection for, 37
   introduction to, 29, **29**
   materials for, 41, **41**
   population density information for, **33**
   purpose of, 39
   research question for, 29
   standards addressed in, 2–3, 42–43
   Teacher Notes for, 39–43
      assessment, 41–42
      content and related concepts, 39–40
      curriculum and instructional considerations, 40–41
      recommendations for implementation, 41
   time required for, 41
   writing an argument for, 38
Diffusion
   Cell Size and Diffusion, 181
   Movement of Molecules in or out of Cells, 159
DNA. *See* Genetics

# INDEX

DNA Family Relationship Analysis activity, 55–65
    developing a claim for, 55–56, **56**
    introduction to, 55
    materials for, 63, **64**
    purpose of, 61
    research question for, 55
    results from STR family relationship analysis test for, 57, **57–59**
    standards addressed in, 2–3, 64–65
    Teacher Notes for, 61–65
        assessment, 63–64
        coding and noncoding sequences of DNA, **61**
        content and related concepts, 61–63
        curriculum and instructional considerations, 63
        recommendations for implementation, 63
        results of STR analysis, **62**
    time required for, 63
    writing an argument for, 60

## E

Ecology
    Decline in Saltwater Fish Populations, 81–101
    Misconception About Insects, 321–327
    Misconception About Interactions That Take Place Between Organisms, 301–307
Environmental Influence on Genotypes and Phenotypes activity, 191–201
    argumentation session for, 192–193, **193**
    getting started on, 191–192
    introduction to, 191
    materials for, 191–192, 199, **199**
    purpose of, 195
    recording your method and observations for, 192
    research question and possible explanations for, 191
    standards addressed in, 134–135, 200–201
    Teacher Notes for, 195–201
        allele or alternative versions of a gene, **195**
        assessment, 199–200
        content and related concepts, 195–197
        curriculum and instructional considerations, 197
        options for implementation, 197–199, **198**
    time required for, 197
    writing an argument for, 194
Evaluate Alternatives instructional model, xviii, xxvi–xxx
    activities for, xxxiii, 133
        Cell Size and Diffusion, 181–190
        Environmental Influence on Genotypes and Phenotypes, 191–201
        *Framework* matrix for, 134–135
        Healthy Diet and Weight, 229–237
        Hominid Evolution, 203–218
        Liver and Hydrogen Peroxide, 171–180
        Movement of Molecules in or out of Cells, 159–169
        Plant Biomass, 149–158
        Plants and Energy, 219–228
        Spontaneous Generation, 137–147
        Termite Trails, 239–248
    goals of, xxvi, xxviii

overview of, xxvi
scoring rubric for, xxx, 367
    student samples and, 343–350, **344, 346, 349**
stages of, xxvi–xxx, **xxvii**
    1: introduce phenomenon to be investigated, research question, and alternative explanations, xxvi–xxviii
    2: generation of data, xxvi–xxviii
    3: generation of tentative arguments and counterarguments, xxviii–xxix, **xxix**
    4: argumentation session, xxix–xxx
    5: reflective discussion, xxx
    6: production of final written argument, xxx
teacher's role during, **xxxvii**
writing prompt for, xxx, **xxxi**
Evaluation of scientific argument, criteria for, x–xi, xvi
Evidence, ix–x
    criteria for evaluation of, x–xi, xv, xvi, xvii
    justification of, x, xx
Evolution
    Color Variation in Venezuelan Guppies, 19–27
    Desert Snakes, 29–43
    Evolutionary Relationships in Mammals, 67–79
    History of Life on Earth, 103–110
    Hominid Evolution, 203–218
    Misconception About Life on Earth, 285–290
Evolutionary Relationships in Mammals activity, 67–79
    amino acid sequence for hemoglobin subunit alpha protein 1–20 for, **73**
    amino acid sequence for hemoglobin subunit alpha protein 20–40 for, **73**
    creating a cladogram for, **71**, 71–72
    developing a claim for, 68–70, **69**
    introduction to, 67–68, **69**
        homologous structures in seven different vertebrate limbs, **67**
    materials for, 77, **77**
    purpose of, 75
    research question for, 68
    standards addressed in, 2–3, 78–79
    Teacher Notes for, 75–79
        assessment, 77–78
        content and related concepts, 75–76
        curriculum and instructional considerations, 76–77
        recommendations for implementation, 77
    time required for, 77
    writing an argument for, 74

## F

Field testing of activities, xiii, 363–365
Food chains and trophic levels: Surviving Winter in the Dust Bowl, 113–122
Fossils. *See* Evolution
Fruit Fly Traits activity, 45–54
    developing a claim for, 46, **46**
    information about the results of various fruit fly crosses for, 46–47, **47–48**
    introduction to, 45, **45**

materials for, 52–53, **53**
purpose of, 50
research question for, 45
standards addressed in, 2–3, 54
Teacher Notes for, 50–54
    allele or alternative versions of a gene, **51**
    assessment, 53–54
    content and related concepts, 50–52
    curriculum and instructional considerations, 52
    recommendations for implementation, 52–53
time required for, 52
writing an argument for, 49

### G

Generate an Argument instructional model, xviii–xxvi
    activities for, xxxii–xxxiii, 1
        Characteristics of Viruses, 123–131
        Classifying Birds in the United States, 5–17
        Color Variations in Venezuelan Guppies, 19–27
        Decline in Saltwater Fish Populations, 81–101
        Desert Snakes, 29–43
        DNA Family Relationship Analysis, 55–65
        Evolutionary Relationships in Mammals, 67–79
        *Framework* matrix for, 2–3
        Fruit Fly Traits, 45–54
        History of Life on Earth, 103–110
        options for implementation of, xxxiv, 369
        Surviving Winter in the Dust Bowl, 113–122
    goals of, xviii
    overview of, xviii
    scoring rubric for, xxv–xxvi, 366
        student samples and, 333–342, **334, 335, 342**
    stages of, xviii–xxvi, **xix**
        1: identification of problem and research question, xviii–xix
        2: generation of tentative argument, xix–xxii, **xx–xxi**
        3: argumentation session, xxii–xxiv, **xxii–xxiv**
        4: reflective discussion, xxiv
        5: production of final written argument, xxiv–xxvii, **xxv**
    teacher's role during, **xxxvi**
    writing prompt for, xxv, **xxv**
Generation of data, xxviii
Genetics
    DNA Family Relationship Analysis, 55–65
    Environmental Influences on Genotypes and Phenotypes, 191–201
    Evolutionary Relationships in Mammals, 67–79
    Fruit Fly Traits, 45–54
    Misconception About Inheritance of Traits, 315–320
    Misconception About Plant Reproduction, 309–314

### H

Healthy Diet and Weight activity, 229–237
    argumentation session for, 231, **231**
    getting started on, 230
    introduction to, 229–230
        percentage of population in different countries that is considered obese, **229**
    materials for, 234, **236**
    purpose of, 233
    recording your method and observations for, 230–231
    research question and possible explanations for, 230
    standards addressed in, 134–135, 237
    Teacher Notes for, 233–237
        assessment, 234–237
        content and related concepts, 233
        curriculum and instructional considerations, 233–234
        options for implementation, 234, **235**
    time required for, 234
    writing an argument for, 232
Heredity. *See* Genetics
History of Life on Earth activity, 103–110
    developing a claim for, 103–104, **104**
    information about the number of different families that have been identified in the fossil record for, **105**
    introduction to, 103
        number of families within some common types of organisms, **103**
    materials for, 109, **109**
    purpose of, 107
    research question for, 103
    standards addressed in, 2–3, 110
    Teacher Notes for, 107–110
        assessment, 109–110
        content and related concepts, 107–108
        curriculum and instructional considerations, 108
        recommendations for implementation, 109
    time required for, 109
    writing an argument for, 106
Hominid Evolution activity, 203–218
    argumentation session for, 208–209, **209**
    getting started on, 207–208
    introduction to, 203
        hominid skull fossils by age, **203**
    materials for, 207, 215, **215**
    purpose of, 211
    recording your method and observations for, 208
    research question and possible explanations for, 204, **204–206**
    standards addressed in, 134–135, 217–218
    Teacher Notes for, 211–218
        assessment, 216–217
        content and related concepts, 211–213
        curriculum and instructional considerations, 213
        options for implementation, 213, **214**
        phylogenetic relationships of hominids, **212**
    time required for, 213
    writing an argument for, 210
Human health: Healthy Diet and Weight, 229

### I

Inquiry-based science, xv–xvii
    as component of science proficiency, xv
    construction of good argument as goal of, xvii

# INDEX

definition of, xv
overemphasis on experimentation in, xv
skills and practices for, xv–xvi
Insects
    Fruit Fly Traits, 45–54
    Misconception About Insects, 321–327
    Termite Trails, 239–248
Instructional models, xii, xiii, xvii. *See also specific models*
    Evaluate Alternatives, xviii, xxvi–xxx
    Generate an Argument, xviii–xxvi

## L

Liver and Hydrogen Peroxide activity, 171–180
    argumentation session for, 173, **173**
    getting started on, 172
    introduction to, 171, **171**
    materials for, 172, 176, **178**
    purpose of, 175
    recording your method and observations for, 172
    research question and possible explanations for, 171
    standards addressed in, 134–135, 179–180
    Teacher Notes for, 175–180
        assessment, 178–179
        content and related concepts, 175
        curriculum and instructional considerations, 176
        options for implementation, 176, **177**
    time required for, 176
    writing an argument for, 174

## M

McComas, William, xxx–xxxi
Mendelian inheritance. *See* Genetics
Microbiology
    Characteristics of Viruses, 123–131
    Misconception About Bacteria, 293–299
Misconception About Bacteria writing activity, 293–299
    purpose of, 294
    standards addressed in, 250–251, 294, 298–299
    student instructions for, 293
    Teacher Notes for, 294–299
        assessment, 296–298
        content and related concepts, 294–295
        curriculum and instructional considerations, 295–296
        options for implementation, 296, 370–371
        student sample scored rubric, **297**
    time required for, 296
Misconception About Inheritance of Traits writing activity, 315–320
    purpose of, 316
    standards addressed in, 250–251, 316, 319–320
    student instructions for, 315
    Teacher Notes for, 316–320
        assessment, 317–319
        content and related concepts, 316–317
        curriculum and instructional considerations, 317
        options for implementation, 317, 370–371
        student sample scored rubric, **318**
    time required for, 317
Misconception About Insects writing activity, 321–327
    purpose of, 322
    standards addressed in, 250–251, 322, 326–327
    student instructions for, 321
    Teacher Notes for, 322–327
        assessment, 324–326
        content and related concepts, 322–323
        curriculum and instructional considerations, 323–324
        options for implementation, 324, 370–371
        student sample scored rubric, **325**
    time required for, 324
Misconception About Interactions That Take Place Between Organisms writing activity, 301–307
    purpose of, 302
    standards addressed in, 250–251, 302, 306–307
    student instructions for, 301
    Teacher Notes for, 302–307
        assessment, 304–306
        content and related concepts, 302–303
        curriculum and instructional considerations, 303
        options for implementation, 304, 370–371
        student sample scored rubric, **305**
    time required for, 304
Misconception About Life on Earth writing activity, 285–290
    purpose of, 286
    standards addressed in, 250–251, 286, 290
    student instructions for, 285
    Teacher Notes for, 286–290
        assessment, 287–290
        content and related concepts, 286
        curriculum and instructional considerations, 286–287
        options for implementation, 287, 370–371
        student sample scored rubric, **288**
    time required for, 287
Misconception About Plant Reproduction writing activity, 309–314
    purpose of, 310
    standards addressed in, 250–251, 310, 312, 314
    student instructions for, 309
    Teacher Notes for, 310–314
        assessment, 311–312
        content and related concepts, 310
        curriculum and instructional considerations, 311
        options for implementation, 311, 370–371
        student sample scored rubric, **313**
    time required for, 311
Misconception About the Methods of Scientific Investigations writing activity, 277–283
    purpose of, 278
    standards addressed in, 250–251, 278, 283
    student instructions for, 277
    Teacher Notes for, 278–283
        assessment, 281–283
        content and related concepts, 278–280

# INDEX

curriculum and instructional considerations, 280–281
options for implementation, 281, 370–371
student sample scored rubric, **282**
two inaccurate depictions of the nature of scientific inquiry, **279**
time required for, 281
Misconception About the Nature of Scientific Knowledge writing activity, 261–268
purpose of, 262
standards addressed in, 250–251, 262, 268
student instructions for, 261
Teacher Notes for, 262–268
assessment, 265–268
content and related concepts, 262–264
curriculum and instructional considerations, 264–265
options for implementation, 265, 370–371
student sample scored rubric, **267**
time required for, 265
Misconception About the Work of Scientists writing activity, 269–274
purpose of, 270
standards addressed in, 250–251, 270, 274
student instructions for, 269
Teacher Notes for, 270–274
assessment, 272–274, 351–360
content and related concepts, 270–271
curriculum and instructional considerations, 271–272
options for implementation, 272, 370–371
student sample scored rubrics, **273, 352, 355, 359**
time required for, 272
Misconception About Theories and Laws writing activity, 253–259
purpose of, 254
standards addressed in, 250–251, 254, 259
student instructions for, 253
Teacher Notes for, 254–259
assessment, 256–259
content and related concepts, 254–255
curriculum and instructional considerations, 255
options for implementation, 255, 370–371
student sample scored rubric, **257**
time required for, 255
Movement of Molecules in or out of Cells activity, 159–169
argumentation session for, **160**, 161–162
conducting a Benedict's test for, 160
introduction to, 158, **158**
materials for, 167, **167**
purpose of, 164
recording your method and observations for, 161
research question and possible explanations for, 159
standards addressed in, 134–135, 169
Teacher Notes for, 164–169
assessment, 168–169
content and related concepts, 164
curriculum and instructional considerations, 164–165
net movement of water into and out of cells in hypertonic, isotonic, and hypotonic solutions, **165**
options for implementation, 165–168, **166**
time required for, 165
writing an argument for, 163

# N

National Research Council (NRC), xvi
Natural selection
Color Variations in Venezuelan Guppies activity, 19–27
Desert Snakes activity, 29–43
Nature of science
Misconception About the Methods of Scientific Investigations, 277–283
Misconception About the Nature of Scientific Knowledge, 261–268
Misconception About the Work of Scientists, 269–274
Misconception About Theories and Laws, 253–259

# O

Osmosis: Movement of Molecules in or out of Cells, 159

# P

Persuasive arguments, xxxi–xxxii
Phenolphthalein agar preparation, 188–189
Phenotypes. *See* Genetics
Photosynthesis
Plant Biomass, 149–158
Plants and Energy, 219–228
Phylogeny. *See* Evolution
Plant Biomass activity, 149–158
argumentation session for, 151, **151**
getting started on, 150
introduction to, 149, **149**
materials for, 150, 155, **157**
purpose of, 153
recording your method and observations for, 150
research question and possible explanations for, 149
standards addressed in, 134–135, 158
Teacher Notes for, 153–158
assessment, 155–158
content and related concepts, 153–154
curriculum and instructional considerations, 154–155
options for implementation, 155, **156**
time required for, 155
writing an argument for, 152
Plants and Energy activity, 219–228
argumentation session for, 221, **221**
getting started on, 219–220
introduction to, 219
materials for, 219–220, 226, **226**
purpose of, 223
recording your method and observations for, 220
research question and possible explanations for, 219
standards addressed in, 134–135, 228

# INDEX

Teacher Notes for, 223–228
    assessment, 226–228
    content and related concepts, 223
    curriculum and instructional considerations, 223–224
    options for implementation, 224, **225**
time required for, 224
writing an argument for, 222

## R

Reflective discussion
    for Evaluate Alternatives, xxx
    for Generate an Argument, xxiv
Refutational Writing, xvii–xviii, xxx–xxxi
    activities for, xxxiii, 249
        *Framework* matrix for, 250–251
        Misconception About Bacteria, 293–299
        Misconception About Inheritance of Traits, 315–320
        Misconception About Insects, 321–327
        Misconception About Interactions That Take Place Between Organisms, 301–307
        Misconception About Life on Earth, 285–290
        Misconception About Plant Reproduction, 309–314
        Misconception About the Methods of Scientific Investigations, 277–283
        Misconception About the Nature of Scientific Knowledge, 261–268
        Misconception About the Work of Scientists, 269
        Misconception About Theories and Laws, 253–259
    options for implementation of, xxxiv, 370–371
    recommendations for, xxxii
    scoring rubric for, xxxii, 368
        student samples and, 351–360, **352, 355, 359**
    writing prompt for, xxxii
Research question(s)
    developing initial answer to, xix–xxii, **xx–xxi**
    identification of, xviii–xix
    introduction of, xxvi–xxviii
    for specific activities
        Cell Size and Diffusion, 181
        Characteristics of Viruses, 123
        Classifying Birds in the United States, 5
        Color Variations in Venezuelan Guppies, 19
        Decline in Saltwater Fish Populations, 82
        Desert Snakes, 29
        DNA Family Relationship Analysis, 55
        Environmental Influence on Genotypes and Phenotypes, 191
        Evolutionary Relationships in Mammals, 68
        Fruit Fly Traits, 45
        Healthy Diet and Weight, 230
        History of Life on Earth, 103
        Hominid Evolution, 204
        Liver and Hydrogen Peroxide, 171
        Movement of Molecules in or out of Cells, 159
        Plant Biomass, 149
        Plants and Energy, 219
        Spontaneous Generation, 138
        Surviving Winter in the Dust Bowl, 114
        Termite Trails, 239
Round-robin format, xxiii, **xxiii–xxiv,** xxx, 6, 19, 30, 46, 55, 68, 83, 104, 114, 124, 139, 151, 161, 173, 182, 193, 208, 221, 231, 240
Rubrics, 331. *See also* Assessment(s)
    for Evaluate Alternatives activities, xxx, 367
        Termite Trails, **344, 346, 349**
    for Generate an Argument activities, xxv–xxvi, 366
        Classifying Birds in the United States, **334, 336, 342**
    for Refutational Writing activities, xxxii, 368
        Misconception About Bacteria, **297**
        Misconception About Inheritance of Traits, **318**
        Misconception About Insects, **325**
        Misconception About Interactions That Take Place Between Organisms, **305**
        Misconception About Life on Earth, **288**
        Misconception About Plant Reproduction, **313**
        Misconception About the Methods of Scientific Investigations, **282**
        Misconception About the Nature of Scientific Knowledge, **267**
        Misconception About the Work of Scientists, **273, 352, 355, 359**
        Misconception About Theories and Laws, **257**

## S

Safety Data Sheet (SDS)
    for Benedict's solution, 168
    for bromothymol blue, 187, 226
    for hydrogen peroxide, 176
    for phenol red, 226
    for vinegar, 187
"Safety in the Science Classroom," xxxiii
Safety notes, xxxiii, 143, 150, 160, 172, 182, 188, 192, 220, 239, 272, 280
Science proficiency, xi, xii, xv
Scientific argument(s)
    assessments of, xxxix, xxxv, 331–360 (*See also* Assessment(s))
    classroom discussions of, xviii, xxxiv
        argumentation session, xxii–xxiv, **xxii–xxiv,** xxix–xxx (*See also* Argumentation session)
        reflective, xxiv, xxx
    construction of, xii, xv, xvi, xvii
    criteria for evaluation of, x–xi
    vs. everyday arguments, ix
    framework for, ix–x, **x**
    generation of, xviii (*See also* Generate an Argument instructional model)
    scientific habits of mind for, xii, xvi, xxii, xxvi
    tentative arguments, xix–xxii, **xx–xxi,** xxviii–xxix, **xxix**
    role in scientific inquiry, xvi
    scoring rubric for, xxv–xxvi, 366
    writing of, xii, xviii, xxiv–xxvi, **xxv** (*See also specific activities*)
        importance of, xxiv–xxv

production of final written argument, xxiv–xxvi, **xxv,** xxx
refutational, xvii–xviii, xxx–xxxii (*See also* Refutational Writing)
writing prompts for, xxv, **xxv,** xxx, **xxxi**
Scientific argumentation
definition of, ix
in *A Framework for K–12 Science Education,* xvi
instructional models to promote student engagement in, xii, xiii, xvii
integration into biology teaching and learning
development of activities for, xiii, xvii–xviii
learning outcomes of, xvii
rationale for, xi–xii
relevance in science education, xvi
Scientific habits of mind, xii, xvi, xxii, xxvi
Scientific investigations
design of, xxviii
identifying research questions for, xviii–xix
Misconception About the Methods of Scientific Investigations writing activity, 277–283
Small-group format, xviii, xix, **xix, xxvii,** xxviii, xxx, xxxiv
Species concept: Classifying Birds in the United States, 5–17
Spontaneous Generation activity, 137–147
argumentation session for, **139,** 139–140
getting started on, 138
introduction to, 137–138
Needham's test of spontaneous generation, 137, **137**
Spallanzani's test of spontaneous generation, 137–138, **138**
materials for, 138, 143, **145**
purpose of, 142
recording your method and observations for, 139
research questions and potential explanations for, 138
standards addressed in, 134–135, 146–147
Teacher Notes for, 142–147
assessment, 145–146
content and related concepts, 142
curriculum and instructional considerations, 142–143
options for implementation, 143, **144**
time required for, 143
writing an argument for, 141
Surviving Winter in the Dust Bowl activity, 113–122
developing a claim for, **114,** 114–115
information about nutritional values and dietary needs for, **116**
introduction to, **113,** 113–114
materials for, 120, **120**
purpose of, 118
research question for, 114
standards addressed in, 2–3, 122
Teacher Notes for, 118–122
assessment, 121–122
content and related concepts, 118–119
curriculum and instructional considerations, 119–120

food chain that consists of four trophic levels, **119**
recommendations for implementation, 120
time required for, 120
writing an argument for, 117

**T**
Teacher Notes, xiii, xxxiii, xxxviii–xxxiv
Assessment section of, xxxix
Content and Related Concepts section of, xxxviii
Curricular and Instructional Considerations section of, xxxiii, xxxviii–xxxix
Recommendations for Implementing the Activity section of, xxxiii–xxxiv, xxxix
for specific activities
Cell Size and Diffusion, 185–190
Characteristics of Viruses, 128–131
Classifying Birds in the United States, 12–17
Color Variations in Venezuelan Guppies, 24–27
Decline in Saltwater Fish Populations activity, 96–101
Desert Snakes, 39–43
DNA Family Relationship Analysis activity, 61–65
Environmental Influence on Genotypes and Phenotypes activity, 195–201
Evolutionary Relationships in Mammals, 75–79
Fruit Fly Traits, 50–54
Healthy Diet and Weight, 233–237
History of Life on Earth, 107–110
Hominid Evolution, 211–218
Liver and Hydrogen Peroxide, 175–180
Misconception About Bacteria, 294–299
Misconception About Inheritance of Traits, 316–320
Misconception About Insects, 322–327
Misconception About Interactions That Take Place Between Organisms, 302–307
Misconception About Life on Earth, 286–290
Misconception About Plant Reproduction, 310–314
Misconception About the Methods of Scientific Investigations, 278–283
Misconception About the Nature of Scientific Knowledge, 262–268
Misconception About the Work of Scientists, 270–274
Misconception About Theories and Laws, 254–259
Movement of Molecules in or out of Cells, 164–169
Plant Biomass, 153–158
Plants and Energy, 223–228
Spontaneous Generation, 142–147
Surviving Winter in the Dust Bowl, 118–122
Termite Trails, 243–248
Teacher's role during activities, xxxiv–xxxv
for Evaluate Alternatives, **xxxvii**
for Generate an Alternative, **xxxvi**
Termite Trails activity, 239–248
argumentation session for, 240–241, **241**
getting started on, 239–240
introduction to, 239, **239**
materials for, 239, 244, **246**

# INDEX

purpose of, 243
recording your method and observations for, 240
research question and possible explanations for, 239
standards addressed in, 134–135, 247–248
Teacher Notes for, 243–248
    assessment, 246–247, 343–350
    content and related concepts, 243
    curriculum and instructional considerations, 243–244
    options for implementation, 244, **245**
    student sample scored rubrics, **344, 346, 349**
time required for, 244
writing an argument for, 242

## W

Writing, xii
    expository, xxxii
    of final argument, xxiv–xxvi, **xxv,** xxx
    importance of, xxiv–xxv
    persuasive, xxxi–xxxii
    refutational, xvii–xviii, xxx–xxxii, xxxiii (*See also* Refutational Writing)
Writing prompts
    for Evaluate Alternatives, xxx, **xxxi**
    for Generate an Argument, xxv, **xxv**
    for Refutational Writing, xxxii